S0-DZC-254

Toward a New Climate Agreement

Climate change is one of the most pressing problems facing the global community. Although most states agree that climate change is occurring and is at least partly the result of humans' reliance on fossil fuels, managing a changing global climate is a formidable challenge. Underlying this challenge is the fact that states are sovereign, governed by their own laws and regulations. Sovereignty requires that states address global problems such as climate change on a voluntary basis, by negotiating international agreements. Despite a consensus on the need for global action, many questions remain concerning how a meaningful international climate agreement can be realized.

This book brings together leading experts to speak to such questions and to offer promising ideas for the path toward a new climate agreement. Organized in three main parts, it examines the potential for meaningful climate cooperation. Part I explores sources of conflict that lead to barriers to an effective climate agreement. Part II investigates how different processes influence states' prospects of resolving their differences and of reaching a climate agreement that is more effective than the current Kyoto Protocol. Finally, part III focuses on governance issues, including lessons learned from existing institutional structures.

The book is unique in that it brings together the voices of experts from many disciplines, such as economics, political science, international law, and natural science. The authors are academics, practitioners, consultants, and advisors. Contributions draw on a variety of methods, and include both theoretical and empirical studies. The book should be of interest to scholars and graduate students in the fields of economics, political science, environmental law, natural resources, earth sciences, sustainability, and many others. It is directly relevant for policy makers, stakeholders and climate change negotiators, offering insights into the roles of uncertainty, fairness, policy linkage, burden sharing, and alternative institutional designs.

Todd L. Cherry is a Professor of Economics at Appalachian State University, USA, and at CICERO Center for International Climate and Environmental Research – Oslo, Norway. He currently holds the Rasmuson Chair of Economics at the University of Alaska Anchorage, USA.

Jon Hovi is a Professor of Political Science at the University of Oslo, and at CICERO Center for International Climate and Environmental Research – Oslo, Norway.

David M. McEvoy is an Associate Professor in the Department of Economics at Appalachian State University, USA.

Routledge Advances in Climate Change Research

Local Climate Change and Society
Edited by M. A. Mohamed Salih

Water and Climate Change in Africa
Challenges and community initiatives in Durban, Maputo and Nairobi
Edited by Patricia E. Perkins

Post-2020 Climate Change Regime Formation
Edited by Suh-Yong Chung

How the World's Religions are Responding to Climate Change
Social Scientific Investigations
Edited By Robin Globus Veldman, Andrew Szasz and Randolph Haluza-DeLay

Climate Action Upsurge
The Ethnography of Climate Movement Politics
Stuart Rosewarne, James Goodman and Rebecca Pearse

Toward a Binding Climate Change Adaptation Regime
A Proposed Framework
Mizan R. Khan

Transport, Climate Change and the City
Robin Hickman and David Banister

Toward a New Climate Agreement
Todd L. Cherry, Jon Hovi and David M. McEvoy

The Anthropology of Climate Change
An Integrated Critical Perspective
Hans A Baer and Merrill Singer

Toward a New Climate Agreement

Conflict, resolution and governance

Edited by

Todd L. Cherry, Jon Hovi and David M. McEvoy

LONDON AND NEW YORK

First published 2014
by Routledge
2 Park Square, Milton Park, Abingdon, Oxon OX14 4RN

and by Routledge
711 Third Avenue, New York, NY 10017

Routledge is an imprint of the Taylor & Francis Group, an informa business.

British Library Cataloguing in Publication Data
A catalogue record for this book is available from the British Library

Library of Congress Cataloging in Publication Data
Toward a new climate agreement : conflict, resolution and governance /
[edited by] Todd L. Cherry, Jon Hovi, David M. McEvoy.
pages cm. -- (Routledge advances in climate change research)
Summary: "This book examines the challenges of sustaining meaningful cooperation among countries striving to manage global climate change through international environmental agreements. Through the perspectives of leading international scholars from multiple disciplines, readers of the book will gain an understanding of how agreements are negotiated, the strength and weaknesses of previous climate agreements and how a more effective future climate agreement can be designed"-- Provided by publisher.
Includes bibliographical references and index.
1. Climatic changes--Government policy--International cooperation.
2. Environmental policy--International cooperation. I. Cherry, Todd L., editor of compilation.
II. Hovi, Jon, 1956- editor of compilation. III. McEvoy, David M. (David Michael), editor of compilation.
QC903.T69 2014
363.738'74561--dc23
2013031853

ISBN: 978-0-415-64379-5 (hbk)
ISBN: 978-0-203-08000-9 (ebk)

Typeset in Times
by Saxon Graphics Ltd, Derby

Contents

Figures

Tables

Contributors

Steinar Andresen, Fridtjof Nansen Institute, Lysaker, Norway

Guri Bang, Center for International Climate and Environmental Research – Oslo, Oslo, Norway

Scott Barrett, Columbia University, New York, USA; Princeton University, New Jersey, USA

Thomas Bernauer, ETH Zürich, Zürich, Switzerland

Todd L. Cherry, Appalachian State University, North Carolina, USA; Center for International Climate and Environmental Research – Oslo, Oslo, Norway; University of Alaska Anchorage, Alaska, USA

Aaron Cosbey, International Institute for Sustainable Development, Winnipeg, Canada

Astrid Dannenberg, Columbia University, New York, USA; University of Gothenburg, Gothenburg, Sweden

Aart de Zeeuw, Tilburg University, Tilburg, The Netherlands

Michael Finus, University of Bath, Bath, UK

Carolyn Fischer, Resources for the Future, Washington DC, USA

Robert Gampfer, ETH Zürich, Zürich, Switzerland

Michael D. Gerst, Dartmouth College, New Hampshire, USA

Michael Hoel, University of Oslo, Oslo, Norway

Jon Hovi, University of Oslo, Oslo, Norway; Center for International Climate and Environmental Research – Oslo, Oslo, Norway

Richard B. Howarth, Dartmouth College, New Hampshire, USA

Shanshan Huang, University of Wyoming, Wyoming, USA

Torbjørg Jevnaker, Fridtjof Nansen Institute, Lysaker, Norway

Steffen Kallbekken, Center for International Climate and Environmental Research – Oslo, Oslo, Norway

Florian Landis, Centre for European Economic Research, Mannheim, Germany

David M. McEvoy, Appalachian State University, North Carolina, USA

Matthew McGinty, University of Wisconsin-Milwaukee, Wisconsin, USA

Matthew E. Oliver, Georgia Institute of Technology, Georgia, USA

Jamison Pike, University of Wyoming, Wyoming, USA

Pedro Pintassilgo, University of Algarve, Faro, Portugal

Jason F. Shogren, University of Wyoming, Wyoming, USA

Jon Birger Skjærseth, Fridtjof Nansen Institute, Lysaker, Norway

Tora Skodvin, University of Oslo, Oslo, Norway; Center for International Climate and Environmental Research – Oslo, Oslo, Norway

Detlef F. Sprinz, Potsdam Institute for Climate Impact Research, Potsdam Germany; University of Potsdam, Potsdam, Germany; Center for International Climate and Environmental Research – Oslo, Oslo, Norway

John K. Stranlund, University of Massachusetts-Amherst, Massachusetts, USA

Håkon Sælen, Center for International Climate and Environmental Research – Oslo, Oslo, Norway

Geir Ulfstein, University of Oslo, Oslo, Norway

Arild Underdal, University of Oslo, Oslo, Norway; Center for International Climate and Environmental Research – Oslo, Oslo, Norway

Christina Voigt, University of Oslo, Oslo, Norway

Jørgen Wettestad, Fridtjof Nansen Institute, Lysaker, Norway

Oran R. Young, University of California-Santa Barbara, California, USA

Foreword

At the 2009 Copenhagen climate change negotiations, a colleague on the US delegation walked into the so-called "shared vision" text break-out session, which was one of about a dozen topic-specific negotiations occurring late in the second week of the talks. One of the co-chairs of the "shared vision" text addressed the room full of staff-level negotiators. He noted that the draft text began with an open bracket and ended with a close bracket. Thus the entire text was bracketed, meaning that after nearly two years of negotiations since the Bali conference, there was agreement on exactly zero words describing the international community's common objectives in combating climate change. The co-chair noted that once a phrase is bracketed, there is no value associated with adding a second pair (or even third pair) of brackets around the phrase. One pair of brackets around a phrase signifies no agreement, and thus additional pairs of brackets are redundant in representing the absence of agreement. So the co-chair asked negotiators to identify potential areas of agreement and requested them to make specific suggestions for removing brackets around phrases, sentences, and paragraphs. The co-chair then called on the next country in the queue to make an intervention. That country expressed displeasure with a specific draft paragraph and called for inserting brackets – a second pair of brackets – around the paragraph in question. The co-chair struggled to suppress his exasperation.

Since this breakout session, there have been few signs of a "shared vision" in combating the risks posed by global climate change. Later that week in Copenhagen, heads of state from about 25 nations, representing every major economy and every UN negotiating bloc, rolled up their sleeves and negotiated line-by-line the Copenhagen Accord. In the hours after heads of state and many ministers left town, lower-level negotiators challenged this agreement in the conference plenary session, requiring an unusual and awkward "taking note" of what the world's leaders personally negotiated. At the 2011 Durban conference, countries agreed to a negotiating path forward, open to meaningful participation by all nations, thereby breaking the anachronistic bifurcation of mitigation effort – through the Annex I and Non-Annex I classifications – that had inhibited progress in tackling climate change for two decades (Aldy and Stavins 2012), only to have some developing nations begin walking back the deal immediately after the talks. At the 2012 Doha conference, the negotiations ran more than a day after the scheduled closing, over disputes on a second commitment period of the Kyoto Protocol, which would cover less than 15 percent of global emissions and simply replicate emission targets already agreed to by European and a few other nations at the 2009 and 2010 talks.

An optimist, however, may point to the foundation of the international climate policy architecture, the UN Framework Convention on Climate Change (UNFCCC), and the agreed-upon ultimate objective, which is the "stabilization of greenhouse gas concentrations in the atmosphere at a level that would prevent dangerous anthropogenic interference with the climate system" (Article II). In Copenhagen in 2009 and Cancun in 2010, the international community elaborated this ultimate objective by calling for warming to be limited to no more than 2°C above pre-industrial levels. While this long-term goal may be shared among members of the international community, there is little agreement and little progress to date on how to realize that goal. The recent experience with climate-related outcomes – emissions and atmospheric concentrations – likely weakens the enthusiasm of even the most ardent climate policy optimist.

Since the 1992 Earth Summit produced the UNFCCC, global carbon dioxide emissions from fossil fuel combustion have increased by more than 50 percent. Global emissions grew more than twice as fast in the decade after the 1997 Kyoto conference than in the decade before that negotiation (Aldy 2013). This growth is occurring in the developing world: in 2011 the developed world's carbon dioxide emissions were about 6 percent below 1990 levels, while the developing world's emissions had grown more than 160 percent since 1990.[1] The developing world's share of fossil carbon dioxide emissions has increased from 40 percent to 60 percent over the past two decades, and this share is expected to grow over time. As a result of the rapid growth in emissions, atmospheric greenhouse gas concentrations in 2013 exceeded 400 parts per million for the first time in human history. Several recent studies have assigned very low probabilities of limiting warming to less than 2°C, even with much more ambitious emission abatement than currently underway (Anderson and Bows 2011; Webster *et al.* 2012).

The track record with process and outcomes likely explains, at least in part, the interest in alternative means of combating global climate change and the recent fragmentation of the international climate policy dialogue.[2] For example, the Major Economies Forum on Energy and Climate serves as a venue for ministers and heads of state of the largest developed and developing economies to discuss topics in the climate negotiations as well as move forward on clean energy action plans. In 2009, the Group of 20 Leaders, representing the largest developed and developing nations, agreed to phase out fossil fuel subsidies and encouraged other nations to do the same, which could lower global carbon dioxide emissions by about 6 percent by 2020 (IEA 2010). In 2009, Canada, Mexico, and the United States proposed to phase down hydrofluorocarbons, high global-warming potential substitutes for ozone-depleting substances such as chlorofluorocarbons and hydrofluororcarbons, under the Montreal Protocol on Substances that Deplete the Ozone Layer. In 2012, a number of developed and developing countries established the Climate and Clean Air Coalition to Reduce Short-Lived Climate Pollutants, with an aim of mitigating black carbon, methane, and hydrofluorocarbon emissions. In 2010, the Nagoya conference of the Convention on Biological Diversity agreed to a decision on the governance of potential research experimentation and deployment of geoengineering.

Given this rather dismal take on the status quo process, its outcomes, and the revealed preference for a multiplication of negotiating fora, how can the international community move forward in its search for a robust, effective international climate policy architecture? In what follows, I suggest a few ideas on how to advance the global effort to combat climate change. These ideas, and many others, resonate in the chapters included in this volume.

First, public policy should explicitly explore all possible means of addressing climate change. While this may appear obvious, the Kyoto Protocol is effectively silent on adaptation, and there have been virtually no discussions of geoengineering in the UN-sponsored climate talks. The debate over future climate policy would benefit from a recognition that we can reduce the risks posed by climate change by *preventing* climate change (emission abatement), *getting used* to climate change impacts (adaptation), and *fixing* the problem (geoengineering and air capture). The incentives differ across these types of risk mitigation, and efforts along one dimension could substitute for or complement efforts along another dimension, depending on the details of implementation. An effective risk mitigation program would likely account for these interactive impacts and seek out low-cost opportunities along each dimension.

Second, the international climate regime reflects a series of actions followed by opportunities for reflection before taking subsequent actions (e.g., 1992 UNFCCC, 1997 Kyoto Protocol, 2009 Copenhagen Accord, etc.). A significant literature in decision sciences and economics supports this kind of act-learn-act approach to climate policy. It suggests, however, that the international community should design and embrace institutions to facilitate such learning. Collecting, analyzing, and disseminating information on the environmental effectiveness and economic consequences of climate change policies can improve the prospects for more successful follow-on efforts to mitigate climate change risks. A system of policy surveillance, not unlike that in other multilateral contexts (such as under the World Trade Organization and the International Monetary Fund), would facilitate learning and ensure that as more countries take on more significant efforts to abate emissions, adapt to climate change, or design geoengineering options, they can benefit from the experience and the best policy practices of those countries that have acted before them.

Third, nations can move forward with policy options for which there is little doubt that their domestic benefits significantly exceed their domestic costs. Consider two straightforward examples. In the first case, black carbon pollution, which is effectively soot and fine particulates, can be mitigated through an array of commercial control technologies and an array of policy instruments (technology mandates, technology subsidies, emission taxes, cap-and-trade, etc.). In addition to contributing to warming, black carbon causes premature mortality. On public health grounds alone, the benefits likely exceed the costs for significant reductions in black carbon pollution. In the second case, eliminating fossil fuel subsidies makes sense on fiscal, economic, and public health grounds, before accounting for climate benefits as well. Most fossil fuel subsidies are poorly targeted and thus the vast majority of the benefits do not accrue to the least well-off in many developing countries. Replacing fossil fuel subsidies with a means-tested cash transfer program, such as Indonesia implemented in 2005, could address redistribution, and energy, economic, and climate goals. In each of these examples, the international community could design mechanisms to facilitate policy transfer and support domestic institutional reforms to deliver reductions in black carbon and in fossil fuel subsidies.

Finally, domestic and international climate policy development occurs within a complicated political economy and system of international relations. A pragmatic approach to climate policy that draws insights from economics, ethics, politics, etc. will likely prove more resilient than an ideologically narrow approach. Some of the stakeholders in the international climate policy process hold strong views of what constitutes the perfect climate policy architecture. I don't believe these perfect models are possible in the current world (or any foreseeable state of the world). Movement toward a meaningful international climate agreement requires a broader perspective that draws pragmatic insights from multiple disciplines, as provided by the collection of chapters in this volume. In terms of the economics, there will not be a single, universal price on carbon. In terms of equity, there will not be consensus on a single, common principle of fairness from which countries could then derive their obligations, rights, and actions under an international climate agreement. In terms of the environment, there will be warming in excess of what some advocates believe is safe. This is not to say that we should not aspire to cost-effective, fair, environmentally ambitious international climate policy agreements. But that we should do so while accounting for the political economy constraints that will produce messy negotiations and messy agreements.

The participants in the international climate negotiations have long benefitted from academic scholarship, especially from research that reflects lines of inquiry from a variety of disciplines. Since the 2011 Durban conference, there has been significant interest in new ideas for international climate policy design to inform the talks leading up to 2015 and beyond. This timely volume, drawing from many top scholars throughout the social sciences, will inform the policy process.

Joseph E. Aldy
Harvard Kennedy School
Cambridge, MA

Notes

1 All emission statistics presented here are for carbon dioxide emissions from fossil fuel combustion, drawing from the Global Carbon Project (www.globalcarbonproject.org/carbonbudget). As another sign of the inadequacy of the status quo regime, the UNFCCC does not collect nor does it require all parties to regularly report their greenhouse gas emissions. For example, until late 2012, the most recent UNFCCC

estimate for greenhouse gas emissions in China was for 1994. Now, the most recent estimate is for 2005. Some efforts to design a system of international consultations and analysis based on the leaders' agreement in Copenhagen aim to remedy this failing.

2 See Keohane and Victor (2010) for a further discussion of the fragmented international climate policy regime.

References

Aldy, J.E. (2013). Designing a Bretton Woods institution to address global climate change. In R. Fouquet (ed.), *Handbook on Energy and Climate Change.* Cheltenham, UK: Edward Elgar, pp. 352–74.

Aldy, J.E. and R.N. Stavins (2012). Climate negotiators create an opportunity for scholars. *Science* 337 (6098), 1043–44.

Anderson, K. and A. Bows (2011). Beyond "dangerous" climate change: emission scenarios for a new world. *Philosophical Transactions of the Royal Society A,* 369 (1934), 20–44.

International Energy Agency (2010). *World Energy Outlook 2010.* Paris: IEA.

Keohane, R.O. and D.G. Victor (2010). The regime complex for climate change. Discussion paper 10-33, Harvard Project on International Climate Agreements, Harvard University, Cambridge, MA.

Webster, M., A.P. Sokolov, J.M. Reilly, C.E. Forest, S. Paltsev, A. Schlosser, C. Wang, D. Kicklighter, M. Sarofim, J. Melillo, R.G. Primm and H.D. Jacoby (2012). Analysis of climate policy targets under uncertainty. *Climatic Change* 112 (3–4), 569–83.

Introduction

Todd L. Cherry, Jon Hovi and David M. McEvoy

Climate change is one of the greatest challenges facing the global community. There is consensus among international scientific organizations that the climate is changing, and that this change is a result, in part, of anthropogenic greenhouse-gas (GHG) emissions. While there remains great uncertainty regarding the effects on our planet, recent studies suggest that the cost of unchecked climate change may be devastating (Weitzman 2011). It is no surprise then that countries have spent significant resources over the past few decades trying to negotiate a strategy to jointly manage GHG emissions.

The fact that climate change must be managed jointly is an important one. GHGs are those that concentrate in the atmosphere and absorb (and then emit) infrared radiation. In short, these gases allow solar radiation to pass through to the earth, but when the infrared radiation (i.e., heat energy) is reflected back to the atmosphere it is absorbed by the GHGs and then re-emitted back to the Earth's surface. Carbon dioxide, the most significant GHG resulting from human activity, is a stock pollutant that mixes uniformly in the atmosphere. Thus the resulting concentration of carbon emissions in the atmosphere is independent of the source of the emissions. This reality means that every country can contribute to the climate change problem through their emissions. It also means that every country can contribute (at least marginally) to mitigating climate change through emissions abatement activities. In this way, atmospheric concentrations of carbon emissions (as well as other GHGs) constitute an international environmental externality.

When externalities cross borders, unilateral management is not an effective option. The benefits from actions taken by one country to reduce carbon emissions will be trivial if other countries fail to reduce their emissions in tandem (or if they respond by increasing their own). And because the benefits of emissions abatement are enjoyed by all countries, there are then incentives for countries to free ride on the abatement activities of other countries (see the third section). For these reasons unilateral management of GHGs will result in higher than optimal concentrations of these gases in the atmosphere. Moreover, unlike regulating domestic pollutants, sovereign countries do not answer to a global authority that can externally impose and enforce GHG regulations in all countries. Rather, countries must coordinate voluntarily by negotiating international climate treaties.

The roots of international cooperation to manage climate change can be traced at least back to 1979, when the first World Climate Conference was arranged by the World Meteorological Organization. However, efforts to manage climate change began in earnest in the early 1990s through the United Nations General Assembly. An Intergovernmental Negotiating Committee was tasked with conducting the negotiation process, and in less than two years the body drafted the United Nations Framework Convention on Climate Change (UNFCCC). The objective of this convention (or treaty) was to "stabilize greenhouse-gas concentrations in the atmosphere at a level that would prevent dangerous anthropogenic interference with the climate system" (UN General Assembly 1992, Article 2). The UNFCCC, however, did not require objective responsibilities for emissions abatement for the parties. Rather, it established a framework through which countries could convene to negotiate binding "protocols" that set mandatory emissions limits. The UNFCCC entered into force in March 1994 and today 195 countries are ratifying members to the convention.

The Kyoto Protocol was adopted in 1997 during the third annual conference of the parties to the UNFCCC (COP-3). The treaty, the first of its kind, established binding emissions limits for developed (Annex I) countries. Annex I countries agreed to reduce their GHG emissions in the period 2008–2012 by an average of 5.2 percent relative to 1990 levels. Although almost all countries signed the treaty soon after its adoption, it would enter into force only when certain participation thresholds were met. Specifically, at least "55 Parties to the Convention, incorporating Parties included in Annex I which accounted in total for at least 55 percent of the total carbon dioxide emissions for 1990 of the Annex I countries" had to ratify the agreement (Kyoto Protocol, Article 25). When Russia ratified in late 2004, the treaty satisfied these participation requirements, which triggered Kyoto to enter into force in February 2005.

As of today, the Kyoto Protocol remains the only international treaty requiring binding limits on GHG emissions. Unfortunately, the treaty failed to motivate the two biggest greenhouse-gas emitters to commit to reducing, or even stabilizing, their emissions. The United States, at the time the biggest emitter (now second biggest), chose not to ratify the treaty. One of the fundamental reasons for the defection was that the United States felt their participation would be futile without having China, then the second biggest emitter (now biggest), also take on emissions abatement responsibilities. While China is a signatory to the protocol, its classification as a developing country under the UNFCCC precludes it from any emissions abatement responsibilities. Kyoto is now in its second commitment period, this time with even fewer key players. In early 2012, Canada withdrew from Kyoto before the first commitment period ended, whereas Belarus, Japan, New Zealand, Russia, and Ukraine have declined to participate further. The Kyoto Protocol, by any metric, has failed to stabilize global GHG emissions.

Why the climate change problem is so difficult to solve

A combination of several factors makes the climate change problem particularly challenging.[1] First, climate mitigation is a global public good, meaning that no country can be excluded from sharing the benefits of climate mitigation, irrespective of whether it contributes to such mitigation. This fact has two important implications. One is that most of the benefits from a country's own mitigation efforts go to other countries, so that governments primarily concerned with their own citizens' welfare have little incentive for action. The other is that any one country's contributions to climate mitigation make (at best) only a small difference for the global climate. Of course, the latter feature is particularly salient for minor countries. For example, even if Norway were to eliminate all of its emissions of GHGs (which is of course unthinkable in practice), global emissions would become only about 0.15 percent smaller than if Norway were to continue business as usual. The effect on global warming would thus be negligible. But even a mid-sized country's emissions have rather modest influence on the climate. For example, even if Canada (the world's eighth largest emitter in 2011) were to eliminate all of its emissions of GHGs (again, this is of course unthinkable), global emissions would drop by less than 2 percent.

Second, whereas the costs of implementing GHG emissions reductions in the present are incurred immediately, the benefits from climate mitigation are realized far into the future. Indeed, these benefits will likely take effect only after several decades. In contrast, policy makers tend to have relatively short time perspectives, and thus prefer measures that accrue immediate benefits with delayed costs. Moreover, policy makers will always have good reasons for delaying action on a long-term problem such as climate change; in particular, they will want to prioritize more immediate societal challenges (e.g., the financial crisis, domestic infrastructure, health care, seniors' welfare, etc.). Policy makers' incentives for delay will likely be reinforced by the fact that the people bearing the costs of climate action today will not be able to enjoy the full extent of the benefits.

Because of these factors, managing global climate change involves a free-rider problem: actors have strong incentives to refrain from contributing to climate change mitigation themselves, while nevertheless enjoying the possible benefits resulting from other actors' mitigation efforts. At best, therefore, climate change mitigation will likely be provided only in suboptimal quantities – despite the fact that some mitigation measures (often called no-regrets options) may involve direct or indirect benefits (such as reduction of local air pollution) that offset the costs of implementing those measures (e.g., Heltberg *et al.* 2009).

Another factor that makes climate cooperation so difficult is that GHG reductions on the scale required to have a real impact on global warming are very costly. The reason is that almost all economic activities produce GHG emissions. Thus the free-rider problem is much more severe for climate change than for other environmental challenges, such as ozone depletion (Barrett 1999). The point is well illustrated by Canada's Environment Minister, Peter Kent, who in December 2011 justified Canada's withdrawal from the Kyoto Protocol in the following way: "The transfer of $14 billion from Canadian taxpayers to other countries – the equivalent of $1600 from every Canadian family – with no impact on emissions or the environment. That's the Kyoto cost to Canadians."[2]

The climate change problem is also difficult for political reasons. Domestic politics tend to advantage those who oppose implementing climate mitigation measures. Because the benefits from unilateral mitigation for domestic constituents are small and highly dispersed, they provide little incentive for collective political action. In contrast, costs are typically highly concentrated and will thus be more likely to motivate concerted efforts to influence policy makers (e.g., Hovi *et al.* 2009).

Finally, several strong asymmetries further exacerbate the climate change problem. Countries vary significantly in their historical responsibility for causing the climate change problem, in terms of their vulnerability and their ability to mitigate and adapt to climate change. These asymmetries are part of the reason why debates about international burden sharing, justice, or equity in relation to climate change mitigation are often contaminated by "self-serving bias": countries tend to invoke distributional principles that favor themselves. For example, the United States highlights the importance of a level playing field in world trade, China points to the moral imperative of historical responsibility, and Norway emphasizes the significance of reducing mitigation costs by permitting countries to cut emissions abroad rather than at home. These asymmetries present barriers to reaching a consensus on what a just climate agreement would look like. At the same time, they make justifying free-riding behavior easy.

Considering all these factors, it is hardly surprising that the first commitment period of the Kyoto Protocol was associated with several types of free riding: first, the United States did not ratify. Second, Canada ratified but subsequently withdrew from the agreement. Third, many countries in Eastern Europe (including Belarus, Russia, and Ukraine) ratified only with very lenient emissions control commitments. Fourth, developing countries (including large countries such as Brazil, China, and India) ratified without *any* emissions control commitments. Finally, it is possible that some countries participated without fully meeting their commitments, which was not yet entirely clear as this volume was being published.

Note that Kyoto imposed only relatively moderate emissions limitation targets on Annex I countries (and no targets at all on other member countries). The unpleasant reality is therefore that a more ambitious future agreement might well further strengthen the incentives to free ride. Thus a new and more effective climate agreement will likely have to include three types of incentives: incentives for ratification with deep commitments, incentives against withdrawal, and incentives for countries to comply with their commitments.

The purpose of this volume

Although the climate change problem is plagued by great uncertainty, most countries agree on three important premises. First, they agree that climate change is occurring and is the result, in part, of the actions of humans. Second, they agree that climate change is a problem worth addressing. Third, they agree that the problem cannot be managed unilaterally. Confirmation of these shared premises can be observed each year at the conference of the parties to the UNFCCC, during which the sovereign countries of the world voluntarily spend valuable time and resources to discuss and negotiate potential solutions. This is the good news.

The bad news is that there is disagreement on almost every other aspect of the climate change problem. Countries approach the problem in different stages of development and from different development paths, and thus with different perspectives. The industrialized countries are responsible for the vast majority of historical GHG emissions, and the wealth of these countries is not independent of their production of carbon emissions. It is also clear that without cooperation on GHG emissions control by large developing countries the problem will persist.

The most recent efforts towards managing global climate change are continued attempts to improve upon the existing Kyoto architecture. At COP-17 in South Africa in 2011, the UNFCCC parties arrived at the so-called Durban platform. This platform, together with the decisions made at COP-18 in Doha, Qatar, extends the Kyoto Protocol to a second commitment period, which will last from 2013 to 2019 (referred to as Kyoto 2). As of today, 31 countries with emissions reduction targets are participating in Kyoto 2: The 27 EU member countries, Australia, Iceland, Lichtenstein, and Norway. Meanwhile, Canada withdrew from Kyoto in 2011, and Belarus, Japan, New Zealand, Russia, and Ukraine have declined to accept new binding targets until countries such as the United States, China, and India also accept binding targets.

The Durban platform also provides a "roadmap" for future negotiations aiming to develop "a protocol, other legal instrument, or an agreed outcome with legal force."[3] The goal is to conclude negotiations on this new "protocol," "instrument" or "outcome" by 2015 and that it should enter into force no later than 2020. However, while the formulation chosen in Durban enabled the parties to avoid a dramatic breakdown in the closing moments of COP-17, it remains unclear what the formulation actually means.

This is where we stand. We are at a point at which all countries agree that the existing international agreement on climate change is insufficient, but without a clear common understanding on how to improve upon it (or depart from it). The purpose of this volume is to present and synthesize recent thoughts and ideas from leading experts on how to move forward in a productive manner. The book is not designed to be a step-by-step approach to resolving the climate change debate, but is rather a sounding board for discussing new ideas and approaches to managing climate change. These ideas range from small tweaks to the existing Kyoto architecture to fundamental shifts in the design of international institutions. Some chapters are narrowly focused on specific aspects of international cooperation. Others cover big picture ideas. Regardless of their particular focus, each chapter offers new insights on strategies to manage the global climate.

The volume brings together the voices of experts from many disciplines engaged in the climate debate. The book contains collaborations between and across economists, political scientists, natural scientists and lawyers. The authors are academics, practitioners, consultants and advisors. Even within disciplines, the book contains a wide range of research methodologies used to approach these problems, consisting of both theory and empirics. Amongst all this diversity in thinking, however, there is much common ground and overlap. The premise that the current generation must bear some (if not most) of the burden of reducing harmful climate change for future generations is one that resonates throughout the volume. There is a shared understanding among contributors that if countries are to successfully manage an inter-temporal, international externality such as GHG emissions, it will require a departure from traditional ways of thinking about the problem. Perhaps at the forefront, there is a shared understanding that concerns about equity and fairness are as real and important as pecuniary ones.

The book is targeted to an audience as diverse as its contributors. In academia, the book should be of interest to graduate students concerned with climate change in the fields of economics, political science, environmental law, natural resources, earth sciences, sustainability, and many others. The book will also serve as a reader for undergraduate courses that concentrate on the climate change problem. The material is also directly relevant for policy makers and stakeholders in the process of international climate negotiations. The book shares insights into the role of uncertainty, fairness, issue and policy linkage, burden sharing, and alternative institutional designs that will be valuable to negotiators.

Overview of chapters

The collection of chapters is organized by three themes: *conflict*, issues that create barriers to an agreement; *resolution*, approaches that offer better prospects for an effective agreement; and *governance*, structures that enhance the effectiveness of an agreement. Though the themes are not mutually exclusive and many chapters contribute to more than one theme, the organization provides an initial framework for the reader.

Part I

The first part includes six chapters that explore sources of conflict. In Chapter 1, Steffen Kallbekken provides a fascinating first-person account of the UNFCCC climate negotiations in Durban, South Africa. Relying on his personal observations during the meetings, he highlights two areas of contention – the firewall between developed and developing countries, and the unwillingness to incorporate climate initiatives external to the UNFCCC process. Kallbekken provides ideas on how to make the process more flexible, break down the firewall, and open the process to external initatives.

Oran Young follows with a chapter that examines the challenge of incorporating concerns of fairness in the climate negotiations. He explores the reasons that make fairness relevant in international agreements and suggests a more equitable allocation scheme for total allowable emissions. Young suggests that the way forward is to allocate allowable emissions globally on an equal per capita basis, with governments managing permits as trustees for their citizens, and allowing trading in emissions permits on a global scale.

In Chapter 3, Michael Finus and Pedro Pintassilgo investigate the role that uncertainty and learning plays in the potential success of climate negotiations. Previous work reported the somewhat counter-intuitive result that the more we learn about climate change, the worse the outcomes from strategic behavior and the more difficult it will be to form climate agreements. Finus and Pintassilgo extend this work by developing a stylized coalition formation model that considers different types of uncertainty and different types of learning. They find, contrary to the literature, that learning is generally conducive to the success of climate negotiations, and that under special and unlikely conditions resolving uncertainty only undermines success. Further, they show that asymmetry among countries can encourage success if the gains from cooperation are shared in a way that makes participation attractive, which implies that both cost-effectiveness and distributional considerations matter for the prospects of a climate agreement.

The distribution of rights and obligations in global climate agreements has resulted in various considerations, such as culpability, economic capacity, growth prospects, and vulnerability. In Chapter 4, Thomas Bernauer, Robert Gampfer and Florian Landis explore the influence of such considerations on the burden sharing problem facing international climate agreements. They outline the most pressing burden sharing criteria and develop specific burden sharing formulas that incorporate the possibility of transfer payments. They highlight the importance of drawing upon both macro and micro level research in developing a more integrated approach when addressing the burden sharing challenge.

In Chapter 5, Scott Barrett and Astrid Dannenberg remind us that the impacts from climate change may be "abrupt and catastrophic," rather than linear and smooth, once greenhouse gas concentrations in the atmosphere exceed a certain threshold. Employing experimental methods, they examine whether the distinction between "gradual" and "dangerous" climate change affects cooperative behavior. Contrary to theory, they find that cooperation is greater with an uncertain prospect of "catastrophic" rather than "gradual" climate change, which informs how framing the problem can influence negotiations.

In Chapter 6, the final chapter of the conflict part, Guri Bang and Tora Skodvin consider how the new abundance of natural gas in the United States might alter the prospects of the United States implementing federal climate policy and participating in an international climate agreement. By reviewing 12 key hearings on energy policy that took place in the Senate Committee on Energy and Natural Resources during the 112th Congress, they find that the shale gas revolution has not changed positions on federal or international climate policy among lawmakers, experts and stakeholders. Their analysis indicates that the fundamental and enduring political barriers to federal policy remain intact and that the likelihood of US participation in an international climate agreement in the near future remains very low.

Part II

The second part focuses on the ways that the international community can more successfully reach an effective climate agreement. Chapter 7 examines a common feature of international environmental agreements – minimum participation requirements. David McEvoy, Todd Cherry and John Stranlund examine whether these requirements facilitate cooperative behavior. Results from their laboratory experiments indicate that when the optimal agreement size requires that all players join, the minimum participation mechanism works very well. However, when optimal agreement sizes allow for free riders, it is far less efficient. In these cases, optimal agreements are blocked to avoid free riding and reduce inequities.

In Chapter 8, Matthew Oliver, Jamison Pike, Shanshan Huang and Jason Shogren explore whether international climate action can be made more successful by focusing on clusters of homogeneous groups that might create more tightly-knit collectives to facilitate trust, commitment and reciprocity. Their experimental study considers such polycentric ordering with and without a complementary coordination mechanism. They find that a polycentric approach alone is not sufficient to support global climate coordination, but the effectiveness of the coordination mechanism is enhanced. The results reinforce the limitations of voluntary action and the potential benefit of combining polycentric ordering with strong institutional arrangements.

Over the past decade, considerable theoretical work has examined international environmental agreements. Matthew McGinty's Chapter 9 presents some insights from this literature, as he discusses the relationship between asymmetry and transfers and provides a review of the models and results. Among the key insights, McGinty points out the need to set abatement targets that account for asymmetry among countries, and the promise of transfers to facilitate voluntary participation in an international agreement.

Richard Howarth and Michael Gerst, in chapter 10, tackle a fundamental issue underlying any international climate agreement – is it worth it? To explore this rate-of-return criterion, they discuss two key factors: the appropriate discount rate and the provision of risk reductions. They conclude that valuing risks based on the preferences people reveal in real-world financial markets supports relatively aggressive climate change policies, and that demanding high rates of return is unreasonable for actions that reduce substantial risks.

Analysts and policy makers alike have suggested that more exclusive approaches to deal with climate change may be more effective than the all-inclusive UN approach. The argument presumes that it is easier to reach agreement within a more confined group of actors. In Chapter 11, Steinar Andresen reviews the experiences of various exclusive approaches to climate governance and concludes that they do not represent a panacea towards a more effective approach to meaningful climate policy. The chapter suggests that many of the most contentious political conflicts are not necessarily easier to confront in more confined forums.

In the final chapter of this section, Jon Hovi, Detlef Sprinz and Arild Underdal provide a comparison of two general approaches – the current "top-down" treaty design that targets a comprehensive solution and an alternative "bottom-up" approach in which countries make commitments based on what they see as economically, politically and administratively feasible at home. They probe the pros and cons of each approach, and also consider the prospects for combining elements of both bottom-up and top-down approaches. While neither approach emerges as a dominant solution, the discussion highlights the distinct advantages of each, which will inform future efforts that seek to combine top-down and bottom-up elements for a more effective approach.

Part III

The final section of the volume addresses issues related to governance, including lessons learned from existing structures. In the first chapter of the governance section, Geir Ulfstein and Christina Voigt examine the legal form and principles of a new agreement. They argue that the Durban Platform allows states to choose between distinctive options, including the adoption of a new protocol, amendments to the UNFCCC, or a combination that includes decisions by the Conference of the Parties. Ulfstein and Voigt discuss the challenges of evaluating the legal form of the different elements of the agreement and developing a more dynamic architecture that overcomes the binary distinction of developed and developing states.

In Chapter 14, Michael Hoel and Aart De Zeeuw explore the notion that international climate agreements can target technology development rather than emissions abatement. They point out that, for sufficiently low abatement costs, many countries might undertake significant emission reductions even without an international agreement on emissions reduction. They show that if some countries value emission reductions, a coalition of countries will undertake research and development to lower abatement costs, leading to technology gains that reduce the emissions from both cooperative and non-cooperative countries. The implication is that international agreements might benefit from a focus on technology development.

Placing a price on carbon emissions is popular among economists, but there are concerns that international trade will undermine the effectiveness of carbon pricing, the competitiveness of domestic industries, and domestic political support. In Chapter 15, Aaron Cosbey and Carolyn Fischer explore the potential for border carbon adjustments to lower some of these barriers. A border carbon adjustment is applied to traded products that seek to make their prices in destination markets reflect the costs they would have incurred had they been regulated under the domestic market's greenhouse gas emissions regime. They explore the complexities of border carbon adjustments, including meeting international legal obligations while remaining environmentally effective and administratively feasible. They also provide one permutation of such a regime.

In Chapter 16 Håkon Sælen asks when enforcement mechanisms are necessary and sufficient for effective international environmental agreements. Focusing on decentralized sanctions that are carried out by individual parties at a cost to themselves and the defector (e.g., trade sanctions), he develops an agent-based model to investigate the conditions under which meaningful international cooperation is likely in the presence of strong reciprocity – a predisposition to cooperate with others, and to punish those who violate the norms of cooperation. The results correspond with rational choice theory's argument that both participation and compliance must be enforced, but they also indicate that a modest number of countries behaving as strong reciprocators can induce broader cooperation.

The European Union Emissions Trading System (EU ETS) is the first international cap-and-trade system to target industry. In Chapter 17, Jon Birger Skjærseth provides a review of the development, implementation and consequences of the EU ETS and identifies potentially important insights from the experience. He finds that the EU ETS has considerable potential as a climate policy instrument, but that it has proven vulnerable to market changes and volatility, and political constraints impede solutions to supply–demand imbalances.

Since the introduction of the EU Emissions Trading System, several complementary trading initiatives have emerged around the world. Jørgen Wettestad and Torbjørg Jevnaker, in the final chapter, explore the ambition and progress of integrating the national and regional trading schemes to create a global carbon market. They review the efforts of the EU, particularly the European Commission, and highlight the limited possibilities for integration. The chapter concludes that linking trading schemes is not an easy fix for the stalled global climate regime.

The climate change problem is daunting, but the contents of this volume suggest that there is reason to be hopeful. It is clear that countries are increasingly engaged on the topic of managing climate change. And if it is true that countries can collectively benefit from mitigating climate change, it must be possible to design a set of rules that can make participation in such an agreement beneficial. The hope, and the driving force behind creating this book, is that such a treaty will be created in time to thwart the most costly effects of a changing climate.

We would like to thank the authors and the anonymous referees for their hard work and timely contributions to this volume, and we greatly appreciate the support provided by Louisa Earls and Charlotte Russell at Routledge. We wish to extend special thanks to CICEP at the Center for International Climate and Environmental Research – Oslo (CICERO) and Appalachian State University for providing support for a 2012 workshop at the University of Oslo that brought together most of the contributors for a few days to present and discuss early versions of their work. Lastly, we wish to thank our friends and families for their support.

Notes

1 This subsection draw extensively on Hovi, Skodvin, and Aakre (2013).
2 Statement by Minister Kent, issued by Environment Canada, 12 December 2011. Available from: www.ec.gc.ca/default.asp?lang=En&n=FFE36B6D-1&news=6B040 14B-54FC-4739-B22C-F9CD9A840800 (accessed: 9 June 2013).

3 A third element in the Durban Platform is an institutional framework for a Green Climate Fund. This fund would provide a means for developed countries to fund adaptation to climate change in developing countries. While the goal is to raise US$100 billion per year from 2020, it remains unclear what the sources of this funding will be. Both public and private sources have been proposed, but the funding issue remains unsettled.

References

Aakre, S. and J. Hovi (2010). Emission trading: Participation enforcement determines the need for compliance enforcement, *European Union Politics* 11 (3): 427–45.

Barrett, S. (1994). Self-enforcing international environmental agreements, *Oxford Economic Papers* 46: 878–94.

——(1999). Montreal versus Kyoto. International cooperation and the global environment. In I. Kaul, I. Grunberg and M.A. Stern (eds.), *Global Public Goods. International Cooperation in the 21st Century*. Oxford: Oxford University Press.

——(2008) Climate treaties and the imperative of enforcement, *Oxford Review of Economic Policy* 24: 239–58.

Barrett, S. and R. Stavins (2006). Increasing participation and compliance in international climate change agreements, *International Environmental Agreements* 3: 349–36.

Breitmeier, H., O.R. Young, and M. Zürn (2006). *Analyzing International Environmental Regimes. From Case Study to Database*. Cambridge, USA: MIT Press.

Downs, G.W., D.M. Rocke and P.N. Barsoom (1996). Is the good news about compliance good news about cooperation? *International Organization* 50: 379–406.

Heltberg, R., P. Bennett Siegel and S. Lau Jorgensen (2009). Addressing human vulnerability to climate change: Toward a 'no-regrets' approach. *Global Environmental Change* 19(1): 89–99.

Hovi, J., T. Skodvin and S. Aakre (2013). Can climate change negotiations succeed? In *Politics and Governance* 1(2): 138–50.

Hovi J., D.F. Sprinz and A. Underdal (2009). Implementing long-term climate policy: Time inconsistency, domestic politics, international anarchy, *Global Environmental Politics* 9(3): 20–39.

UN General Assembly (1992). United Nations Framework Convention on Climate Change. FCCC/INFORMAL/84, http://unfccc.int/resource/docs/convkp/conveng.pdf (accessed: August 2013).

Weitzman, Martin (2011). Fat-tailed uncertainty in the economics of climate change. *Review of Environmental Economic Policy* 5(2): 275–92.

Part I

Conflict

Barriers to a new agreement

1 Observations from the climate negotiations in Durban, South Africa[1]

Steffen Kallbekken

On Monday December 14 2009 I was sitting inside the plenary negotiations hall at the Bella Centre in Copenhagen. I was attending the 15th Conference of the Parties (COP15) to the United Nations Framework Convention on Climate Change (UNFCCC). This was the eye of the storm. The world had come to Copenhagen: delegations from more than 190 nations, represented by 119 heads of state, with 5,000 journalists attending. Adding to this an unprecedented number of observers, there were in total more than 30,000 people attending. This made COP15 one of the largest summits in world history. For a few crucial days, the attention of the world was focused on the conference in Copenhagen.

By December 14 time was starting to run out as negotiators were far from reaching agreement on a draft text to presents to their heads of state, who would start arriving in a couple of days. You would expect hectic activity inside the plenary hall at such a critical juncture. But it was quiet inside the eye of the storm. The podium, where the COP president should be sitting, chairing the debate, was empty. Ahead of me, inside the massive hall where the delegates should have been seated behind their country name plates, were hundreds of empty chairs.

The negotiations had stalled. Delegates from the G77 and China refused to resume the talks. They claimed the developed countries were trying to escape their obligation to reduce their emissions by introducing a new draft agreement. By introducing a draft not originating from the process, the presidency had failed to respect the rules of the game.

This anecdote is a reflection of a broader picture. We are running out of time if we are to "prevent dangerous anthropogenic interference with the climate system" (UN 1992). The pace of emissions growth is far outstripping the pace of progress at the UNFCCC negotiations. The UNFCCC process is hampered by a range of institutional barriers, and it is allowing itself to be thus hampered.

Actual climate negotiations appear nothing like the rational, organized and stylized discussions in our academic papers. Our abstractions are necessary and useful in order to allow the analysis to penetrate the fog of more or less irrelevant details. Consequently, researchers abstract away from a lot of frustrating – and sometimes interesting – details that take place within the walls of the negotiation rooms. This first chapter provides an inside view of the UNFCCC negotiations, and therefore a useful reality check for the remaining chapters.

Many of the reasons why the UNFCCC process is unlikely to produce an ambitious global agreement can be analyzed and understood without in-depth data or observation. Some other challenges can perhaps only be fully understood by attending the negotiations and seeing how they actually play out. Observing the negotiations reveals some barriers that reduce the prospects of reaching an ambitious and effective agreement, barriers that have taken specific forms that are not inherent to any global climate negotiations. They are, rather, institutional barriers produced by the history and culture of these specific UNFCCC climate negotiations.

I will not focus on the challenges inherent to any global climate negotiations, such as those discussed by, for instance, Barrett (2006) and Victor (2011), but on specific institutional barriers that have emerged and become entrenched within the UNFCCC process. I will consider two particular barriers that it might be possible to overcome: the firewall between developed and developing countries, and the unwillingness to relate to initiatives external to the UNFCCC process.

The end of history: freezing the definition of developed and developing countries

> Historical responsibility is not, and never can be, a dynamic concept.
>
> Saudi Arabian Negotiator at COP17 in Durban

As Francis Fukuyama proclaimed the end of history in 1989, you might be surprised to hear that it ended in 1992. You might be more surprised to learn that climate negotiators take this claim most seriously: effort sharing in international climate agreements should be based on historical responsibility, and history only extends as far as 1992. Furthermore, effort sharing means that developed countries should take the lead and reduce their emissions, while developing countries should not take on any binding commitments. The definition of who is developed and who is developing is based on the world in 1992, and should not be revised. At least, this is the world according to some parties.

A brief review of the history of the UNFCCC effort sharing principles will help make it clear how this situation emerged, why it is so problematic, and how it can potentially be remedied.

The CBDR/RC-principle

The United Nations Framework Convention on Climate Change, adopted in Rio in 1992, states that "the Parties should protect the climate system for the benefit of present and future generations of humankind, on the basis of equity and in accordance with their common but differentiated responsibilities and respective capabilities." To any neutral observer not involved in the negotiations, this would probably seem a reasonable principle. The problem is how the principle, usually referred to by its acronym CBDR, came to be put into operation, and how this was cemented over time.

The sentence following on from the one cited above states that "the developed country Parties should take the lead in combating climate change and the adverse effects thereof" (UN 1992). Annex I to the convention contains a list that defines which countries are "developed country parties." All other countries are by default defined as developing country parties (sometimes referred to as non-Annex I countries). The convention imposes commitments on the Annex I parties, including that they shall "adopt national policies and take corresponding measures on the mitigation of climate change." Importantly, the objective of the Annex I mitigation effort is to "demonstrate that developed countries are taking the lead in modifying longer-term trends in anthropogenic emissions."

At COP1 in Berlin in 1995, the division into developed and developing countries was taken one step further with the adoption of the Berlin mandate. The mandate acknowledges the fact that "the largest share of historical and current global emissions of greenhouse gases has originated in developed countries" (UNFCCC 1995). The mandate agrees to begin a process which will aim to "set quantified limitation and reduction objectives within specified time-frames" for Annex I parties. Crucially, the mandate then adds that the process will "not introduce any new commitments for Parties not included in Annex I"(UNFCCC 1995).

The Berlin mandate was the start of a process that resulted in the adoption of the Kyoto Protocol at COP3 in 1997. In the Kyoto Protocol the Annex I parties agreed to "quantified emission limitation or reduction commitments," while non-Annex I parties did not take on similar commitments. This cemented the creation of a sharp division between developed and developing country parties in the UNFCCC process.

This brief history illustrates how, over time, the CBDR principle has come to mean that developed countries should take on binding commitments to reduce their emissions, while developing countries should not. Furthermore, almost all attempts to revise the list of Annex I countries, and thereby redefine who is developed and who is developing, have been blocked.

Acknowledging that developing countries need to prioritize their domestic social and economic development, and that developed countries must take the lead in reducing emissions, should not be a problematic idea. What is problematic is the division into two groups of countries, where one group has commitments whereas the other does not, and the inflexibility with respect to defining who is a developed country party, and who is a developing country party.

Unchanging definitions in a changing world

The world has changed significantly in twenty years. By 2010 developing countries were responsible for 60 percent of global emission of CO_2 from fossil fuels and cement (GCP 2012). Any emissions path consistent with reaching the 2°C target must by mathematical necessity include emissions reductions in all major economies. Rogelj *et al.* (2011), for instance, estimate that in order to have a likely (greater than 66 percent) chance of limiting warming to 2°C, global emissions should peak between 2010 and 2020, and be reduced to 45 percent (35-

55 percent) below 1990 levels by 2050. Emissions reductions of this scale are not possible without the active participation of developing countries. This does not mean that developing countries must necessarily implement and finance domestic mitigation measures. The mathematical necessity only dictates that some of the mitigation measures need to be *implemented* in developing countries.

The principle that those countries defined as developed countries should reduce their emissions first, while some developing countries are allowed to prioritize domestic social and economic development, seems morally flawed for those cases where non-Annex I countries are far wealthier than Annex I countries. The most recent comparable data from the World Bank shows that 69 non-Annex I countries are richer than Ukraine, which is the poorest Annex I country (GNI per capita, Atlas method). There are also 46 non-Annex I countries that are richer than Belarus, and 44 that are richer than Bulgaria. Qatar, Kuwait, Singapore and South Korea are among the richest countries in the world, yet they are defined as developing countries.

In terms of emissions per capita, 7 of the 10 countries with the highest per capita emissions are non-Annex I countries (mostly countries with significant oil and gas extraction). In terms of absolute emissions, China, a non-Annex I country, has the highest emissions, currently contributing around 28 percent of global CO_2 emissions (GCP 2012).

Figure 1.1 shows a plot of GNI per capita (logarithmic scale) against emissions per capita. The figure shows that there is significant overlap between Annex I and non-Annex I countries in terms of both income and emissions per capita. The figure highlights how flawed the idea is that the world can be separated into two distinct groups with widely different obligations. However, the figure also shows that on the whole most non-Annex I countries have significantly lower income than most Annex I countries, and the same holds for emissions per capita, making some form of differentiation of obligations reasonable.

Vested interests in keeping the CBDR

The way the CBDR principle is currently interpreted effectively establishes what is known as the "firewall" between developed and developing countries: the first group is expected to take on commitment in UNFCCC agreements, while the second is not. This interpretation of the principle is often invoked and rarely disputed.

There is no doubt that retaining the CBDR principle is a key strategic target for the developing countries in the negotiations. A senior Indian negotiator explained to a journalist at the Rio+20 summit that the CBDR will be the country's principal negotiating weapon in the years to come, and is quoted saying that "we can use this [the CBDR] to fend off almost any demand from the West" (Chaudhuri 2012).

The pressure from developing countries to retain the CBDR principle in any future negotiating text was seen clearly at COP18 in Doha. The previous year, at COP17 in Durban, many observers argued that the key achievement was that the parties agreed to a process for negotiating a new global climate agreement by

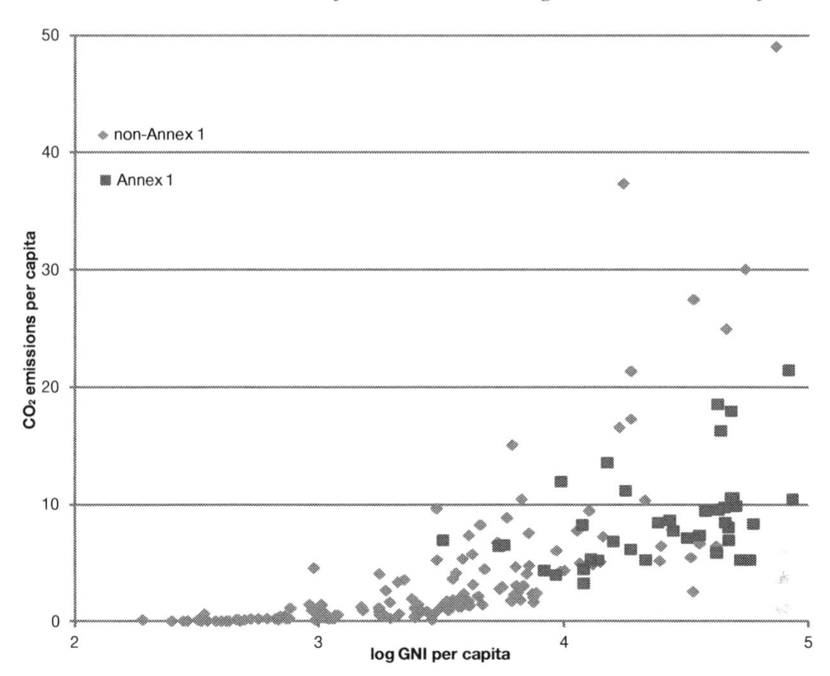

Figure 1.1 GNI per capita and CO_2 emissions per capita for Annex I countries (squares) and non-Annex I countries (diamonds) in 2008

2015 without any reference to the CBDR principle (see for instance Bodansky 2011). Importantly, the Durban Platform instead states that the new agreement should be "applicable to all parties" (UNFCCC 2011).

During COP18 attempts by developing country parties to make a more explicit reference to the CBDR principle within the Durban Platform process dominated parts of the negotiations. In the end the developing countries succeeded in including the words "[the process] shall be guided by the principles of the Convention" into the decision text on advancing the Durban Platform (UNFCCC draft decision CP.18). This is significant as the "principles of the Convention" include the CBDR principle. At the same time the debate might seem superfluous as the platform text itself refers to the principles of the convention several times, which only highlights how contentious the issue has become.

The CBDR principle is at the same time used by the USA to justify inaction. In 1997 the US Senate adopted the Byrd-Hagel resolution, stating that the US should not be a signatory to any climate agreement which would "mandate new commitments … for the Annex I Parties, unless the protocol or other agreement also mandates new specific scheduled commitments to limit or reduce greenhouse gas emissions for Developing Country Parties." Given the current state of climate policy in the US, it is unlikely the country would agree to ambitious commitments in any agreement even if it included commitments by developing country parties, but the lack of commitments by China and other significant economies still plays a role in justifying the lacking US willingness to commit.

There is every reason to believe that the debate over the CBDR principle will be at the heart of the negotiations until COP21 in Paris in 2015, which is supposed to conclude with a decision to adopt a new global climate agreement. If this is not to result in the whole process derailing, or in an ineffective agreement with (weak) commitments for Annex I countries and none for the other parties, how can the issue of equity be approached?

Tear down this firewall (by reinterpreting the CBDR principle)

Equity-based effort sharing is not as such a problematic concept. In fact, most countries subscribe to some or other equity-based principle that respects social and economic differences between countries. The CBDR principle itself, if interpreted without the burden of twenty years of negotiations history, seems to offer a reasonable approach. The problem lies in how the principle was first made operational at COP1, and since then has been interpreted in a very specific and narrow manner.

The primary problem with making the CBDR principle operational is that it is neither sufficiently dynamic nor broad enough. It needs to be more dynamic with respect to countries' responsibilities and capabilities. It also needs to be broader with respect to its conception of effort sharing. A dichotomous world where to commit or not to commit is the only question limits the scope of potential agreements, and ignores that there are multiple dimensions along which mitigation efforts can potentially be defined. One way to differentiate commitments, ruled out by the interpretation of the CBDR, is to allow different types of commitments for different groups of countries.

One possible way forward is to argue for a reinterpretation of the principles of the convention. The preamble to the convention acknowledges "that the global nature of climate change calls for the widest possible cooperation by all countries ... in accordance with their common but differentiated responsibilities and respective capabilities and their social and economic conditions" (UN 1992). Three key concepts are left out of today's established interpretation of the principle: that the nature of the problem calls for "the widest possible cooperation by all countries," and that this should happen in accordance with their "respective capabilities and their social and economic conditions." These concepts align well with the need for a broader and more dynamic approach to effort sharing.

A broader and more dynamic conception of effort sharing

There are numerous aspects to mitigation, including reporting emissions, implementing mitigation measures, providing financial support for mitigation, and taking on emissions targets of different types and levels of ambition. If "effort sharing" is to reflect the total scope of international climate policy, and not only mitigation, it should also include adaptation, climate finance, technology cooperation, capacity building, and the costs of climate change impacts (loss and damage).

It is possible to argue that "the widest possible cooperation" should mean not only global participation, but also the broadest possible conception of what constitutes *effort*, i.e., which policies and measures should be included – and counted towards a country's effort.

The salient point is that different countries have different "lines in the sand" they will not cross. A broader conception of which efforts are to be shared, combined with greater flexibility with respect to which types of commitments are accepted, would give negotiators more levers to pull in the negotiations: Brazil might not accept a market mechanism for reduced deforestation, but might be willing to commit to internationally verified emissions cuts. Norway would probably not accept an emission reductions commitment defined solely in terms of domestic mitigation efforts, but would accept a (more ambitious) commitment if international emissions trading is allowed.

National interests will not change simply by redefining the term "effort." However, a narrow conception of effort makes it more difficult to reach agreement because ideological, cultural and economic interests vary substantially across countries. A type of effort acceptable to one country is not necessarily acceptable to another: cutting emissions by 10 percent through a cap-and-trade scheme might be unacceptable to a country, while at the same time pledging a sum of money to support renewable energy investments and energy efficiency improvements – which would reduce emissions by 10 percent – might be acceptable to the same country.

Respective capabilities and social and economic conditions are largely neglected elements of the CBDR/RC principle. Placing greater emphasis on countries' capability to reduce emissions, adapt to climate change, raise climate finance, and develop and deploy new technologies could open up a much more dynamic approach to effort sharing than the current focus only on historical responsibility. Capability is a dynamic concept, and it can be defined with respect to a range of relevant factors (see for instance Metz *et al.* 2002).

Trying to reinterpret the principles of the convention in a broader and more flexible manner could potentially help move the negotiations substantially forward. Any attempt to ignore the principles of the convention in the new (2015) climate agreement would be met with fierce opposition, but a thorough discussion on what the principles mean might offer a road of less resistance. Allowing a broader range of commitments would, however, produce new challenges, such as how to allow greater flexibility while retaining stringency and a high level of ambition; how to compare effort involved in complying with different types of commitments; and how to account for non-mitigation efforts in effort sharing.

The unwillingness to relate to external initiatives

> Impatience with the proceedings does not justify seeking solutions in fora outside the UNFCCC.
>
> Indian negotiator at the intersessional meeting in Bonn 2009 (TWN 2009)

Despite slow progress and frequent stalemates, the parties to the UNFCCC process have remained remarkably loyal to it, and at the same time unwilling to bring initiatives from other fora into the process.

At COP15 the Danish presidency was spectacularly unsuccessful at introducing ideas from a parallel discussion on climate policy into the UNFCCC negotiations. The so-called Greenland Dialogue was a series of informal roundtables on climate policy with a specially invited group of climate and environment ministers from both developing and developed countries (Meilstrup 2010). The roundtables were initiated by Connie Hedegaard (later to become COP15 president), and were used to discuss and test ideas for a potential agreement at COP15. In the build-up to the negotiations, the presidency prepared a compromise text through bilateral consultations with leaders from both developed and developing countries. The initial draft was revised based on the input from these consultations.

This might seem an effective approach to try to identify a possible win set before engaging in cumbersome full-scale negotiations. However, when this proposal, referred to as "the Danish text," was leaked via a newspaper during the negotiations, the reception was anything but welcoming. China stressed that "the only legitimate basis" for a decision at COP15 was an outcome based on the process itself, and that the presidency could not "put forward text from the sky" (IISD 2009). The Sudanese negotiator who represented G77 and China put it less diplomatically: "It has become clear that the Danish presidency – in the most undemocratic fashion – is advancing the interests of the developed countries at the expense of the balance of obligations between developed and developing countries" (Black 2009).

The impact of the leakage upon the fragile trust between developing and developed countries was highly adverse. Yvo de Boer, then executive secretary of the UNFCCC, wrote to his staff that "the Danish letter presented at an informal meeting a week before the COP destroyed two years of effort in one fell swoop" (Meilstrup 2010: 129).[2]

Closed to initiatives external to the UNFCCC process

There is a strong focus on ensuring an inclusive, transparent and party-driven process, particularly after the events at COP15. In practice this means that it is not possible to introduce draft decisions from other fora, or build on the achievements of other climate-related agreements. For instance, at the intersessional in Bonn in 2009, the US argued that the UNFCCC process was not moving ahead at sufficient speed, and that discussions at alternate fora should be advanced. Ideas from these fora should then be brought to the UNFCCC. The response from developing countries was overwhelmingly negative (perhaps not surprising given the USA's continued unwillingness to agree to binding commitments). Saudi Arabia, for instance, responded that "the UNFCCC is the only negotiating body and the meetings of the G8, Major Economies Forum etc. are not binding on the negotiations" (TWN 2009).

Given the many problems associated with the UNFCCC process, why try to reform it? Why not focus on potentially more effective approaches, such as sectoral approaches, bilateral agreements or supply side agreements? One key reason is that most countries, and especially frontrunners like the EU countries, are politically committed to the UNFCCC process. This could change, but the frontrunners are not likely to champion any other alternative initiative in the short to medium term. Too much political capital has been committed to the process, and no other process could have the same global political legitimacy. This is not to say that no other initiatives are being advanced. A regime complex is already emerging (Keohane and Victor 2010). To some extent these other policy initiatives concern issues outside the defined scope of the UNFCCC process, such as emissions from international shipping and aviation (e.g., within the IMO and ICAO), or short-lived climate forcers (e.g., the black carbon component of the Climate and Clean Air Coalition). Other initiatives are directly complementary to the process and relate to formal initiatives within the UNFCCC, such as the various REDD+ activities. To a lesser extent there are alternative fora for discussing international climate policy, such as the G20 or the Major Economies Forum on Energy and Climate Change, but none of these have attempted to bring forward effective and comprehensive agreements. Not yet at least.

We are currently in a situation where a regime complex is emerging, in which the frontrunners are still committed to the UNFCCC process, and that process is reluctant to engage with the other initiatives. This prompts the pertinent question: what reforms can be made to the UNFCCC process to make sure that the interaction between the UNFCCC process and the emerging regime complex is a constructive one?

Complementary initiatives

One possible starting point is to ask why the UNFCCC process, already broad in scope and global in participation, should relate to other initiatives. The answer is, first, because we need to acknowledge that climate policy can never exist in a political vacuum. Climate policies overlap and interact with green growth strategies, energy policy, transport policy, health, education, and a number of other policy domains. To succeed, climate policy needs to effectively influence key policies within all these other domains. This links to the second and main reason: the benefits relating to climate policy in isolation (impacts avoided) will probably not be sufficient to support an ambitious climate agreement. In order to succeed with ambitious international climate policies, they need to be aligned with positive incentives within other policy domains. For instance, climate policies need to provide benefits in terms of improved health, improved energy security, job creation, food security, or technological innovation. Only by ensuring that such co-benefits are realized does it become realistic to mobilize industry, voters and NGOs to a sufficient extent to make ambitious climate policies politically feasible. This is why the UNFCCC process, if it is to succeed in creating or facilitating ambitious climate policies, needs to relate much more directly to climate policy initiatives stemming from fora external to the UNFCCC process.

A large number of potentially complementary initiatives already exist. These initiatives include the phasing out of fossil fuel subsidies (e.g., the G20), addressing short-lived climate forcers (e.g., the Climate and Clean Air Coalition), and developing and deploying energy efficient technologies (e.g., Sustainable Energy for All). In terms of actors, the various initiatives involve the private sector, other non-state actors (such as cities), other agreements (e.g., the Montreal Protocol), other international institutions (e.g., the IMO and ICAO), and other UN-based initiatives (e.g., Sustainable Energy for All).

Blok *et al.* (2012) analyzed 21 major initiatives, all independent of the UNFCCC process, that could help bridge the emissions gap.[3] Together these initiatives are estimated to have the potential to reduce emissions by around 10 Gt CO_2e by 2020 (compared to annual global CO_2 emissions of around 35 Gt). The initiatives include green financial institutions, major cities initiatives, energy efficiency measures, boosting renewable energy, phasing out subsidies for fossil fuels, reducing deforestation, and enhancing measures to reduce air pollutants. Many of these initiatives also have substantial co-benefits. As an example, introducing efficient cook stoves in rural areas would reduce emissions of greenhouse gases (CO_2 and black carbon), improve local air quality, and reduce the pressure on forests.

The UNFCCC as facilitator in the regime complex

With a strong trust deficit between developed and developing countries, and with many negotiators highly loyal to the UNFCCC process, how might it be possible to ensure a more open and inclusive process?

First of all, it is important that the UNFCCC does not co-opt other policy initiatives, as this could jeopardize their effectiveness. In other words, it's important that well-functioning policy initiatives do not become trapped by the politics and institutional challenges of the UNFCCC. At the same time, the UNFCCC can potentially contribute to the success of other initiatives by giving them visibility, ensuring greater transparency (for example, by inclusion in national communications or other reports to the UNFCCC), and lending them some of the legitimacy that a UN process brings. Furthermore, there is a potential for the UNFCCC process to help scale up smaller initiatives by providing funding (for instance, through the Green Climate Fund), or by giving them greater political attention.

An example that illustrates both the potential pitfalls and benefits of relating external initiatives more closely to the UNFCCC process is the history of international efforts to reduce emissions from deforestation and forest degradation (REDD) since COP11 in 2005. Linking to the UNFCCC offers the potential to deliver substantial funding through carbon markets and increased transparency and legitimacy, but it can also slow progress and lead to less effective mechanisms through mission creep (the adoption of multiple non-climate objectives). Bilateral REDD initiatives are currently moving ahead at a faster pace than the negotiations on the topic are progressing, yet they seek to complement the UNFCCC process, in part by adopting the same principles (Angelsen *et al.* 2012).

A "soft" first step towards greater openness to, and inclusion of, external initiatives might be to use the UNFCCC to facilitate an exchange of information about successful climate policies, in particular policies with substantial co-benefits. Several negotiators have called for a forum for sharing experiences with climate policies, and informing other parties about their own efforts. This might also serve to foster more rapid progress. Underdal *et al.* (2012) argue that "in a setting in which at least some parties adopt strategies of conditional cooperation, a move by one party to contribute [more] can, in fortunate circumstances, cause a chain of positive and mutually reinforcing responses." To the extent that parties make conditional commitments, a more continuous process of discussion and knowledge sharing about national policies could potentially lead to more rapid progress than a system where large agreements are negotiated every 10–15 years. A more continuous process offers more frequent opportunities for conditionalities to be met, and will perhaps trigger new commitments more frequently and rapidly.

In some cases the UNFCCC could function as a coordinating device. There are, for instance, many emissions trading schemes in operation or under planning, including in China, the EU, New Zealand, South Korea, and in parts of the USA (RGGI and WCI). Linking emissions trading schemes would yield large efficiency gains. While direct linkage is highly demanding politically, legally and technically, indirect linkage is far more feasible in the short term (Tuerk *et al.* 2009), and would realize most of the efficiency gains of direct linkage (Anger 2008). The UNFCCC could facilitate indirect linkage, for instance by ensuring that the new market-based mechanism (UNFCCC 2011: Article E.83) can function as an offset mechanism accepted by all existing and emerging emissions trading schemes. The Clean Development Mechanism (CDM) to some extent fulfils this function today, but is limited both by high transaction costs and by legal restrictions on its use.

Another idea acknowledges as its starting point that the UNFCCC is unlikely to deliver a very ambitious agreement (the law of the least ambitious is more likely to prevail, Underdal 1980). The argument is that parties should not aim for the UNFCCC process to be the frontrunner in the international climate regime, but rather use the UNFCCC process to establish "minimum commitments." These minimum commitments can then be supplemented by commitments under initiatives complementary to the UNFCCC process, and where the UNFCCC process provides incentives for parties to take on such additional commitments. For developing countries, for instance, the incentives for adopting additional commitments could be linked to support through the Green Climate Fund, or related to technology transfers.

Concluding remarks

I have attempted to identify some approaches that could help break down the firewall between developed and developing countries, and make the UNFCCC process more open to external initiatives. There is a strong political link between the two issues. An important reason why many countries do not want the UNFCCC process to relate to external initiatives is that these are not governed by the CBDR principle, but by other principles that give them less power.

If the CBDR principle could be reinterpreted to place greater emphasis on "the widest possible cooperation" as well as "respective capabilities," it might be possible to approach effort sharing in a more comprehensive and dynamic way. A key question is: which efforts are to count? Allowing a broader range of commitments (policies and measures) to count towards a country's effort would open the door to acknowledging efforts made under external initiatives, without necessarily requiring that these efforts have to be negotiated under the UNFCCC. Broadening the range of measures and policies to be negotiated would provide "more levers to pull" and could reduce some of the inflexibilities of the current process.

There is a caveat to the main argument in this chapter: the challenging UNFCCC process should not be allowed to co-opt other policy initiatives, as this would risk its problems spilling over into other and more successful processes. The implication is that the UNFCCC process should relate to (learn from, facilitate, or help coordinate) other policy initiatives, without these initiatives being brought into the UNFCCC negotiations. Alongside the vital research on alternatives to the UNFCCC process, an applied and potentially very constructive research agenda would be to explore the potential and limitations of the role of the UNFCCC process in facilitating a more effective climate regime complex.

Notes

1 The chapter concerns observations from the negotiations. As there is no official record of the negotiations, the chapter relies in large part on my personal notes from the negotiations, as well as grey literature (for instance the reporting by the Earth Negotiations Bulletin). Where no reference follows a quotation, the quotation is taken from my personal notes. The observations also rely extensively on my official, but non-published, report from a Chatham House Rules workshop on the climate negotiations, organized by the Nordic Working Group for Global Climate Negotiations (NOAK) in 2012.

2 Yvo de Boer refers to a letter sent from Denmark to Russia, China and the USA containing an alternate version of "the Danish text" (but his assessment of the effect concerns the same series of events as described above).

3 "Emissions gap" is the term used to describe the difference between expected emissions in 2020 and the emissions that would be consistent with being on a pathway towards reaching the 2°C target.

References

Angelsen, A., M. Brockhaus, W.D. Sunderlin, L.V. Verchot and T. Dokken (2012). *Analysing REDD+: Challenges and choices.* Bogor, Indonesia: Center for International Forestry Research (CIFOR).

Anger, N. (2008). Emissions trading beyond Europe: Linking schemes in a post-Kyoto world. *Energy Economics* 30(4), 2028–49.

Barrett, S. (2006). *Environment and Statecraft: The Strategy of Environmental Treaty-Making.* New York: Oxford University Press.

Black, R. (2009). *Copenhagen climate summit negotiations 'suspended'.* Available at: http://news.bbc.co.uk/2/hi/science/nature/8411898.stm (accessed: August 2013).

Blok, K., N. Höhne, K. v. d. Leun and N. Harrison (2012). Bridging the greenhouse-gas emissions gap. *Nature Climate Change* 2, 471–74.

Bodansky, D. (2011). *The Durban Platform Negotiations: Goals and Options.* Harvard Project on Climate Agreements Policy Brief.

Chaudhuri, P. P. (2012). 4-word principle India's big green victory at Rio. *Hindustan Times,* 21/6/2012.

GCP (2012). *Global Carbon Budget 2012.* Available at: www.globalcarbonproject.org/carbonbudget/12/files/CarbonBudget2012.pdf.

IISD (2009). *Earth Negotations Bulletin COP15* no. 10. International Institute for Sustainable Development.

Keohane, R.O. and D.G. Victor. (2010). *The regime complex for climate change.* Harvard Project on Climate Agreements Discussion Paper. Harvard Kennedy School.

Meilstrup, P. (2010). The runaway summit: the background story of the Danish presidency of COP15, the UN Climate Change Conference. In H. Hvidt and H. Mouritzen (eds.), *Danish Foreign Policy Yearbook 2010.* Copenhagen: Danish Institute for International Studies.

Metz, B., M. Berk, M. den Elzen, B. de Vries and D. van Vuuren (2002). Towards an equitable global climate change regime: compatibility with Article 2 of the Climate Change Convention and the link with sustainable development, *Climate Policy* 2(2), 211–30.

Rogelj, J., W. Hare, J. Lowe *et al.* (2011). Emission pathways consistent with a 2 °C global temperature limit, *Nature Climate Change* 1, 413–18.

Tuerk, A., M. Mehling, C. Flachsland and W. Sterk (2009). Linking carbon markets: concepts, case studies and pathways, *Climate Policy* 9(4), 341–57.

TWN (2009). Bonn News Update 10. Third World Network.

UN (1992). United Nations Framework Convention on Climate Change. New York: United Nations.

Underdal, A. (1980). *The Politics of International Fisheries Managements: The Case of the Northeast Atlantic.* Oslo: Scandinavian University Press.

Underdal, A., J. Hovi, S. Kallbekken and Tora Skodvin (2012). Can conditional commitments break the climate change negotiations deadlock? *International Political Science Review* 33(4), 475–93.

UNFCCC (1995). Report of the Conference of the Parties on its first session, held at Berlin from 28 March to 7 April 1995 – Addendum ("Berlin mandate"). Bonn: United Nations Framework Convention on Climate Change.

——(2011). Resolution 1/CP.17 Establishment of an Ad Hoc Working Group on the Durban Platform for Enhanced Action. Bonn: United Nations Framework Convention on Climate Change.

Victor, D.G. (2011). *Global Warming Gridlock: Creating More Effective Strategies for Protecting the Planet,* Cambridge, UK: Cambridge University Press.

2 Does fairness matter in international environmental governance?

Creating an effective and equitable climate regime

Oran R. Young

Introduction

Provisions pertaining to fairness or equity are prominent features of international environmental agreements. The 1992 United Nations Framework Convention on Climate Change (UNFCCC), for instance, states that "[t]he Parties should protect the climate system for the benefit of present and future generations of humankind, on the basis of equity" (UNFCCC 1992 Art. 3.1). The 1997 Kyoto Protocol to the UNFCCC seeks to put this provision into operation through the adoption of the principle of common but differentiated responsibilities, which distinguishes between developed (UNFCCC Annex I) and developing (UNFCCC non-Annex I) states with regard to targets and timetables for reducing greenhouse gas (GHG) emissions, and through the inclusion of a commitment to devising measures to protect particularly vulnerable societies, such as small island developing states (Kyoto Protocol 1997). Even the 2009 Copenhagen Accord, widely regarded as a political statement cobbled together following a failure of the Conference of the Parties to the UNFCCC to agree on the terms of a new legally binding agreement on climate change, states clearly that "[w]e shall … on the basis of equity and in the context of sustainable development, enhance our long-term cooperative action to control climate change" (Copenhagen Accord 2009).

There is nothing unusual about the case of climate change in these terms. Similar statements appear in agreements relating to other large-scale environmental problems. For their part, practitioners devote considerable attention to the challenge of crafting provisions that the parties to such agreements are prepared to accept as fair or equitable. Yet social scientists, and especially those whose expertise is rooted in political science and economics, regularly argue that the influence of such considerations is marginal in the creation of such agreements. A typical formulation comes from David Victor, a prominent commentator on issues relating to climate change, who asserts that "for most states most of the time, the decision making process is mainly a selfish one. Consequently, there exists very little evidence that fairness exerts a strong influence on international policy decisions" (Victor 1996: 3; see also Victor 2011).

A particularly striking feature of this view is that it is shared for the most part by analysts who espouse neo-realist, neo-liberal, and constructivist perspectives on international relations. Neo-realists think that what happens in the world is all about power in the material or structural sense. Neo-liberals focus on the interests of the relevant actors and direct attention to processes of bargaining or negotiation among self-interested actors seeking deals that maximize net benefits for themselves. Constructivists direct attention to the role of ideas in the form of discourses or conceptual frameworks. But this does not mean they expect considerations of fairness or equity to loom large in these terms. The influential idea of Gramscian hegemony suggests that prevailing discourses generally reflect the views or preferences of the rich and powerful who are not prone to espousing considerations of fairness or equity. Thus, none of the main streams of thinking about international relations suggests that considerations of fairness or equity constitute a major force to be reckoned with in creating and administering international environmental governance systems that prove effective in solving problems.

What should we make of this gap? Is the emphasis on fairness or equity in international environmental negotiations largely a matter of rhetoric, in the sense that drafters of documents like the UNFCCC and the Copenhagen Accord confine references to fairness and equity to general statements that have little operational content? Are negotiators cynical representatives of self-interested states trying to deceive others – including members of the attentive public – about what really goes on in efforts to address issues like climate change? Are they deceiving themselves, trying to make believe that their efforts involve something more than the pursuit of hard-core national interests and that they themselves are enlightened individuals who care about more elevated concerns like the plight of the world's poor, the destruction of small island developing states, or the fate of Earth's climate system?

In this chapter, I offer an alternative perspective on this complex of issues. Using the case of climate change as a source of concrete examples, I argue that there is an identifiable and significant class of environmental issues in which we should expect those endeavoring to solve problems to take considerations of fairness or equity seriously. This argument does not depend on assumptions about the influence of altruism or the existence of a strong sense of community that produces deep feelings of social solidarity among the members of international society. Rather, I take the view that states have instrumental reasons to think hard about matters of fairness or equity when it comes to addressing a well-defined class of environmental problems. The problem of climate change, I argue, belongs to this class.

My argument proceeds as follows. I begin by exploring conditions that characterize situations in which it makes sense to think hard about matters of fairness or equity. I then turn to climate change as an illustrative case and develop what I regard as a compelling way to address issues of fairness or equity in the effort to create an effective climate regime. In the concluding section, I draw on the preceding analysis to distill some recommendations arising from this way of thinking that should appeal to those responsible for efforts to reach agreement on ways forward in coming to terms with the problem of climate change in the post-

Copenhagen era. Although these recommendations may seem unrealistic to those laboring in the trenches of the climate negotiations, I argue that the actual impacts of climate change may alter this equation sooner rather than later.

When, why, and how does fairness matter?

I see little evidence that the members of international society or the agents who act on their behalf in international negotiations are so deeply socialized regarding considerations of fairness that they respond to such concerns out of some sense of social obligation. Rather, my argument is that such considerations become important in instrumental terms in situations that exhibit a cluster of identifiable features. The most prominent of these features are: (i) the inability of key states to pressure or coerce others into accepting their preferred solutions: (ii) the limited usefulness of utilitarian calculations like various forms of cost/benefit analysis; and (iii) the need to foster buy-in or a sense of legitimacy regarding the solutions adopted in order to achieve effective implementation and compliance over time with the prescriptions embedded in any agreements reached. Together, these conditions produce situations in which states have good reasons to pay attention to considerations of fairness or equity.

Consider these conditions as they arise in the case of climate change. No individual state or cluster of states (e.g., the European Union) has the capacity to impose a solution to the climate problem on other key members of international society. Between them, the United States and China account for well over 40 percent of current emissions of greenhouse gases (GHGs).[1] Any solution to the problem would require active and committed participation on the part of both states. There is no prospect whatsoever that either of these countries can impose its preferred solution on the other. Other major players in this realm include the European Union, Brazil, India, Indonesia, Japan, and Russia. Together, these eight actors account for well over two-thirds of current emissions of GHGs. An agreement among them to launch a forceful attack on the problem of climate change would go a long way toward coming to terms with this problem, though it is already late in the day to be launching a serious effort to tackle the problem in such a way as to fulfill the UNFCCC's goal of avoiding "dangerous anthropogenic interference with the climate system" (UNFCCC Art. 2). But the politics of this situation make it clear that none of these actors is in a position to operate as a hegemon forcing the others to accept the terms of a climate regime they do not like. An effective climate regime must be acceptable to all the key players, at least in the sense that these players feel that they are getting a fair deal.

By itself, this condition may simply provide an incentive to engage in vigorous bargaining in search of a mutually agreeable deal. But add to this the fact that ordinary calculations of benefits and costs are not particularly helpful in addressing the problem of climate change. This is not just a matter of debates about discount rates or the relative merits of different policy instruments, such as cap-and-trade systems, carbon taxes, and conventional command-and-control regulations (Aldy and Stavins 2007). The limitations go deeper than that; they center on fundamental problems in calculating both the likely costs of allowing climate change to run its course and the costs of taking the steps required to avoid or at least to ameliorate the impacts of climate change. Given the continuing debate about the biophysical effects of projected concentrations of carbon dioxide and other GHGs in Earth's atmosphere, not to mention the role of short-lived climate pollutants like carbon soot, calculations regarding the magnitude of the costs associated with climate change are little more than guesswork. It is not surprising, therefore, that projections regarding these costs range from almost nothing to tens of trillions of dollars. And this is before we even come to questions about the incidence of these costs and about the prospect that climate change could produce winners as well as losers. Nor are we in much better shape when it comes to estimating the costs of tackling the climate problem effectively. Optimists reckon that we could solve the problem by investing 1–2 percent of gross world product (GWP) in the near term and that the impact of this effort on the growth of GWP by the year 2050 would be slight (Stern *et al.* 2007; Stern 2008). Responsible pessimists (as distinct from climate deniers), on the other hand, anticipate that serious efforts to tackle the climate problem, including adaptation as well as mitigation, would run into hundreds of billions or even trillions of dollars and constitute a significant drag on conventional economic growth (Nordhaus 2007). In the face of uncertainties of this magnitude, the way forward is to adopt a kind of Rawlsian perspective, seeking a solution to the climate problem that will seem fair to all parties concerned regardless of whose predictions about costs turn out to be correct (Rawls 1971).[2]

A third condition centers on matters of implementation and compliance; it reinforces the other two in providing good reasons to pay attention to fairness or equity. Agreement on the terms of an international regime means little unless those subject to these terms make a good faith effort to implement them and to comply with the requirements and prohibitions that form the nucleus of the agreement. This is especially true in a case like climate change where implementation will require substantial changes from business as usual and commitments that will remain in place over an indefinite period of time. Implementation often emerges as the Achilles heel of innovative policies, even at the national level where governments have some capacity to enforce the rules or to provide key actors with incentives to alter their behavior in ways needed to conform to new arrangements. The challenge of implementation is considerably greater at the international level, where the member states themselves are the subjects of the rules and there is no higher authority capable of compelling or persuading these actors to abide by the rules.

As the case of the Montreal Protocol Multilateral Fund in the ozone regime suggests, there are cases in which incentive mechanisms operating at the international level can help to bring on board actors that are willing to comply but lack the capacity to do so in the absence of outside assistance. But this is an unusual – perhaps even unique – case. Providing a similar mechanism in the case of climate change would require the international community to mobilize tens to hundreds of billions of dollars per year over a considerable period of time. The 2009 Copenhagen Accord, which envisions mobilizing $100 billion a year by 2020 to address mitigation and adaptation, offers some basis for hope regarding this issue. But no one expects the UNFCCC Annex I countries to provide "new and additional" funding of the sort needed to address issues of mitigation and adaptation effectively during the foreseeable future. This means that there is no way to ensure the implementation over time of the provisions of the climate regime unless those subject to its provisions feel a clear sense of obligation or responsibility to fulfill the commitments they make in the agreements. What is needed to meet this condition is a sense on the part of individual members that both the basic terms of the agreement and the process of agreeing on their content are fair.

What are the implications of this line of reasoning for solving problems like climate change? If you cannot force actors to accept and abide by the provisions of international agreements, and if you cannot rely on utilitarian reasoning to convince them that the benefits of compliance outweigh the costs, the only way forward is to devise governance systems that members feel obligated to abide by because they were developed through procedures regarded as fair and because their major provisions add up to what they can accept as an equitable deal. This is not a matter of the deontological status of principles of fairness in international society. My argument is that states will have instrumental reasons to take considerations of fairness or equity seriously in cases where these conditions obtain.

Nor is the view I have presented an attempt to refute the arguments of those who assert that climate change is a "diabolical problem" (Steffen 2011). I do not dispute the claim that we may fail to arrive at the terms of a governance system that can solve the problem of climate change. Rather, I want to drive home the proposition that there is an important class of problems arising at the international/global level with regard to which taking considerations of fairness or equity seriously is necessary (though not always sufficient) to arrive at effective solutions. Climate change is a prominent member of this class of problems.

What are the elements of a fair or equitable climate regime?

There is a vast literature on ways to think about fairness or equity in efforts to address the problem of climate change. It is not the purpose of this chapter to review, much less to critique, the approaches that various authors have proposed (Gardiner *et al.* 2010). But one fundamental distinction in this realm separates approaches that treat the problem as a matter of governing the use of a common pool resource from approaches that deal with it as a matter of finding a way to supply a public or collective good. In this analysis, I focus on the common pool resource branch of this literature. The chapter by Bernauer, Gampfer, and Landis in this volume directs more attention to the public goods branch of the literature.

Allocating total allowable emissions

Suppose we are able to reach agreement regarding the concentration of GHGs in the atmosphere (measured in CO_2e) that is compatible with the UNFCCC's commitment to avoiding "dangerous anthropogenic interference with the climate system." We could then adopt a budget approach, calculating total allowable emissions (TAEs) on an annual basis and developing a procedure for ratcheting down TAEs from one year to the next. The next step would be to devise a mechanism governing the allocation of these TAEs each year and computing reductions in allowable emissions from one year to the next in such a way as to avoid exceeding the agreed upon target, whether it is set at 350, 450, 550ppmv or some other level of concentration of CO_2e in Earth's atmosphere.

Is there a way to address this budget problem that key players would regard as constituting a fair deal and agree to abide by in practice? This challenge belongs to a well-known class of issues involving the allocation of use rights or permits to use valuable natural resources (e.g., fish stocks) that are commons in the sense that they are not subject, at least initially, to recognized use rights (Raymond 2003). The valuable resource in this case is the capacity of Earth's atmosphere to serve as a repository for wastes in the form of GHG emissions, so long as concentrations in the atmosphere do not reach a level expected to cause "dangerous anthropogenic interference with the climate system." If we start from the premise that Earth's atmosphere is a part of the common heritage of humankind, and if we assume that all humans are equal with regard to their entitlement to use this common heritage – as seems only reasonable – then we should adopt as a point of departure the proposition that emissions permits should be allocated in a manner that is proportional to human population and that annual reductions in TAEs should be imposed in the form of equal across-the-board percentage cuts in these initial allocations. Assuming that emissions permits (much like catch shares in some marine fisheries) are tradable and that "major" producers of goods and services worldwide are required to obtain permits to cover their GHG emissions – as also seems reasonable – the eventual distribution of TAEs in any given year will depart substantially from the initial allocation.[3] If the resultant trading system is global in scope and not subject to severe market failures, this arrangement should yield an efficient distribution of TAEs in the sense that permits end up in the hands of those who value them most highly.[4]

This is, in essence, a way to avoid the pitfall of the tragedy of the commons on a global scale. As is true of all such situations, we can proceed either by adopting a strategy of privatization, granting permits to individuals, or by opting for a strategy featuring a role for public authorities, entrusting permits to governments acting on behalf of their citizens (Hardin 1968; Baden and Noonan 1998; Ostrom *et al.* 2002). Privatization in this case would amount to allocating an equal number of emissions permits to each human being alive on the planet in any given year. Turning to public authorities, by contrast, would entail relying on governments to act as trustees, managing emissions permits calculated on a per capita basis for all their citizens. Those familiar with the literature on the tragedy of the commons will recognize these options as examples of standard responses to this problem in a variety of settings.[5]

There are good reasons to prefer the public option over the private option in this case, even if we focus on efficiency and favor situations featuring a large role for markets. Privatizing TAEs would lead to a situation fraught with problems. Many individuals would be either unable to understand the nature of these assets or be motivated to liquidate them immediately in order to mobilize the resources required to address urgent needs (e.g., malnutrition). They would become easy marks for unscrupulous manipulators with money to purchase their assets at cut rates. Such an arrangement would also make it impossible – or at least very difficult – for governments to use these assets to mobilize the resources needed to create social infrastructure and to provide the incentives needed to energize green development. Of course, many governments are corrupt. There is no denying the prospect that powerful individuals in control of some governments would find ways to exploit publicly held or managed emissions permits without taking the actions needed to improve the lot of the poor or to provide a broader basis for sustainable development (Acemoglu and Robinson 2012). Nevertheless, the public option would open up opportunities for governments to take the steps needed to use these assets in a progressive and socially beneficial manner.

Some will dismiss this approach to allocating emissions permits as politically naïve or highly unrealistic, regardless of its virtues from the perspective of fairness or equity. But let me suggest some reasons why this conclusion may be premature. No doubt, powerful actors that are used to getting their way in today's world (e.g., major oil companies) would advocate some sort of grandfathering procedure that would provide them with generous shares of the TAEs free of charge.[6] And experience in such areas as allocating individual transferable quotas (ITQs) or catch shares in marine fisheries makes it clear that political considerations often play an influential role regarding allocation mechanisms (Raymond 2003).

Nevertheless, there are important considerations that cut the other way in the case of climate change. Once integrated into the system, producers of all sorts of goods and services would incorporate the cost of emissions permits into their cost functions and pass this cost along to their customers. So long as the system is global, no subset of producers would enjoy an unfair advantage in this regard. For their part, consumers are used to the fact that prices for most products include the cost of waste disposal. They should not find it hard to understand the unfairness of a system in which producers of many products must internalize the cost of waste disposal, while producers of products generating GHG emissions are not required to include the cost of disposing of these wastes in their cost functions. So long as they are able to pass these costs along to their customers, corporations will not suffer. So long as the resultant price increase for most products is not extreme, consumers should not only be able to understand the logic of this arrangement but also be prepared to acknowledge its essential fairness. Even though this approach seems radical in light of the current negotiations regarding climate change, the impacts of climate change – especially in the form of sharp climate shocks – may change this assessment in short order.

As is the case with all allocation mechanisms linked to population, it will be necessary to establish a procedure to prevent countries being rewarded for taking steps to promote population growth as a means of upping their shares of the TAEs. But it is easy to exaggerate this concern. While individuals might be motivated to enlarge their families as a way to receive additional emissions permits, it is hard to envision situations in which such considerations would become important drivers of national policies regarding population. Moreover, to the extent that national governments manage shares of the TAEs in trust for their citizens, it should be possible for the global climate regime to avoid addressing a suite of complex problems that arise when permits are distributed to individuals, including the use of permits as collateral for loans, the treatment of permits in connection with divorce, the transfer of permits to heirs, and so forth.[7]

Equally important is the fact that this allocation system has the potential to solve or at least alleviate several problems that have proven to be major sticking points in negotiations aimed at strengthening the climate regime. So long as no significant producer of goods and services is allowed to emit GHGs without obtaining valid permits to cover its emissions, this system would lead to a substantial transfer of wealth from emitters to governments acting as trustees for their citizens in managing shares of the TAEs. The scale of these transfers would depend on a number of things, including the number of permits distributed at the outset, the schedule for ratcheting down TAEs over time, the ability of emitters to come up with inexpensive ways to reduce their emissions, and the rate of economic growth in various parts of the world. But it is reasonable to suppose that, at least at the outset, the transfers in question would amount to several hundred billion dollars per year, and that the bulk of these transfers would go to populous developing countries where a large proportion of the world's poor people reside. To the extent that this is the case, it would obviate the highly contentious debates about financial transfers that have emerged as a sticking point in efforts to reach agreement on ways to finance mitigation and adaptation under the terms of the climate regime.

Note also that this system can circumvent the problems of selecting a discount rate and finding an explicit solution to the problem of fairness to future generations. Presented as a conventional matter of establishing a (social) rate of discount with regard to future benefits and costs, the discount rate problem appears intractable. On the other hand, if we can agree on a TAE for year one and on a schedule for ratcheting down TAEs in subsequent years, there will be no need to tackle the problem of discounting directly. Of course, this procedure does require the adoption of a method of addressing the problem of time. If we agree on an overall target (e.g., 350 or 450ppmv) and on a date by which we must stabilize atmospheric concentrations at this level, we will have agreed de facto to a position regarding the costs of anthropogenic disturbance of the climate system over time. Nevertheless, this way of addressing the problem may prove less contentious than setting a precise discount rate, which is essential to the cost/benefit approach, or reaching some mutually agreeable position regarding the rights of members of future generations, which is a hard proposition to sell in most political arenas.

Taking the broader context seriously

Many observers have noted that we should not become so obsessed with the climate problem that we lose track of all other issues on the international policy agenda. The climate problem is one of the great issues of our times. A failure to address this problem effectively could lead in the coming decades to serious losses of social welfare, not least among residents of less developed countries already struggling with multiple threats to human well-being. Still, climate is not the only game in town. Among those living on $1–2 a day and struggling to deal with urgent problems of health, education, and welfare, it is understandable that the climate problem may not seem paramount (Collier 2007). If you are preoccupied with the challenges of staying alive today and tomorrow, the disruptions that climate change may cause a few decades from now or by the end of this century are likely to seem distant and even irrelevant. For shorthand purposes, we can think of this problem as the issue of trade-offs between addressing the climate problem and focusing our energy on efforts to fulfill the Millennium Development Goals (MDGs) or some similar set of welfare goals. To the extent that efforts to strengthen the climate regime marginalize efforts to fulfill the MDGs, we would be faced with major concerns about fairness (Stern 2009).

How serious is this problem, and is it likely to emerge as a deal breaker in efforts to negotiate agreement on the terms of a strengthened climate regime? This is not an easy question to answer. Much depends on the cost of dealing with climate change and the extent to which we can structure efforts to address climate change in such a way as to address the MDGs or similar goals simultaneously. If optimists like Stern are correct, we should be able to solve the climate problem and make progress toward fulfilling the MDGs at the same time. A particularly interesting prospect in these terms is that efforts to move toward the creation of the green economies needed to reduce GHG emissions may be organized in such a fashion that they also lead to job creation and poverty reduction. There is no basis for expecting that this will happen automatically. But if those who think about these issues in terms of the idea of resource productivity are right, there may well be opportunities to take steps that lead simultaneously to reductions in GHG emissions and the growth of employment opportunities in emerging industries in such a way as to alleviate poverty (von Weizsäcker *et al.* 2010). If we can succeed in devising strategies of this sort, what may seem initially like an unfair trade-off between climate change and the MDGs could turn into a synergistic interaction producing win-win outcomes.

An important point in terms of this analysis arises from the fact that the allocation mechanism discussed in the preceding subsection should produce significant financial transfers, and that a sizable proportion of these funds would be entrusted to the governments of countries with large populations of poor people (e.g., Bangladesh, India, Nigeria). There is no doubt that corruption or bad policies would divert some of these funds to unproductive uses. But such governments could use these funds as a means of financing various forms of low carbon development that would simultaneously address the climate problem and lift sizable numbers of people out of poverty. This argument assumes that there is merit in the collection of ideas dealing with what is loosely described as a green economy. There are legitimate questions about the feasibility of making progress in this realm. To the extent that the obstacles to moving in this direction are financial and institutional rather than behavioral and technological, however, the strategy for dealing with the climate problem under consideration here could make a real difference.

What does climate fairness demand?

If we take the argument of the preceding section seriously, what advice can we offer negotiators endeavoring to strengthen the existing climate regime or to replace it with a more effective alternative? First and foremost, this analysis suggests approaching the problem of climate change as a matter of managing the use of a global commons under conditions in which demand for the resource (construed as the capacity of the climate system to absorb additional GHGs without experiencing disruptive change) is outstripping the supply. This means we should address this challenge through the perspective we have come to think of as governing the commons (Ostrom 1990).

From the perspective of fairness or equity, the crucial step is to adopt a budget approach, calculating TAEs on an annual basis, focusing on methods for allocating emissions permits, and devising a procedure for ratcheting down the TAEs year over year. I have made the case that fairness requires allocating TAEs on a basis proportional to Earth's human population and designating national governments to act as trustees in managing the emissions permits allocated to their citizens. Of course, some governments are corrupt, and it would be naïve to assume that such a management regime would operate smoothly in all cases. Nonetheless, the alternatives – some form of private property rights for 7–9 billion people or some sort of common property regime of the sort we can observe in some small-scale societies – seem unworkable in this case.

This system for allocating TAEs has a number of virtues. Assuming that all "major" producers are required to obtain permits to cover their emissions, the outcome of such a system would be substantial transfers of financial resources to developing countries, thereby circumventing many current debates about funding mechanisms as elements of the climate regime. It would also provide resources to allow developing countries to take vigorous steps toward introducing green economies, thereby alleviating some of the current concerns about unfairness arising from the fact that today's advanced industrial systems were able to rely on cheap energy derived from fossil fuels to drive economic growth.

In my judgment, the arrangements described in the preceding paragraphs should go far toward alleviating the issues often described as addressing intergenerational equity and leveling playing fields. To the extent that we create a workable procedure to calculate TAEs on an annual basis (including the year-over-year reductions required to fulfill the objective of Article 2 of the UNFCCC) and to allocate them fairly, we can bypass hot-button issues regarding reparations for past injustices and setting aside funds specifically to protect the interests of members of future generations. This approach should also level the playing field for major producers worldwide with regard to terms of trade, and provide funds needed by developing countries to move toward the creation of green economies without passing through all the stages of energy intensive development that today's advanced industrial economies have experienced. While all these concerns about fairness are entirely understandable, it seems likely that approaching them through the procedures described in this chapter has a better chance of succeeding than an approach that features an effort to tackle these contentious issues directly.

On the other hand, I see no way to avoid the issues described in the preceding section under the heading of taking the broader context seriously. Not only does fairness demand a concerted effort to address the issues grouped under the heading of the MDGs; it also seems highly unlikely that major developing countries, like Brazil, China, India, Indonesia, Nigeria, and South Africa, will accept meaningful obligations regarding reductions in GHG emissions that deflect attention from efforts to fulfill the MDGs or siphon off resources required to pursue these goals vigorously. This means that the only way to make real progress in addressing the climate problem is to come up with procedures that link the two sets of issues – solving the climate problem and fulfilling the MDGs or similar goals – together in a synergistic fashion, so that policy makers will not approach this set of concerns as a matter of making painful trade-offs between solving the climate problem and promoting sustainable development. This will not be easy. But I believe it can be done, especially if resources are available to allow developing countries to move directly toward the establishment of green economies.

Some will regard the approach I have described as politically naïve, especially in the context of the current climate negotiations. But is this really the case? As a number of observers have pointed out, solving the twin problems of climate change and poverty reduction constitutes the grand challenge of our era (Stern 2009). Our lives and those of our children and grandchildren may literally depend on achieving success in efforts to meet this challenge. One or more severe climate shocks may make options that appear far-fetched today seem far more realistic tomorrow. There is no guarantee that we will find a solution to the climate problem. But I am convinced that provisions that take fairness or equity seriously will form a necessary element of any effective response to this problem.

Acknowledgements

I am grateful to Thomas Bernauer, Jon Hovi, Tora Skodvin, and one anonymous reviewer for helpful comments on an earlier version of this chapter.

Notes

1 The latest calculations from the Global Carbon Project put China at 28 percent, the US at 16 percent, the EU 27 at 11 percent, and India at 7 percent (Global Carbon Project 2012).

2 On the effects of uncertainty, see also the chapter by Barrett and Dannenberg in this volume.

3 The assumption here is that it will be easier to control the actions of producers of goods and services that generate GHG emissions than those of a much larger number of

consumers of these goods and services. Still, this formula begs the question of what constitutes a "major" producer. There are various ways to address this issue, none of which is obviously superior to the others. But the key point here is that the requirement regarding permits should extend to all producers in the relevant categories, wherever they are located.

4 The German Advisory Council on Global Change arrives at somewhat similar conclusions using what it calls a "budget approach" (WBGU 2009).

5 I am assuming here that international society is not a setting conducive to the evolution of the sort of informal rules among appropriators of common-pool resources discussed by analysts who study "governing the commons" in small-scale societies (Ostrom 1990; Ostrom *et al.* 2002).

6 This has already happened in the case of the European Emissions Trading Scheme. The sorts of proposals discussed in the US often feature similar arrangements, although the cap-and-trade system that California has put in place has provisions for auctioning permits.

7 For a discussion of these issues as they arise in the case of the Alaska limited-entry permit system for fisheries under state management, see Young 1983.

References

Acemoglu, D. and J. Robinson (2012). *Why Nations Fail: The Origins of Power, Prosperity, and Poverty.* New York: Crown Business.

Aldy, J.E. and R.N. Stavins (eds.) (2007). *Architectures for Agreement: Addressing Global Climate Change in the Post-Kyoto World.* Cambridge, UK: Cambridge University Press.

Baden, J.A., and D.S. Noonan (eds.) (1998). *Managing the Commons.* Bloomington: Indiana University Press.

Collier, P. (2007). *The Bottom Billion: Why the Poorest Countries Are Failing and What Can Be Done about It.* Oxford: Oxford University Press.

Gardiner, S., S. Caney, D. Jamieson and H. Shue (eds.) (2010). *Climate Ethics: Essential Readings.* Oxford: Oxford University Press.

Global Carbon Project (2012). *The Carbon Budget 2012*, available at www.globalcarbonproject.org/carbonbudget (accessed: August 2013).

Hardin, G. (1968). The tragedy of the commons. *Science* 162, 1243–48.

Kyoto Protocol (1997). *Kyoto Protocol to the United Nations Framework Convention on Climate Change.* Available at: http://unfccc.int.

Nordhaus, W.D. (2007). A review of the Stern review on the economics of climate change. *Journal of Economic Literature* 45, 686–702.

Ostrom, E. (1990) *Governing the Commons: The Evolution of Institutions for Collective Action*, Cambridge, UK: Cambridge University Press.

Ostrom, E., T. Dietz, N. Dolsak, P. Stern, S. Stonich and E. Weber (eds.) (2002). *The Drama of the Commons.* Washington, DC: National Academies Press.

Rawls, J. (1971). *A Theory of Justice.* Cambridge: Harvard University Press.

Raymond, L. (2003). *Private Rights in Public Resources: Equity and Property Allocation in Market-Based Environmental Policy.* Washington, DC: Resources for the Future.

Steffen, W. (2011). Climate change: A truly complex and diabolical policy problem. In *Oxford Handbook of Climate Change and Society.* Oxford: Oxford University Press, pp. 21–37.

Stern, N. (2007). *The Economics of Climate Change: The Stern Review*, Cambridge, UK: Cambridge University Press.

——(2008). The economics of climate change. *American Economic Review Papers & Proceedings* 98(2), 1–37.

——(2009). *The Global Deal: Climate Change and the Creation of a New Era of Progress and Prosperity.* New York: Public Affairs.

UNFCCC (1992). *United Nations Framework Convention on Climate Change*. Available at: http://unfccc.int (accessed: August 2013).

Victor, D.G. (1996). The regulation of greenhouse gases: Does fairness matter? Unpublished paper.

——(2011). *Global Warming Gridlock: Creating More Effective Strategies for Protecting the Planet.* Cambridge, UK: Cambridge University Press.

von Weizsäcker, E., C. Hargroves, M.H. Smith and C. Desha (2010). *Factor Five: Transforming the Global Economy Through 80% Improvements in Resource Productivity.* London: Earthscan.

WBGU (German Advisory Council on Global Change) (2009). *Solving the Climate Dilemma: The Budget Approach.* Berlin: Special report, Secretariat WBGU.

Young, O.R. (1983). Fishing by permit: Restricted common property in practice. *Ocean Development and International Law* 13(2), 121–70.

3 Formation of climate agreements

The role of uncertainty and learning

Michael Finus and Pedro Pintassilgo

Introduction

Climate change is a major challenge to international cooperation, as emphasized for instance by the Stern Review and various IPCC reports (Stern 2006 and IPCC 2007) but also many others. One of the main problems of achieving cooperation under international climate agreements is free-riding. Countries have an incentive to adopt a non-cooperative behavior. Emission reduction constitutes a public good. No country can be excluded to benefit from the emission reduction of other countries. Moreover, by not contributing to emission reduction, a country saves on abatement cost. The literature on the formation of self-enforcing international environmental agreements (SEIEAs) studies the underlying incentive structure in detail, by considering various assumptions related to the behavior of countries and their objectives, the cost–benefit structure and many other economic features affecting the incentive structure of governments to join climate treaties, but also pointing to possibilities to mitigate the free-rider incentives. For surveys, see for instance Barrett (2003) and Finus (2003, 2008).

Uncertainty is also an important element in climate change and so determines the formation of international agreements. In fact, despite intensive research, there are still large uncertainties regarding the impact of greenhouse gases on the climate system and on caused environmental damages. In addition, predictions about abatement costs are difficult (IPCC 2007). These uncertainties may well have an impact on the governance of global climate change. For instance, the former US President George Bush used uncertainty (as an excuse?) as one argument for his decision to withdraw from the Kyoto Protocol. In a letter to senators, dated March 13, 2001, as quoted by Kolstad (2007), he wrote: "I oppose the Kyoto Protocol … we must be very careful not to take actions that could harm consumers. This is especially true given the incomplete state of scientific knowledge."

A recent strand of literature has analyzed the formation of international agreements on climate change in the context of uncertainty (Kolstad 2007; Kolstad and Ulph 2008, 2011; Na and Shin 1998). The main conclusion is that the "veil of uncertainty" is conducive to the success of international climate agreements. That is, in a model that captures the strategic interaction between countries in climate change, more information through learning can lead to worse outcomes. Outcomes are measured in terms of aggregate welfare (i.e., the sum of welfare levels in all countries) in the equilibria of a coalition formation game under various assumptions about the degree of learning. This result is certainly puzzling as it runs counter to the general wisdom that more/better information can never harm. This result is also somehow disturbing because it basically implies that all scientific research aimed at reducing the uncertainty around climate change would be counter-productive. This leads to three research questions which we would like to address in this chapter.

1 What are the driving forces that generate the "negative result" about learning?
2 How general is this negative result?
3 Can the problem be fixed?

In answering these questions, it is helpful to point out that the papers cited above use stylized models and make a couple of simplifying assumptions to derive their results. Hence, one route to addressing our questions could be to set up a model with "more realism." This route has been pursued in Dellink *et al.* (2008) and Dellink and Finus (2012), who use a calibrated climate change model with twelve world regions and determine stable coalitions based on a large set of Monte Carlo Simulations under various assumptions about the distribution of the key parameters. From their results, it appears that the negative impact of learning is less evident than in the purely theoretical papers. In fact, in most cases, full learning leads to better outcomes than partial or no learning. Another route is to modify and generalize previous theoretical models. This approach has been taken by Finus and Pintassilgo (2012, 2013) of which we summarize the main findings in a non-technical way in this chapter. The advantage of the theoretical approach is that driving forces can be isolated and one can analyze the generality of the previous results in a systematic way. Nevertheless, we are well aware of a couple of interesting extensions, which could add more realism to our model. These extensions are briefly discussed in the last section.

The model

Coalition formation is usually modeled using game theory – a branch of mathematics that studies the strategic interaction between decision makers ("players") by using various equilibrium concepts to predict the outcome of these interactions (Finus 2001). In the following, we first introduce the coalition formation game, then explain the three types of uncertainty, and finally lay out the three scenarios of learning.

Coalition formation game

International environmental agreements are typically "single agreements," meaning that countries decide either to join a treaty (in which case they are a member of the "coalition") or to abstain (in which case they act as singletons). Moreover, participation is voluntary and membership is open to all, i.e., a country can neither be forced into nor excluded from participation. Therefore, we model coalition formation as a two-stage open membership single coalition game.[1] In the first stage, players (in our context countries) decide whether to join an agreement (in our context a climate treaty) or remain an outsider as a singleton. In the second stage, players choose their policy levels (in our context abatement). The game is solved backward, assuming that strategies in each stage must form a Nash equilibrium, i.e., they are mutual best replies.

This simple game has also been called cartel formation game, with non-members called fringe players. It originates from the literature in industrial organization (d'Aspremont *et al.* 1983) and has been widely applied in this literature (for surveys see Bloch 2003; Yi 1997, 2003) but also in the literature on self-enforcing international environmental agreements (for surveys see Barrett 2003; Finus 2003).

In the first stage, players' membership decisions lead to a coalition structure, $K=\{S, 1_{n-m}\}$, which is a partition of players, with n being the total number of players, m the size of coalition S, $m \leq n$, 1_{n-m} denotes $n-m$ the singletons and N the set of players, $S \subseteq N$. Due to the simple structure of this coalition formation game, there can be at most one non-trivial coalition, with "non-trivial" referring to a coalition of at least two players. Hence, we can simply talk about coalition S with the understanding that all players that are not in S are singletons, i.e., single players.[2] Typically, we will denote a member of S by i and call it a signatory, and a non-member of S by j and call it a non-signatory.

In the second stage, given that some coalition S has formed in the first stage, players choose their abatement levels q_i. The decision is based on the following payoff function:

$$(1) \quad \Pi_i = B_i\left(\sum_{k=1}^{n} q_k\right) - C_i(q_i),\ i \in N$$

where $B_i(\bullet)$ is country i's concave benefit function (i.e., benefits increase at a constant or decreasing rate) from global abatement (in the form of reduced damages, e.g., measured against some business-as-usual-scenario), with global abatement being the sum of all abatement and $C_i(\bullet)$ its convex abatement cost function (i.e., costs increase at a constant or increasing rate) from individual abatement. The global public good nature of abatement is captured by the benefit function which depends on the sum of all abatement contributions. For a start, we assume that all functions and their parameters are common knowledge and introduce uncertainty later on.

Working backward, we assume that the optimal economic strategies in the second stage are the Nash equilibrium of the game between coalition S, with its m members, and the $n-m$ singletons. The equilibrium is derived by assuming that coalition members maximize the aggregate payoff of their coalition, whereas all singletons maximize their own payoff. That is, coalition members cooperate and the coalition internalizes the externality among its members. In contrast, singletons behave selfishly, ignoring the externality they impose on others. The simultaneous solution of these maximization problems leads to the equilibrium abatement levels of signatories $q_i^*(S)$ and of non-signatories $q_j^*(S)$. The abatement levels depend on coalition S. If, say, a non-member k joins coalition S such that $S \cup \{k\}$, then the equilibrium abatement level of coalition members will increase and those of non-members will decrease or remain constant, i.e., $q_i^*(S) < q_i^*(S \cup \{k\})$ and $q_j^*(S) \geq q_j^*(S \cup \{k\})$. It can be shown that total abatement will increase if a non-signatory k joins coalition S such that $S \cup \{k\}$ forms. Hence, non-members are better off under $S \cup \{k\}$ than under S, as benefits from total abatement will be higher and their abatement costs will remain constant or will drop. That is, non-members benefit from the greater cooperation of others, which explains the strong free-rider incentive, which typically shows up as only small coalitions being stable.

If equilibrium abatement levels of signatories $q_i^*(S)$ and of non-signatories $q_j^*(S)$ are inserted in the payoff function (see (1) above), in the second stage of the coalition formation game we derive equilibrium payoffs, which are denoted by $\Pi_{i \in S}^*(S)$ and $\Pi_{j \notin S}^*(S)$ respectively, given that coalition S has formed.

In the first stage, stable coalitions are determined by invoking the stability concept of internal and external stability, which is de facto a Nash equilibrium in membership strategies.

(2) Internal stability: $\Pi_i^*(S) \geq \Pi_i^*(S \setminus \{i\}) \; \forall \, i \in S$
(3) External stability: $\Pi_j^*(S) > \Pi_j^*(S \cup \{j\}) \; \forall \, j \notin S$.

That is, no signatory should have an incentive to leave coalition S to become a non-signatory and no non-signatory should have an incentive to join coalition S. In order to avoid knife-edge cases, we assume that if players are indifferent in choosing between joining coalition S or remaining outside of S, they will join the coalition. Coalitions which are internally and externally stable are called stable. In case there is more than one stable coalition, we apply the Pareto-dominance selection criterion. That is, we delete from our set of stable coalitions those stable coalitions where at least one player could be made better off and no player worse off by moving to another stable coalition.

Up to now, we have assumed the absence of transfers. However, given the assumption of joint welfare maximization of coalition members and the fact that we allow for asymmetric payoff functions, it is perceivable that coalition members share their total payoff $\Pi_S^* = \sum_{i \in S} \Pi_i^*(S)$ through transfers t_i such that the "corrected" payoffs are $\Pi_i^*(S) + t_i$ with $\sum_{i \in S} t_i = 0$. If t_i is positive, a coalition member receives transfers from other members and if it is negative a member makes a contribution to other members. The sum of transfers is zero, implying that there are no outside sources to pay for transfers.

In the case of transfers, many schemes are perceivable, which typically lead to different sets of stable coalitions. In order to avoid this sensitivity, we follow the concept of an almost ideal sharing scheme (AISS) proposed by Eyckmans and Finus (2009) with similar notions discussed, for instance, in Fuentes-Albero and Rubio (2009), McGinty (2007) and Weikard (2009). They basically argue that if, and only if,

(4) Potential internal stability: $\sum_{i \in S} \Pi_i^*(S) \geq \sum_{i \in S} \Pi_i^*(S \setminus \{i\})$

(5) Potential external stability: $\sum_{j \notin S} \Pi_j^*(S) > \sum_{j \notin S} \Pi_j^*(S \cup \{j\})$ for all $j \notin S$.

hold, then there exists a transfer system that makes S internally and externally stable. For instance, potential internal stability means that the total payoff of coalition S exceeds the sum of free-rider payoffs. In other words, the surplus $\sigma(S) = \sum_{i \in S} \Pi_i^*(S) - \sum_{i \in S} \Pi_i^*(S \setminus \{i\})$ is positive. Eyckmans and Finus (2009) show that such a transfer scheme is the almost ideal sharing scheme (AISS) which gives each coalition member his/her free-rider payoff $\Pi_i^*(S \setminus \{i\})$ plus a share of the surplus $\sigma(S)$. For our purposes, the following properties of AISS are important. First, among those coalitions that are potentially internally stable, the coalition with the highest aggregate welfare is stable, i.e., internally and externally stable. This explains the word "ideal." Second, this coalition with the highest aggregate welfare may not be the grand coalition because free-rider incentives may be too strong (i.e., the surplus is negative). This explains the word "almost." Finally, for the stable coalition that generates the highest aggregate welfare, its size will never be lower with transfers than it is without transfers, but most likely higher. That is, a smart transfer scheme like AISS will normally improve outcomes.

Three types of uncertainty

We now turn to the assumption about the uncertain parameters of the payoff functions, which are summarized in three types of uncertainty. Due to the complexity of coalition formation, the consideration is required of a particular payoff function, as well as the parameters that are uncertain and their distributions. In order to avoid the exclusive focus on the binary equilibrium choices "abate" or "not abate" in the second stage, as for instance in Kolstad (2007) and Kolstad and Ulph (2008), and to capture the first and second stage strategic effect from learning, we consider a strictly concave payoff function (used for instance by Barrett 2006; Na and Shin 1998; and many others) that is still simple enough to derive analytical results:

(6) $$\Pi_i = b_i \sum_{k=1}^{n} q_k - c_i \frac{q_i^2}{2}, i \in N, b_i > 0, c_i > 0$$

where b_i is a benefit parameter, $b_i \sum_{k=1}^{n} q_k$ is the benefit from global abatement, c_i is a cost parameter, and $c_i \frac{q_i^2}{2}$ is the abatement cost from individual abatement.

Generally, the benefit as well as the cost parameters could be uncertain. In this chapter, we focus on benefit parameters and report briefly on results obtained in Finus and Pintassilgo (2012) for the cost parameters used herein. Hence, we simplify the model by dividing payoffs by the cost parameter c_i and defining the benefit–cost ratio by $\gamma_i = b_i / c_i$, and hence the payoff function reads:

(7) $$\Pi_i = \gamma_i \sum_{k=1}^{n} q_k - \frac{q_i^2}{2}, i \in N, \gamma_i > 0.$$

Henceforth, we call γ_i the benefit parameter. If this parameter is uncertain, then it is represented by the random variable Γ_i, with associated distribution f_{Γ_i}.

For all three types of uncertainty, uncertainty is symmetric. That is, all players know as much or little about their own as about their fellow players' payoff functions. We first lay out the specific assumptions and then provide a wider interpretation. An overview is displayed in Table 3.1 and the formal description as well as generalizations are provided in Finus and Pintassilgo (2012, 2013). All results and assumption are exactly those in Finus and Pintassilgo (2012).

Type 1: uncertainty about the level of benefits

Uncertainty of type 1, which the authors call systematic uncertainty as it relates to a *common parameter*, is considered in Kolstad (2007) and Kolstad and Ulph (2008). All players have the same expectations ex-ante, and once uncertainty is resolved, all countries have the same benefit parameter ex-post, which we call symmetric realization. Important is that this type of uncertainty is de facto about the *level of the benefits* from global abatement.

Table 3.2 illustrates the implication with a simple example, which assumes only two players and a particular distribution of the benefit parameter, which takes two possible values with equal probability. Ex-post, all players have either a low

Table 3.1 Three types of uncertainty about the benefit parameters

Type of Uncertainty	*Interpretation of Parameters*	*Ex-ante Expectations of Parameters*	*Ex-post Realizations of Parameters*
1 pure uncertainty about the level of benefits	common	symmetric	symmetric
2 pure uncertainty about the distribution of benefits	individual	symmetric	asymmetric
3 simultaneous uncertainty about the level and the distribution of benefits	common and individual	symmetric	symmetric and asymmetric

Table 3.2 Ex-post realization: example

Type of Uncertainty	*Possible Ex-Post Realizations*	*Feature*
1 pure uncertainty about the level of benefits	(1,1), (2,2)	symmetric
2 pure uncertainty about the distribution of benefits	(1,2), (2,1)	asymmetric
3 simultaneous uncertainty about the level and the distribution of benefits	(1,1), (2,2), (1,2), (2,1)	symmetric or asymmetric

value of $\gamma_i = 1$ or a high value $\gamma_i = 2$, but both have the same value and hence, ex-post, the sum of marginal benefits is either $\sum_{i=1}^{2} \gamma_i = 2$ or $\sum_{i=1}^{2} \gamma_i = 4$. The ex-ante expectation is $E(\Gamma_i) = 1.5$ and $\sum_{i=1}^{2} E(\Gamma_i) = 3$. This can be viewed as an urn with tickets representing all possible values of the benefit parameter γ_i. One player is chosen randomly and draws a ticket; this value applies to all.

Type 2: uncertainty about the distribution of benefits

Uncertainty of type 2 is considered in Na and Shin (1998) and relates to individual parameters. Though expectations about the benefit parameters are again symmetric, their realizations are asymmetric among players. More specifically, assume that there are n different tickets γ_i in an urn with values from 1, 2, ..., n; each player selects a ticket without replacement. Hence the realization of the benefit parameter is asymmetric. Table 3.2 illustrates this for our simple example. If player 1 draws $\gamma_1 = 1$, then player 2 will have $\gamma_2 = 2$ and vice versa. This implies that the sum of marginal benefits is known ex-ante, $\sum_{i=1}^{2} E(\Gamma_i) = \sum_{i=1}^{2} \gamma_i = 3$ but not the individual marginal benefits, γ_i, with expected value $E(\Gamma_i) = 1.5$.

Different from Na and Shin (1998), we consider an arbitrary number of players and not only three players. We assume a discrete uniform distribution of the benefit parameters over all permutations of vector $(1, 2, ..., n)$. Consequently, the sum of marginal benefits is fixed and its variance is zero. Hence uncertainty is purely about the distribution of the benefits from global abatement as the level of global benefits is known and constant $\sum_{i=1}^{n} E(\Gamma_i) = \sum_{i=1}^{n} \gamma_i = \frac{n^2 + n}{2}$. Note that the benefit vector can be viewed as different shares of the global benefits from abatement. Because the realization of the benefit parameters are asymmetric ex-post, it will turn out that transfers are useful and make a difference to the outcome when players learn this value (which is the case under our scenario of full learning, which is explained below).

Type 3: uncertainty about the level and distribution of benefits

Uncertainty of type 3 is a combination of the previous two types of uncertainty and hence there is uncertainty about common and individual parameters. This translates in our setting into uncertainty about the level and distribution of the benefits from global abatement. This is captured by assuming that, for each player, the benefit parameter follows a discrete uniform distribution over the values of vector $(1, 2, ..., n)$, which, contrary to type 2, is independent from the benefit parameters of the other players. This can be viewed as players drawing tickets from an urn with different tickets with values 1, 2, ..., n, but with, not without, replacement. Hence the sum of marginal benefits is uncertain, with positive variance that is larger than for uncertainty of type 2, but smaller than for uncertainty of type 1. Again, Table 3.2, which shows that symmetric but also asymmetric vectors of the benefit parameter may come about ex-post, is useful for illustrative purposes. On average, the asymmetry across players is larger than for uncertainty of type 1 but smaller than for uncertainty of type 2.

Interpretation of the three uncertainty cases

All three types of uncertainty capture important aspects of the uncertainty surrounding climate change. There is much uncertainty about the absolute level of the benefits from reduced damages but also much debate about their regional distribution: which countries will be suffering more from climate change? Hence, uncertainty of type 3 represents the most comprehensive assumption, but type 1 and 2 are useful benchmarks in order to isolate effects.

Three Learning Scenarios

Following Kolstad and Ulph (2008), we assume risk-neutral agents as players are governments and not individuals, and distinguish three simple learning scenarios: (1) full learning, (2) partial learning, and (3) no learning. The impact for coalition formation is illustrated in Table 3.3.

Full Learning (abbreviated FL) can be considered as a benchmark case in which players learn about the true parameter values before taking the membership decision in the first stage. Hence uncertainty is fully resolved at the beginning of the game. This scenario could also be called certainty. For Partial Learning (abbreviated PL) it is assumed that players decide about membership under uncertainty but know that they will learn about the true parameter values before deciding upon abatement levels in the second stage. Hence the membership decision is based on expected payoffs, under the assumption that players will take the correct decision in the second stage. We could also call this scenario partial uncertainty. Finally, under No Learning (abbreviated NL) the abatement decision also has to be taken under uncertainty. That is, players derive their abatement strategies by maximizing their expected payoffs. The membership decisions are also taken based on expected payoffs, though these expected payoffs differ from those under partial learning, given that less information is available under no learning in the second stage. We could also call this scenario uncertainty.

Full learning is certainly an optimistic, and no learning a pessimistic, benchmark about the role of learning in the context of climate change. Partial learning approximates (because beliefs are not updated in a Bayesian sense) the fact that information becomes available over time. For instance, between the signature of

Table 3.3 Three scenarios of learning

Stage 1: Membership		
NL expected Π_i	PL expected Π_i	FL true Π_i
Stage 2: Abatement Decision		
NL expected γ_i	PL true γ_i	FL true γ_i

the Kyoto Protocol in 1997, its entry into force in 2005, and the target period of implementation 2008–12 (commitment period), more information has emerged, as documented by various updated issues of IPCC reports.

Results

Preliminaries

In this section, the main results of the coalition formation game are presented and discussed. Detailed derivations are found in Finus and Pintassilgo (2012). Note that for a sensible comparison across different scenarios, we measure aggregate welfare in terms of expected values, i.e., ex-ante. For FL, with no uncertainty, each possible realization is assumed to be equally likely and thus expected welfare is de facto an average welfare.

In order to understand the intuition of the subsequent results, it is useful to define what we call the first and second stage effect from learning. We say that the first stage effect from learning is positive (negative) if the size or the composition of the coalition leads to higher (lower) global welfare the more players learn. In our setting, a larger average coalition size usually results in higher global welfare. However, in some special cases it is not the average size of stable coalitions which matters but the composition of coalition members. For instance, consider a coalition S of size m and an asymmetric realization of the benefit parameter γ_i. The highest global payoff is obtained if the m countries are those with the highest γ_i -values: the higher the γ_i-values of members, the higher their implemented abatement level and hence global abatement. Finally, we say that the second stage effect from learning is positive (negative) if for a given generic coalition S, global welfare is higher (lower) the more players learn, which is the result of the choice of equilibrium abatement levels.

The idea is to separate effects in both stages though, for the overall result, effects in both stages matter. Three comments are in order. First, it is evident from Table 3.3 that partial and full learning are identical in the second stage. Second, due to backwards induction, the first stage effect cannot be completely isolated from what players do in the second stage. Nevertheless, thinking about first and second stage effects is useful when discussing the intuition of our results. Third, proofs use information about first and second stage effects. If a learning scenario performs (weakly) better with respect to both stages, it is straightforward to conclude that it performs (weakly) better overall. If effects in both stages go in opposite directions, the relative importance of effects needs to be weighted.

Main results

We now turn to our main results, which are summarized in three Propositions according to the three types of uncertainty.

Proposition I: uncertainty of type 1 (uncertainty about the level of benefits)
For uncertainty of type 1, under the full, partial, and no learning scenario, equilibrium expected total payoffs are ranked as follows: FL = PL > NL
Thus, if there is only uncertainty about the level of the benefits from global abatement, "learning is good" in terms of aggregate welfare.[3] A detailed analysis would reveal that the first stage effect from learning is neutral, the size of stable coalitions for all three scenarios of learning is the same, namely $m^* = 3$. So what matters is the second stage effect from learning, which is positive for full and partial learning compared to no learning. Under no learning, abatement levels are chosen based on expected benefit parameter values. If the realized benefit parameters are all high (low), total expected abatement is too high (low) compared to the true parameter values. In other words, under no learning, there is systematic over- or undershooting of abatement, which is costly and leads to lower aggregate welfare than under full and partial learning.

Proposition II: uncertainty of type 2 (uncertainty about the distribution of benefits)
For uncertainty of type 2, under the full, partial and no learning scenario, equilibrium expected total payoffs are ranked as follows:

No Transfers $\begin{cases} NL = PL > FL & \text{if} \quad n = 3 \\ NL > PL > FL & \text{if} \quad n \geq 4 \end{cases}$

Transfers $\begin{cases} FL = PL = NL & \text{if} \quad n = 3 \\ NL > FL = PL & \text{if} \quad 4 \leq n \leq 8 \\ NL > FL > PL & \text{if} \quad n = 9 \\ FL > NL > PL & \text{if} \quad n \geq 10 \end{cases}$

The main conclusion one can draw from Proposition II is that the more players learn, the lower will be global welfare if there are no transfers. This result can somehow be reversed with transfers. The intuition is the following.

First, assume no transfers. Consider the first stage of coalition formation in which players choose their membership. For uncertainty of type 2, players are ex-ante symmetric, though ex-post asymmetric. For partial and no learning this does not affect coalition formation compared to uncertainty of type 1 because players take their membership decisions based on expected payoffs which are symmetric (see Table 3.3). Hence $m^* = 3$. This does not apply to full learning. Members of a coalition receive different payoffs. As we assumed symmetric abatement cost function, cost-effectiveness requires that all coalition members contribute the same abatement level. However, members benefit to a different extent from this joint action as they have different benefit parameters γ_i. Those with high γ_i-values benefit more on average and those with low γ_i-values less on average. Thus the low γ_i-value countries have an incentive to leave the coalition. This implies smaller stable coalitions than under symmetry. Hence, $m^* = 1$ if $n = 3$ and $m^* = 2$ if $n \geq 4$. In other words, the first stage effect from learning is negative for full learning but neutral for partial learning. Hence, regarding the first stage and assuming no transfers, the ranking is $NL = PL > FL$.

Consider now the second stage of coalition formation in which players choose their abatement levels. Consider the simplest case where all players are singletons and hence all behave non-cooperatively. For payoff function (7) the first order conditions, implying that individual marginal benefits are equal to marginal costs, delivers $q_i^* = \gamma_i$ under full and partial learning and $q_i^* = E[\Gamma_i]$ under no learning. Let us simplify things even further and consider only two countries and the realized parameter values as listed in Table 3.2. Then under full and partial learning either $q_1^* = 1$ and $q_2^* = 2$ or $q_1^* = 2$ and $q_2^* = 1$ whereas under no learning $q_1^* = 1.5$ and $q_2^* = 1.5$. The average or total abatement level is the same under all scenarios of learning but the individual abatement levels are asymmetric under full and partial learning but symmetric under no learning. For symmetric abatement cost function, cost-effectiveness requires symmetry, and hence total abatement costs under full and partial learning are higher (and hence total welfare lower) than under no learning. Thus, in a strategic setting, more information can lead to worse outcomes overall. Specifically, the second stage effect from learning is negative.

In our context, this simple example illustrating the second stage effect of learning generalizes in the following sense. First, the same result holds if we consider not only two but any number of countries. Second, the result holds not only if all players are singletons but for any coalition structure where there is a coalition S with m members and $n - m$ single players. If, and only if, S is the grand coalition, then the second stage effect from learning is neutral. Hence, regarding the second stage, the ranking is $NL \geq PL = FL$, with strict inequality if the coalition is not the grand coalition.

Thus the overall result summarized in Proposition II, under the assumption of no transfers, with ranking $NL \geq PL = FL$ with strict inequality if $n \geq 4$, $PL > FL$ is (exclusively) due to the negative first stage effect from learning and $NL > PL$ is (exclusively) due to the negative second stage effect from learning ($n = 3$ is an exception because under PL and NL the grand coalition forms, in which case the second stage effect from learning is neutral, which explains $NL = PL$ in this particular case). Consequently, we can conclude that the ranking $NL > FL$ is due to a negative effect from learning in the first and second stage.

The remaining question is: what do transfers change? They do not change the choice of equilibrium abatement levels and hence cannot change the second stage effect from learning. So, regarding the second stage, the ranking $NL \geq PL = FL$ still holds, with strict inequality; if not, the grand coalition forms. However, transfers change the first stage effect from learning provided this is due to asymmetry. Hence transfers will make no difference to no and partial learning and consequently the ranking between these two scenarios will not change. We recall, regarding the first stage, this means $NL = PL$. This is different for full learning. First, transfers imply for full learning at least the same coalition size than for no and partial learning. Second, if asymmetries between players are pronounced enough, which increases with the number of players for the considered distribution (and hence results in Proposition II depend on n), even larger coalitions than under no and partial learning can be stable.[4] That is, the first stage effect from learning is neutral or even positive. According to Proposition II, the first stage effect from learning is strictly positive for $n \geq 9$, which explains the overall ranking $FL > PL$ if $n \geq 9$, and it is sufficiently strong for $n \geq 10$ to compensate the negative second stage effect from learning compared to no learning, which explains the overall ranking $FL > NL$ if $n \geq 10$. Thus asymmetry is a burden without transfers but becomes an asset if a smart transfer scheme is used to balance asymmetries. The intuition is that the relative gains from cooperation accruing to coalition members, compared to no cooperation, increases with asymmetry, making it more attractive to join the coalition than remain an outsider.

Taken together, we generalize the negative result of Na and Shin (1998) about the role of learning by considering more than three players and including the intermediate case of partial learning in the analysis. Even more important, we qualify their conclusion by considering transfers and showing that this conclusion can be reversed, at least for full learning.

Finally, as for uncertainty of type 2, so for uncertainty of type 3: players are ex-ante symmetric but ex-post asymmetric. The average degree of asymmetry ex-post is positive, therefore larger than for uncertainty of type 1, but smaller than for type 2. Not surprisingly, this improves upon the relative performance of full learning compared to uncertainty of type 2, but weakens it compared to type 1 if there are no transfers. With transfers, like in case 2, heterogeneity becomes an asset under full learning. Like for uncertainty of type 1, the second stage effect from learning is positive, which explains why partial learning always performs better than no learning and that this may even be true for full learning. Even without transfers, the first stage effect from learning under full learning can be positive compared to partial and no learning because not only the size but also the composition of coalition members matters.

Proposition III: uncertainty of type 3 (uncertainty about the level and distribution of benefits)

For uncertainty of type 3, under the full, partial, and no learning scenario, equilibrium expected total payoffs are ranked as follows:

$$\text{No Transfers} \quad \begin{cases} PL > NL > FL & \text{if} \quad n < 29 \\ PL > FL > NL & \text{if} \quad 29 \leq n < 32 \\ FL > PL > NL & \text{if} \quad n \geq 32 \end{cases}$$

$$\text{Transfers} \quad \begin{cases} FL = PL > NL & \text{if} \quad n = 3 \vee n = 4 \\ FL > PL > NL & \text{if} \quad n \geq 5 \end{cases}$$

If we view uncertainty of type 3 as the most relevant case of actual negotiations, because at the same time it captures uncertainty about both the level and the distribution of the benefits from cooperation, both relevant in climate change, then our results come to a far less negative conclusion than the previous literature. Even in a strategic context, more information must not necessarily be detrimental to the self-enforcing provision of a public good. However, the larger the uncertainty about the distribution of the gains from cooperation, the more important it is to hedge against free-riding through an appropriate transfer scheme.

Extensions

So far the analysis has focused on uncertainty about the benefits from abatement. A general conclusion was that the second stage effect from learning is only negative if there is pure uncertainty about the distribution of benefits (uncertainty of type 2). Already, if there is some uncertainty about the level of benefits (uncertainty of type 3), this effect became positive and it was also positive if there was pure uncertainty about the level of benefits. If we were to assume the same three types of uncertainty about costs instead of benefits, it can be shown that the second stage of learning is always positive.

Another conclusion was that the first stage effect of learning can be negative for full learning compared to no and partial learning if the ex-post realization of the benefit parameters is sufficiently asymmetric under no transfers, but that this was just reversed under transfers. For uncertainty about the cost parameters, a similar result holds. Asymmetric costs can lead to smaller coalitions under full learning if not accompanied by transfers. Transfers neutralize this negative effect but, unlike uncertainty about benefits, cannot turn it into a positive effect.

Overall, regardless of the type of uncertainty about costs, partial learning leads always to better outcomes than no learning and this is also true for full learning provided transfers are used to balance possible asymmetries.

Conclusion

The results in this chapter challenge the conclusion that the "veil of uncertainty" is conducive to the success of international agreements on climate change, as suggested in the previous literature. We have shown that learning is only bad in a strategic context in very specific situations: there is pure uncertainty about the distribution of the benefits from abatement. However, this is most unlikely in the climate change context. Moreover, should the problem arise, it can be mitigated, fixed or even turned into an asset through an appropriate transfer scheme. Such a transfer scheme could also be replicated through an appropriate allocation of permits under a carbon-trading scheme. Hence our message is not that we deny the possibility that we find in a strategic context: that what is good for single players can turn out to be bad at the aggregate. To the contrary, it is important to be aware of this possibility. However, if this possibility is anticipated, appropriate measures can be taken. Moreover, we derive a very optimistic message: diversity can be an asset if managed well. This is an important message because conventional wisdom would suggest that the more diverse agents are, the more difficult it is to establish cooperation. This wisdom is true if no compensation measures are taken, but can be reversed if the gains from cooperation are shared such that participation is attractive to all. For addressing climate change effectively, this means that more efforts are needed not only to design policies that are cost-effective, but also to be aware of their distributional implications. This is particularly important because the more countries differ regarding costs and benefits, the larger the potential gains of cost-effective climate policies.

The most obvious extensions for future research include the following items. First, in our model, if players learn, there is perfect learning. A more natural assumption would be gradual learning through the update of beliefs. Our assumption about partial learning does not really capture this. Second, learning in our model is exogenous. Therefore, learning-by-doing, learning-by-research and governmental policies that influence the stock of knowledge could be part of the model. Third, the first and second stages of coalition formation are one-shot decisions. In reality, countries can revise their membership decisions as well as their abatement decisions. Fourth, we assumed that all players hold the same information, whereas in reality there is dispute about scientific evidence, with the IPCC, for instance, aiming at moderating this dispute. It would be interesting to analyze the effect of such institutions on the outcome of climate negotiations.

Acknowledgements

This chapter was written while Michael Finus was visiting the Institute Henri Poincaré, Paris in January–February 2013. The hospitality and financial support of the Institute are gratefully acknowledged.

Notes

1 For possible other coalition formation games, see for instance Finus and Rundshagen (2003).
2 Hence, one could also talk about a singleton coalition, i.e. a coalition with just one player. However, we find it easier to reserve the term coalition for a coalition of at least two players and refer to all players not belonging to this coalition as singletons or non-coalition members.
3 This result is in stark contrast to Kolstad (2007) and Kolstad and Ulph (2008). Since the technical details to explain this difference are quite involved, we refrain from discussing this issue here but refer the interested reader to Finus and Pintassilgo (2013).
4 For the uniform distribution of the benefit parameter γ_i, the variance of the realized parameter values increases with the number of players n.

References

d'Aspremont, C., A. Jacquemin, J.J. Gabszewicz and J.A. Weymark (1983). On the stability of collusive price leadership. *Canadian Journal of Economics* 16(1), 17–25.

Barrett, S. (2003). *Environment and statecraft: the strategy of environmental treaty-making*. Oxford and New York: Oxford University Press.

——(2006). Climate treaties and "breakthrough" technologies. *American Economic Review* 96(2), 22–25.

Bloch, F. (2003). Non-cooperative models of coalition formation in games with spillovers. In C. Carraro (ed.), *The endogenous formation of economic coalitions*. Cheltenham, UK: Edward Elgar, pp. 35–79.

Dellink, R. and M. Finus (2012). Uncertainty and climate treaties: does ignorance pay? *Resource and Energy Economics* 34(4), 565–84.

Dellink, R., M. Finus and N. Olieman (2008). The stability likelihood of an international climate agreement. *Environmental and Resource Economic* 39(4), 357–77.

Eyckmans, J. and M. Finus (2009). An almost ideal sharing scheme for coalition games with externalities. University of Stirling: *Stirling Discussion Paper Series, 2009–10.*

Finus, M. (2001). *Game theory and international environmental cooperation*. Cheltenham, UK and Northampton MA: Edward Elgar.

——(2003). Stability and design of international environmental agreements: the case of transboundary pollution. In H. Folmer and T. Tietenberg (eds.), *International yearbook of environmental and resource economics 2003/4*. Cheltenham, UK: Edward Elgar, pp. 82–158.

——(2008). Game theoretic research on the design of international environmental agreements: insights, critical remarks and future challenges. *International Review of Environmental and Resource Economics* 2(1), 29–67.

Finus, M. and P. Pintassilgo (2012). International environmental agreements under uncertainty: does the veil of uncertainty help? *Oxford Economic Papers* 64(4), 736–64.

——(2013). The role of uncertainty and learning for the success of international environmental agreements. *Journal of Public Economics* 103, 29–43.

Finus, M. and B. Rundshagen, B. (2003). Endogenous coalition formation in global pollution control: a partition function approach. In *The Endogenous Formation of Economic Coalitions*. Fondazione Eni Enrico Mattei Series on Economics and the Environment. Cheltenham, UK: Edward Elgar, pp. 199–243.

Fuentes-Albero, C. and S.J. Rubio (2009). Can the international environmental cooperation be bought? *European Journal of Operation Research* 2002, 255–64.

IPCC (2007). *Climate Change 2007: Synthesis Report.*

Kolstad, C.D. (2007). Systematic uncertainty in self-enforcing international environmental agreements. *Journal of Environmental Economics and Managament* 53(1), 68–79.

Kolstad, C.D. and A. Ulph (2008). Learning and international environmental agreements. *Climatic Change* 89(1–2), 125–41.

——(2011). Uncertainty, learning and heterogeneity in international environmental agreements. *Environmental and Resource Economics* 50, 389–403.

McGinty, M. (2007). International environmental agreements among asymmetric nations. *Oxford Economic Papers* 59, 45–62.

Na, S.-L. and H.S. Shin (1998). International environmental agreements under uncertainty, *Oxford Economic Papers* 50(2), 173–85.

Stern, N. (2006). *Stern Review: the economics of climate change.* Report prepared for HM Treasury in the UK.

Weikard, H.-P. (2009). Cartel stability under optimal sharing rule. *The Manchester School* 77, 575–93.

Yi, S.-S. (1997). Stable coalition structures with externalities. *Games and Economic Behavior* 20(2), 201–37.

——(2003). Endogenous formation of economic coalitions: a survey of the partition function approach. In C. Carraro, (ed.), *The endogenous formation of economic coalitions.* Cheltenham, UK: Edward Elgar, pp. 80–127.

4 Burden sharing in global climate governance

Thomas Bernauer, Robert Gampfer and Florian Landis

Introduction

Global climate change politics constitutes a problem of global public goods provision whose solution is hampered by the discounting of future benefits of climate change mitigation and free-riding incentives among the countries that should reduce their greenhouse gas (GHG) emissions (Bernauer 2013). The fact that the more than 190 countries involved in global efforts to set up a governance system for limiting GHG emissions differ strongly along several lines adds another layer of complexity; in particular, they differ in their individual historical responsibility for global warming, their contemporary GHG emissions, their economic growth prospects, and their vulnerability to climatic changes. These differences have led to intense debates about how the global GHG mitigation burden should be structured, in the sense of how much specific countries should contribute to the effort, and in what timeframe.

In this contribution we shall discuss both macro and micro level facets of the burden sharing problem in global climate governance. We start by outlining the most important normative criteria for burden sharing that have emerged from academic and policy debates. The following section translates a selection of these criteria into specific burden sharing formulas. Specifically, it looks at how a given global emissions budget could be allocated to countries or groups of countries, and then adds the possibility of transfer payments. We then move to a discussion of how laboratory experiments and surveys contribute to measuring and explaining individual citizens' burden sharing preferences. Understanding such preferences is important because, in contrast to some other global policies whose implementation occurs outside the purview of the "ordinary" citizen (e.g., monetary policy, arms control), the implementation of climate policies has very direct effects on citizens (e.g., in the form of energy taxes or regulatory standards for energy efficiency). Hence the need to allocate the global GHG mitigation burden in ways that are widely regarded as fair by citizens (voters). The concluding section connects insights from macro and micro level research on burden sharing and argues for a more integrated approach to studying both.

Normative principles

Problems of distributional justice (Shue 1999) in global climate governance can be viewed from two perspectives. The first perspective views the capacity of Earth's climate system to absorb GHG emissions in terms of a common pool resource. The main issue here is how to allocate rights to use this resource. Such rights can then be conceptualized as emission budgets (see Oran Young's contribution to this book in Chapter 2). The second perspective focuses on averting detrimental changes in Earth's climate system and/or on adaptation measures. The main issue here is how to allocate the burden of supplying the public good of climate system protection. While the two perspectives differ in emphasis and should in fact be viewed as complementary, our contribution adopts primarily the second perspective. Furthermore, burden sharing, as discussed in this chapter, applies more directly to mitigation than to adaptation. Since reducing GHG emissions has the same effect on the global climate system no matter where it occurs, any given burden can be allocated freely among countries without implications for the effectiveness of the effort as a whole. This is not true to the same extent for adaptation, which is often specific to local geographic, social, and economic conditions. Still, the burden of paying for adaptation globally could in principle be allocated freely among all countries worldwide.

Article 3 of the 1992 UN Framework Convention on Climate Change (UNFCCC) states that

> The Parties should protect the climate system for the benefit of present and future generations of humankind, on the basis of equity and in accordance with their common but differentiated responsibilities and respective capabilities. Accordingly, the developed country Parties should take the lead in combating climate change and the adverse effects thereof.

This (rather generic) legal provision for burden sharing notwithstanding, debates in academic and policy circles over the past 20 years on how exactly to share the global GHG mitigation burden have not produced a consensus. Still, they have at least resulted in a quite clearly defined set of normative principles that could be used to allocate GHG emission rights to countries and, conversely, allocate mitigation obligations (Ringius *et al.* 2002; Boston and Gregory 2008), making those principles relevant for both perspectives on global climate governance noted above.

Table 4.1 summarizes what we consider the most important normative principles. Like Ringius *et al.* (2002), we focus on substantive fairness principles, as opposed to principles of procedural justice that describe aspects of the governance process such as inclusiveness, transparency, and representation (see e.g., Bernauer and Gampfer 2013).

Table 4.1 omits one important principle that is often listed in the scientific and policy literature as well: vulnerability to climate change (e.g., Baer 2013). This principle is primarily relevant for climate change adaptation, whereas we are

Table 4.1 Normative principles for sharing the global GHG mitigation burden

Principle	*Description*	*Burden sharing rule (examples)*
Individual equality	Each individual has an equal right to emit GHGs and to be protected from the consequences of GHG emissions.	Reduce GHG emissions in a way that produces convergence in per capita emissions at a fixed amount.
Economic equality	Each country has the right to grow economically and impose an equal amount of externalities on other countries per unit of economic output.	Reduce GHG emissions in a way that produces convergence in emissions per unit of GDP at a fixed level. Reduce GHG emissions in a way that results in equal marginal abatement costs per GDP across countries (equalize net welfare change across countries).
National grandfathering	Countries are sovereign entities and have acquired a historical right to emit GHGs.	Reduce GHG emissions of all countries proportional to their contemporary emissions (e.g., keeping national shares in total global emissions constant while reducing their absolute levels).
Historical responsibility / polluter pays	Those who have contributed more to the stock of GHG emissions in the atmosphere should invest more in CHG mitigation.	Allocate GHG reduction targets proportional to a country's accumulated (historical) contribution to climate change.
Current responsibility / polluter pays	Those who are currently contributing more to the flow of GHG emissions should invest more in CHG mitigation.	Allocate GHG reduction targets proportional to a country's current contribution to climate change
Historical economic benefits from emissions	Those who benefited most (economically) from being able to emit GHGs should invest most in reducing GHG emissions.	Scale GHG reduction targets to historical GHG emissions and associated economic growth.
Economic capacity	Those with a stronger economic capacity to solve the problem should do more.	Allocate GHG reduction targets proportional to economic capacity (e.g., GDP per capita).
Cost–benefit analysis	Countries should adjust their emissions such that their marginal emissions create as much benefit for themselves in the present as they impose harm to the global community in the future.	Global damages may be represented in monetary or utility terms. If damages are represented in monetary terms, countries should behave as if they would be held liable for climate damages.

focusing on GHG mitigation policy. It holds that those who are most likely to suffer from the consequences of climate change should receive most international assistance. International assistance, in turn, could be organized according to burden sharing principles that resemble those for climate change mitigation (e.g., current or historical responsibility, economic capacity; see Dellink *et al.* 2009).

Empirically, the application of different normative burden sharing principles is likely to lead to similar burden sharing outcomes. For example, most of the highly vulnerable countries are also among those with the lowest economic capacity, and most high-capacity countries are those with a large historical responsibility for GHG concentrations in the atmosphere. However, some closely-related principles could also produce contrasting allocations of the global mitigation burden. For instance, responsibility for current emissions implicates a much larger obligation for large emerging economies (e.g., China) than historical responsibility. Such differences in implications between normative principles complicate the choice of appropriate burden sharing rules and suggest that some combination of several principles needs to be applied in order to achieve allocations that are regarded as fair and acceptable by the large majority of stakeholders.

The 1997 Kyoto Protocol to the UNFCCC is the first, and thus far only, global climate treaty that includes specific GHG emission (reduction) targets. The allocation of these targets was not explicitly derived from clearly-defined normative principles. Nonetheless, it implicitly reflects several of the principles listed in Table 4.1, particularly historical responsibility and economic capacity.

The most widely discussed burden sharing approach during the negotiation of the Kyoto Protocol was formally proposed by Brazil (Ringius *et al.* 2002; see also http://themasites.pbl.nl/tridion/en/themasites/fair/applications/060_Brazilian_proposal/index-2.html). This proposal focused mainly on historical responsibility as the guiding normative principle. Its methodology concentrated on establishing how much a country has contributed, through its GHG emissions, to the increase until today in global mean temperature. The approach proposed by Brazil was not in the end used to derive emission targets in the Kyoto Protocol, but was passed on to the Subsidiary Body on Scientific and Technical Advice (SBSTA) of the UNFCCC for consideration.

Another prominent proposal, by India, focuses on per capita emissions. It holds that a country must start mitigation action when its emissions reach global average per capita emissions or its GDP per capita reaches US$20,000 (Baer *et al.* 2007).

Yet another prominent proposal for burden sharing combines several principles: the Responsibility-Capacity Index (RCI) (Baer 2013). It has been strongly promoted particularly by non-governmental organizations and some, mostly developing, countries since the 2009 Conference of the Parties (COP) to the UNFCCC in Copenhagen. The underlying idea has also received some support in the academic literature (e.g., Caney 2010). The RCI is a composite measure that takes into account both countries' contribution to GHG emissions after 1990 and their gross national income above a certain "development threshold" – whose appropriate value is, however, strongly debated.

To the extent that burden sharing principles have been translated into specific allocation proposals, these proposals have concentrated mostly on a one-time allocation of emission reduction paths. However, it seems more realistic to assume that international agreements on this issue will cover no more than two to three decades. If, for instance, the world community agreed to allocate the burden according to a per capita emissions principle, emission targets would probably require minor updates in the regularly recurring negotiations, but would not experience major shifts. But if the burden of reducing emissions were allocated based on current emissions, the distribution of cost could shift significantly over time, and even if historic emissions were used, responsibility shares would probably change significantly (Botzen *et al.* 2008). These caveats notwithstanding, the existing scientific and policy literature now offers a rather exhaustive set of normative criteria that negotiators can resort to when trying to design regional or global climate governance systems. As discussed in the next section, based on these criteria, climate-economic modeling has studied the implications of different allocation rules for the major stakeholders.

Insights from climate-economic modeling

Academics have picked up on the Brazilian and Indian proposals, as well as other proposals, and have explored their implications for burden sharing. For instance, the FAIR model (Framework to Assess International Regimes for burden sharing; see den Elzen and Lucas 2005) has been used to compare allocations derived from the "Brazilian" approach with allocations derived from other normative principles. The main conclusion from this research is that using historical and/or per capita emission principles favors developing countries, whereas including all GHG emissions as well as land-use related emissions favors industrialized countries. According to the FAIR model results, the Brazilian approach would require all countries (including developing countries) to start reducing emissions immediately, though a participation threshold (at a specific level of GDP per capita or per capita emissions) could also be used in order to produce faster convergence of per capita emissions over time.

Other recent work has combined the application of normative burden sharing principles with an emissions budgeting approach (UNEP 2012; Fuessler *et al.* 2012). This approach, which combines the two perspectives on distributional justice mentioned above, starts with a total amount of GHG emissions that would be compatible with the widely supported 2 degrees Celsius target (e.g., Meinshausen *et al.* 2009; Rogelj *et al.* 2011). It then applies specific burden sharing principles to study their implications for total and country or country-group specific emission trajectories until the end of the twenty-first century. A recent study by Fuessler *et al.* (2012) applies three specific principles:

- Indian proposal 2008: a country must start reducing emissions when its emissions reach global average emissions per capita, or when its income level reaches US$20,000.

- Equal cumulative per capita emissions: this approach aims at equal emissions per capita, aggregated for the time period 1990 to 2100.
- Responsibility-Capacity Indicator: each country's share in the global mitigation effort must be proportional to this indicator. The latter is composed of three-quarters "polluter pays" (per capita emissions in the last 10 years) and one quarter "ability to pay" (expected per capita GDP) elements.

Figure 4.1 illustrates the implications of these three approaches for Western Europe and China.

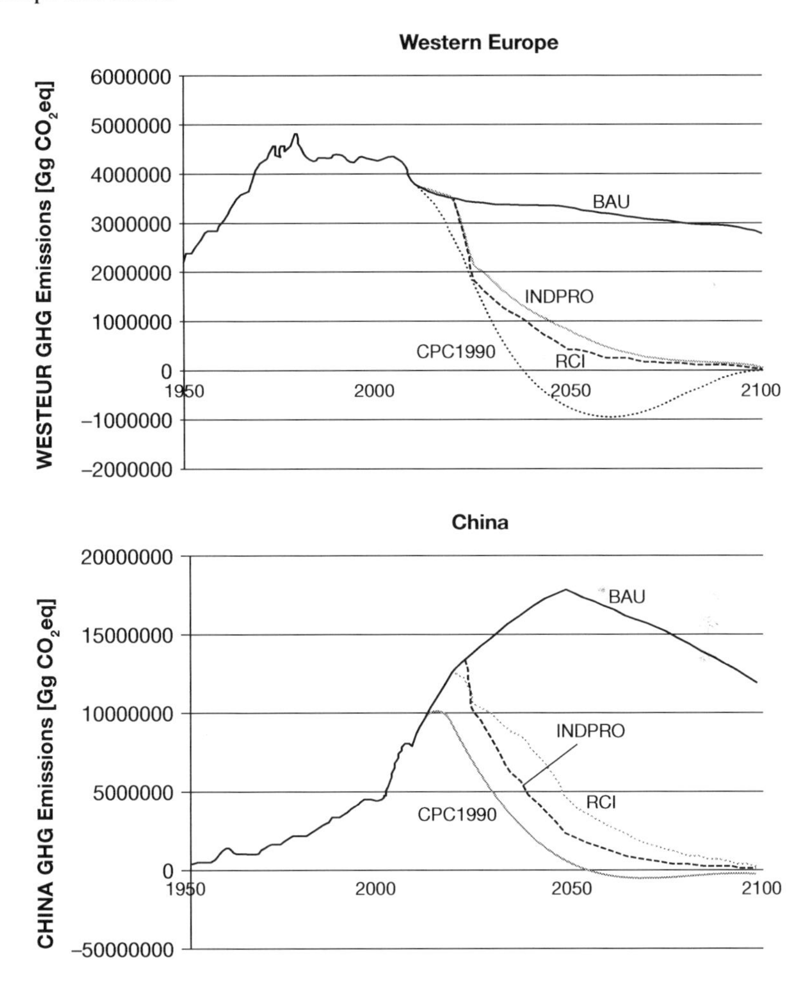

Figure 4.1 Implications of three normative principles

Source: Adapted from Fuessler *et al.* 2012. INDPRO = Indian proposal 2008; CPC1990 = Equal cumulative per capita emissions 1990; RCI = Responsibility-Capacity Indicator; BAU = business as usual (no mitigation effort). In the BAU scenario, Annex I country emissions are assumed to level off, non-Annex I country emissions are assumed to more than double until 2050. Negative allocations means that the respective country would have to fund emission reductions outside its territory or engage in carbon sequestration.

The implications for Western Europe and the United States (not shown in Figure 4.1), as derived from this modeling exercise, are rather similar, though the high historical per capita emissions of the United States produce an even more negative allocation for the US by 2050. Under the Indian and RCI principle, emissions in Western Europe would have to decline by around 80 percent by the year 2050, and even more if the equal per capita emissions principle (CPC1990) is used. The allocation for China would be somewhere in between the allocation for Annex I and non-Annex I countries. All three normative principles would require China to reduce its emissions quite rapidly. In part, this is due to the strong growth of China's emissions since the baseline year 1990.

By and large, the Indian proposal and the RCI approach result in somewhat similar mitigation trajectories, both for Annex I and non-Annex I countries. None of them leads to a negative emissions allocation for Annex I countries as a group. The CPC1990 principle, in contrast, mandates a stronger decline of emissions and leads to a large negative emissions budget (deficit) for Annex I countries in the 2030–2100 period. The reason is that it focuses primarily on historical and present emissions.

Deriving emission reduction trajectories from normative principles, such as the ones listed in Table 4.1, is appealing because these principles are clear-cut and intuitive, and the scientific methods for computing required emissions reductions based on specific normative principles are quite well developed. One disadvantage of this approach, however, is that it does not pay much attention to the different economic growth prospects of countries. Such differences are, in our view, at least in part responsible for the fact that none of the normative principles listed above has so far been explicitly used to define legally binding GHG reduction commitments, neither at the global level nor within the most ambitious regional climate governance system, the one set up by the European Union (see http://climateactiontracker.org/countries.html). Moreover, the studies discussed so far do not tell us much about the uncertainty associated with reduction trajectories derived from normative principles.

A recent study by two of the authors of this chapter (Landis and Bernauer 2012) addresses this challenge. Instead of distributing a given global carbon budget between countries, the two authors study the implications for burden sharing of differences in countries' perception of the social cost of carbon (SCC). Such differences in perception are presumed to result from differences across countries in trading off the cost of emission reductions today against the global benefits from reduced climate change in the future. This trade-off can be based on a utilitarian or a liability principle. The former compares the current country-specific utility loss from GHG abatement to the future global utility gain from climate protection, while the latter compares the same current country-specific loss to a future country-specific utility gain from having to pay less compensation for climate related damages inflicted on the global community. In both cases, a Euro spent or received has a bigger impact on a poor country's utility than it has on a rich country's utility (according to the widely accepted assumption of diminishing marginal utility of consumption). This implies that carbon prices based on SCC are lower for low income countries with prospects of high income growth rates than for rich countries.

The study explores in detail the influence of different assumptions about climate change, damages, and socio-economic futures on the ratios between different world regions' carbon prices. The main finding is that current uncertainty about climate change and the damages it entails seems to influence such ratios far less than uncertainty about the socio-economic future of different countries and the world as a whole. The authors compute regional social costs of carbon (RSCC) for the regions Asia (ASIA), Africa and Latin America (ALM), and countries undergoing economic reform (REF; the former Soviet Union) (see www.nature.com/nclimate/journal/v2/n8/full/nclimate1548.html, especially Fig. 2).

Their results indicate that, in spite of major uncertainty about climate change damages, future economic growth, and other factors, carbon prices in the OECD are systematically higher than carbon prices in the other world regions. This result does not, of course, indicate by how much emissions should be reduced, or by whom. But it shows that distributional matters can be discussed in a utilitarian framework based on RSCCs because the distributional implications remain confined in spite of high uncertainty. This means that the basic idea of a global carbon price, which is very popular among climate-economic modelers because it would efficiently distribute emission reduction efforts across the globe, cannot appropriately reflect the different economic situations in which different countries find themselves. Consequently, if an inter-regional (or global) climate agreement should produce a uniform carbon price, this price would have to represent a compromise between RSCCs. This implied compromise could be designed such that the gains for richer countries from being able to apply a lower-than-RSCC carbon price equal the losses for the poorer world regions from having to use a higher-than-RSCC carbon price.

Transfer payments from industrialized to developing and emerging economies could make such a compromise feasible. Landis and Bernauer derive their argument for transfer payments without explicitly taking into account historical responsibility or development status – though implicitly, development status plays a role in the scenarios' common assumption of an economic catching-up potential for developing countries. Their argumentation is solely based on regional intertemporal optimization, given two different ways of fully internalizing externalities produced by GHG emissions. It is free from arguments of equal access to the global public good that the atmosphere represents and can be kept separate from historic responsibility for pre-agreement GHG emissions. Still, the study arrives at the conclusion that the biggest part of the abatement cost burden should be borne by industrialized countries. To illustrate how mitigation costs could be distributed in an ambitious global climate agreement, Landis and Bernauer assume a globally uniform carbon price of US$35/t$CO_2$ and impute RSCCs that could lead to this carbon price compromise. They find that transfers in the range of US$15–45 billion (depending mostly on assumptions about the utility function of regions and their socio-economic future) would be appropriate. They also find that the abatement costs under the imputed RSCCs justify cost distributions under which industrialized countries would pay 54–75 percent of the global abatement cost if the liability criterion is applied, and around 90 percent or more under the utilitarian criterion.

Insights from surveys and laboratory experiments

Normative burden sharing principles, such as the ones discussed in the previous sections, have emerged from public discourses among stakeholders and scientists around the world. Laboratory experiments and surveys have explored the micro-foundations of such principles; that is, they have examined how individuals, the most fundamental units of any political system, form preferences with respect to burden sharing in climate policy.

The empirical study of micro-foundations of normative principles in climate policy is important. Even though global climate governance arrangements are negotiated between governments rather far away from the purview of the "ordinary" citizen, they need to be implemented locally (e.g., in the form of energy taxes or building codes). This means that effective implementation of climate policies, particularly in democratic countries, requires a high level of public support. Governments thus have a strong incentive to pay close attention to domestic public opinion on climate policies.

Since burden sharing is likely to be a key factor in individuals' perceptions and evaluations of international climate policy, this aspect has received considerable attention in micro level research. Surveys and survey-embedded experiments have served to measure and explain the attitudes, preferences, and behavioral intentions of "ordinary" citizens. Laboratory experiments have been used to observe individual preferences and behavior, and also to simulate interactions among states, with experimental participants proxying for whole nations or their governments.

Existing large-scale multi-national surveys, perhaps surprisingly, do not address burden sharing directly. They do, however, capture public opinion with respect to climate policy more generally, which also provides some indirect insights into fairness perceptions. A World Bank poll in 16 countries, conducted in 2010, asked whether respondents thought their country should commit to reducing GHG emissions even if no international agreement on climate change was reached (World Bank 2010). Large majorities in all except one country opined that they would support such unilateral commitments. This suggests that, on average, individuals want their respective country to accept at least some of the global mitigation burden. A representative survey in the United States found that most respondents supported the Kyoto Protocol as well as unilateral US GHG mitigation actions (Leiserowitz 2006). The current stalemate in global climate negotiations, however, is hard to square with such strong public opinion in favor of ambitious climate policies.

One important problem associated with surveys of this kind is that survey items gauging support for climate policy do not usually indicate the economic burden associated with mitigation measures. The question about responsibility for climate change in the 2010 World Bank survey, for example, reads "Do you think our country has a responsibility to take steps to deal with climate change?" This leaves open what exactly such a "responsibility to take steps" means. Even if surveys make potential economic consequences of mitigation measures explicit, which the World Bank survey in fact does in other parts of the questionnaire whose items are not placed in the context of international burden sharing, it is probably very difficult for respondents to estimate the economic consequences for themselves or their country. Another important problem is social desirability bias. If respondents think that there is a social norm to be "pro-climate" they may respond according to this norm. In both cases, responses may not reflect true individual preferences, and may in fact convey an overly optimistic picture of public support for climate policy.

An additional disadvantage of standard surveys is that they do not allow for rigorous causal inferences concerning relationships between specific climate policy aspects and individual preferences. To overcome this limitation, some recent studies on burden sharing have used survey-embedded experiments. In such experiments, survey respondents are randomly assigned to treatment groups that receive different stimulus texts, or answer different questions. Participants then respond to survey items measuring the dependent variable of interest, and responses between treatment groups (and if appropriate a control group) are compared.

Tingley and Tomz (2011), in a survey experiment with US participants (convenience sample), found that respondents support unilateral emission reductions by their country. Treatment groups were exposed to information about mitigation commitments with different levels of ambition by other countries. The main finding is that respondents favor economic and diplomatic sanctions against countries that violate their mitigation commitments. Implicitly, this finding suggests that people subscribe to the general notion that the global mitigation burden should be shared between countries. But it does not allow for any conclusions concerning burden sharing arrangements that are perceived as fair, or which criteria should be used to structure such arrangements.

Bechtel and Scheve (2013) examine preferences concerning burden sharing in a more direct way, using an experimental conjoint survey that elicits preferences about different characteristics of climate agreements. Their survey is based on representative national samples in four countries. It turns out that costs per se have a strong influence on support for the (hypothetical) climate agreement: an increase of costs equivalent to 0.5 percent of gross domestic product decreases support by 20 percent. Participants were also sensitive to the international distribution of these costs. Both economic capacity and responsibility for greenhouse gas emissions seem to affect preferences: agreements in which countries contribute to mitigation according to their wealth and according to their historical emissions receive more support than agreements in which no such burden sharing principle is applied. Thus this study lends support to two of the main normative burden sharing principles depicted in Table 4.1 above.

Carlsson *et al.* (2010) also use a conjoint survey experiment to explore preferences for different burden sharing rules. The participants in this experiment were from China and the United States, the world's largest GHG emitters. The burden sharing principles participants had to rate were capacity, responsibility in terms of historical or current emissions, and individual equality. The main result was that respondents strongly favored the principle that is likely to be the least costly for their own country. Whereas US participants preferred allocating the burden according to current emissions, Chinese participants mostly supported the principle of historical emissions. Interestingly, individual preferences thus appear to be shaped by both normative (fairness) and utilitarian (economic) considerations. This finding also highlights the need to pay attention to respondents' country-specific background when studying stated preferences concerning international climate policy.

To mitigate the problem of "stated preference" approaches – participant responses usually do not have material consequences for them and might thus not reflect their true preferences – researchers have conducted interactive laboratory experiments based on standard cooperation or coordination games. Monetary incentives make participant decisions costly (at least to some extent), with the final payoff depending on each player's actions and other players' reaction. Individual preferences are thus revealed through behavior in the game.

Critics often raise questions about the external validity of laboratory experiments. Hence we will address this criticism up front before discussing insights from existing research using laboratory experiments. The most important criticism holds that the pool from which participants in these experiments are recruited (usually university students) is not a random sample of any given country's total population or of the country's political decision-makers. In response to this criticism, some researchers argue that laboratory experiments test theories about human interactive behavior in generic choice situations that are considered to be applicable to, for example, international climate governance as well (Morton and Williams 2010: 341). Milinski *et al.* (2008: 2291), for example, interpret their climate-change related public goods experiment as a fundamental "collective-risk social dilemma" that exists in various social choice situations. Proponents of laboratory experiments also point to studies comparing decisions of political elites and students in game-theoretical experiments. These studies have found that the choices of these groups do not differ significantly from each other (e.g. Fréchette *et al.* 2005; Potters and van Windern 2000). If anything, elite decision-makers appear to act more rationally than students. This implies that underlying rationalistic decision-models are likely to hold for political elites, even if they perform less well for student populations (Fatas *et al.* 2007; Fréchette *et al.* 2005). For the time being, however, there are no systematic studies on whether and how results of laboratory experiments concerning climate policy differ between specific types of participants, such as students, government officials, and "ordinary" citizens.

Another type of criticism concerns the realism of laboratory experiment settings. One particular problem is that the stakes for participants are very small compared to the real world consequences of solving or not solving the climate change problem. Even with experimental designs where participants can lose all or a substantial amount of the money they are endowed with at the beginning of the experiment, losses will not extend to their actual private wealth. The fact that laboratory participants can expect to leave the experiment with a net material benefit (at a minimum the so-called show-up fee) is sometimes seen as a factor potentially biasing their behavior. However, there is still no conclusive evidence from methodological studies on whether this is indeed a problem. Observations from some experiments indicate that participants act rather loss-averse with respect to their experimental endowment, suggesting that the potential bias from the impossibility of real losses is limited.

The most common designs of laboratory experiments on burden sharing in climate policy are variations of a public goods game in which players contribute money to a common pool (usually representing the public good of international climate policy, climate change mitigation, or similar concepts). Such a design is, for instance, used by Brekke *et al.* (2012) to examine the effect of differences in economic capacity. Over several periods, participants' contributions must reach a certain threshold for the public good to be provided, in which case total contributions are multiplied and evenly redistributed to the participants. The main result is that participants who were given a high initial amount of money contributed significantly more to the public good than those with a low initial amount. In addition, framing contributions differently affected the contribution behavior of "poor" players. The latter paid very small amounts when the frame emphasized how little they could contribute to reaching the threshold, given their small endowments. This result suggests that the capacity-to-pay principle has an important impact on contributions.

In another public goods game where participants had to reach a certain contributions threshold to avoid "dangerous climate change," Tavoni *et al.* (2011) also allocated different initial endowments to players ("dangerous climate change" meant loss of their entire initial endowment at the end of the game – this can be thought of as a catastrophic climate event that destroys a substantial part of a country's economic wealth or capital stock). They found that in groups where rich players already contributed high amounts in early rounds, overall contributions were higher and the group was more likely to reach the threshold. This result was even more pronounced when rich players were given the opportunity to declare their contribution intentions at the beginning of the game. Explicitly invoking a capacity-to-pay principle for burden sharing was thus reflected by corresponding actual behavior.

Very few laboratory experiments have thus far investigated whether burden sharing rules involving current or historical emissions play a greater role in shaping individual climate policy preferences. In a field experiment with a sample representative of the German voting-age population, Diederich and Goeschl (2011) gave participants the choice between simply taking a random monetary payout from a lottery, or contributing to greenhouse gas emission reductions by using the money to buy and retire emission allowances from the EU Emissions Trading System. Participants who reported being conscious about GHG emissions that their everyday behavior had caused in the past were more likely to choose emission reductions. This result provides some evidence for the historical responsibility principle. However, there is still plenty of scope for more laboratory or field experiments that directly and systematically test the influence of different allocation principles, as well as interactions between them, on individual burden sharing preferences.

Another interesting opportunity to examine individual burden sharing preferences, by means of laboratory and/or survey experiments, are climate policies that are enacted unilaterally by one country or a group of countries, but that have economic consequences for other countries. Unilateralism has been gaining ground in view of persisting stalemate in negotiations on a successor agreement to the Kyoto Protocol. From a standard perspective that views international environmental cooperation as some form of prisoners' dilemma, unilateral GHG mitigation measures appear puzzling – unless one assumes that such measures are virtually costless, or they generate co-benefits that are large enough to offset mitigation costs (both of which are unlikely). Viewed differently, however, one could argue that countries (or groups of countries) may decide to become frontrunners in international climate policy in order to set precedents and entice laggard countries to follow suit (Bernauer 2013). Such a perspective leads to at least two questions about unilateralism. First (from the standard perspective), why would a country be willing to bear some part of the GHG mitigation burden in the absence of reciprocity, and in the presence of strong uncertainty about whether others will follow up with mitigation measures in the future? Second (and departing from the conventional view), can unilateral moves that create positive or negative incentives for other states to eventually participate in burden sharing be successful?

One of the most important examples of this type of unilateralism is the decision of the European Commission to subject emissions from air transport to the EU Emissions Trading System from January 2012 on. This policy applies to all airlines flying in and out of EU airports, and covers the entire flight, including emissions not occurring in the EU's airspace. Politicians and airline managers from many non-EU countries have claimed that with this policy the European Commission is infringing the sovereign rights of other nations to regulate emissions within their respective airspace.

Bernauer, Gampfer, and Kachi (2013) conducted a survey experiment with US and Indian participants – that is, participants from the two largest democracies outside the EU that are most strongly opposed to the new EU regulation on GHG emissions from aircraft. The aim was to find out whether "ordinary" people in the two countries are as strongly opposed to the EU's policy on grounds of cost and sovereignty implications as have been policy makers and airline executives in the two countries. Treatment groups were primed either with different levels of cost implications of the policy (airline ticket price increases), or with stimuli emphasizing political costs (the alleged sovereignty infringement). Whereas treatments priming participants for high cost implications significantly and substantially increased opposition towards the EU's unilateral climate policy, effects of political cost treatments were mostly insignificant in both the Indian and the US samples. This finding suggests that unilateral climate policies that impose an economic burden on third countries will not necessarily lead to popular opposition in those countries (and thus pressure on their governments to oppose and perhaps even actively undermine those policies) as long as negative economic consequences for individuals there remain modest.

Yet another facet of burden sharing, namely whether individuals differentiate between burden sharing for mitigation and for adaptation, has so far received only scant attention. Most studies on burden sharing concentrate on mitigation costs – that is, most experiments and surveys are framed in those terms. One exception is Klinsky *et al.* (2012). They use standardized surveys and semi-structured interviews to elicit fairness preferences, both in the context of mitigation and adaptation. When asked about sharing a mitigation burden, respondents most often mentioned that the countries causing emissions should pay for reducing them. That is, they focused on the current responsibility principle. In the context of adaptation, respondents tended to advocate a burden allocation according to the needs and capabilities of the countries most affected by climate change. Economic capacity and vulnerability to climate change were thus seen by respondents as important criteria in the adaptation context. This suggests that adaptation preferences are influenced by different fairness criteria than mitigation preferences.

Research on individual preferences concerning market-based policy instruments may also contribute to the study of burden sharing. Credit-based mechanisms such as the clean development mechanism (CDM) and emissions trading systems are likely to have profound consequences for burden sharing. These implications remain to be examined theoretically and empirically. Such research could, for instance, explore the conditions under which people consider particular types of market-based climate policy instruments fair and legitimate. Both state-sponsored and private market-based instruments are worthwhile studying in this context. As to the latter, for example, some recent work has examined people's willingness to engage in private carbon offset initiatives (Blasch and Farsi 2013). Since such carbon offset decisions are taken directly by individuals, studying the implications of their use for perceptions of fairness in burden sharing is certainly worthwhile.

Finally, obvious candidates for further research on burden sharing preferences are transfer payments from developed to developing countries, low-carbon technology transfers, and potential international compensation payments for damage from climate change. The many policy proposals that have been put forth on these issues aim at redistributing the mitigation and adaptation burden in a fair manner, and a considerable body of literature has emerged that examines their fairness implications either theoretically or empirically at the macro level (e.g., Tol and Verheyen 2004; see also sections 2 and 3 of this chapter). Individual preferences vis-à-vis such policy instruments remain to be examined.

Conclusion

Effective global climate governance hinges on states' ability to find burden sharing arrangements that are widely regarded as fair and legitimate, and therefore acceptable both to political and economic elites and to citizens more broadly. Various normative principles have been proposed by policy makers and academics, and academic researchers have studied the implications of key principles for GHG emission reduction trajectories and emission allocations to countries or country groups. In practice, none of these principles has been explicitly used by policy makers to define global or regional emission allocations. But the allocations set forth in the Kyoto Protocol, which is part of the first, and thus far only, global climate treaty, and within the European Union are broadly in line with some of these principles, in particular with the principles of historical responsibility and economic capacity.

Comparisons of different normative principles and their implications for the allocation of emission targets to countries and country groups show that all principles tend to result in a larger mitigation burden for developed than developing countries. Nevertheless, remaining variation in these implications still seems large enough to prevent a global climate agreement, in spite of a general consensus that action is necessary. Large emerging economies such as China, and advanced industrialized countries such as the United States, for example, disagree over the relative importance of the contemporary versus historical emissions criterion.

A rather small but growing number of surveys, survey-embedded experiments, and laboratory experiments have examined whether and how normative burden sharing principles, as they appear at the international level, correspond to the preferences and behavioral patterns of individuals (i.e., citizens). On the assumption that policy makers' choices are usually, by and large, in line with what the majority of people in the respective society wants, studies on individual preferences and behavior can provide insights into which of the various burden sharing principles tend to be more appealing to what type of country under what conditions. Moreover, since it is virtually impossible to conduct experiments with international climate negotiators in a real world setting, laboratory experiments offer at least a rough approximation that can generate useful knowledge on what types of burden sharing arrangements could be both acceptable and effective in terms of problem solving under specific conditions. In view of the current stalemate in global climate governance, experimental studies on options for unilateral climate policies and private mitigation initiatives, such as carbon offsets, are also useful.

To date, results from laboratory experiments are generally in line with political outcomes. The current stalemate in global climate negotiations can be explained by disagreement among countries about the relative importance of different burden sharing principles and the tendency for each country to favor principles that minimize the burden for itself. This means that even if countries generally agree that globally coordinated climate policy would be beneficial, they have thus far been unable to agree on a specific burden sharing principle. Similarly, survey findings show that most citizens accept the need for GHG mitigation measures, but tend to favor burden sharing principles that minimize costs for their own country. One silver lining, however, is that surveys find rather strong support for the current practice of unilaterally enacting climate policies, which many observers in fact regard as the most promising approach in view of the current impasse at the global level.

The most general conclusion from this chapter is that both macro and micro level studies of burden sharing in climate policy are important and insightful. Both approaches, which have thus far progressed independently from each other, could indeed benefit from paying greater attention to each others' results. This may help macro level studies to become more policy-relevant and would also provide greater coherence to micro level studies, in the sense of enabling systematic comparison of preferences and behavioral patterns under a wide range of potentially relevant normative burden sharing principles.

Acknowledgements

The authors are very grateful for comments on earlier versions of this chapter by Oran Young, an anonymous reviewer, the editors of this book, and the participants in a book workshop in Oslo in November 2012. Part of the research for this contribution was supported by ERC Advanced Grant 295456 (SourceLeg).

References

Baer, P. (2013). The greenhouse development rights framework for global burden sharing: reflection on principles and prospects. *Wiley Interdisciplinary Reviews: Climate Change* 4(1), 61–71.

Baer, P., T. Athanasiou and S. Kartha (2007). *The Right to Development in a Climate Constrained World, The Greenhouse Development Rights Framework*. Berlin: Heinrich Böll Foundation.

Bechtel, M. and K. Scheve (2013). Mass support for global climate agreements depends on institutional design. *Proceedings of the National Academy of Sciences* 110(34), 13763–68

Bernauer, T. (2013). Climate change politics. *Annual Review of Political Science* 16, 421–48.

Bernauer, T. and R. Gampfer (2013). Effects of civil society involvement on popular legitimacy of global environmental governance. *Global Environmental Change* 23(2), 439–49.

Bernauer, T., R. Gampfer, and A. Kachi (2013). European unilateralism and involuntary burden sharing in global climate politics: a public opinion perspective from the other side. *European Union Politics*, forthcoming.

Blasch, J. and M. Farsi (2013). Context effects and heterogeneity in voluntary carbon offsetting – a choice experiment in Switzerland. *Journal of Environmental Economics and Policy*, forthcoming.

Boston, J. and R. Gregory (2008). *Policy Quarterly – Special Issue on Global Climate Change Policy: Burden Sharing Post-2012*. Wellington: University of Wellington.

Botzen, W.J.W., J.M. Gowdy and J.C.J.M. van den Bergh (2008). Cumulative CO_2 emissions: shifting international responsibilities for climate debt. *Climate Policy* 8(6), 569–76.

Brekke, K.A., J. Konow and K. Nyborg (2012). Cooperation is relative: framing and endowment effects on public goods. Working Paper.

Caney, S. (2010). Climate change and the duties of the advantaged. *Critical Review of International Social and Political Philosophy* 13(1), 203–28.

Carlsson, F., M. Kataria, A. Krupnick, E. Lampi, A. Löfgren, P. Qin, T. Sterner and S. Chung (2010). A fair share. Burden sharing preferences in the United States and China. University of Gothenburg Working Papers in Economics 471.

Dellink, R., M. den Elzen, H. Aiking, E. Bergsma, F. Berkhout, T. Dekker and J. Gupta (2009). Sharing the burden of adaptation financing: an assessment of the contributions of countries. Fondazione Eni Enrico Mattei, Working Paper.

Diederich, J. and T. Goeschl (2011). Giving in a large economy: price vs. non-price effects in a field experiment. University of Heidelberg Economics Discussion Paper Series 514.

Elzen, M. den and P. Lucas (2005). The FAIR model: a tool to analyse environmental and costs implications of regimes of future commitments. *Environmental Modeling & Assessment* 10(2), 115–34.

Fatas, E., M. Neugebauer and P. Tamborero (2007). How politicians make decisions: a political choice experiment. *Journal of Economics* 92(2), 167–96.

Fréchette, G., J.H. Kagel and M. Morelli (2005). Behavioral identification in coalitional bargaining: an experimental analysis of demand bargaining and alternating offers. *Econometrica* 73(6), 1893–937.

Fuessler, J., M. Herren, M. Guyer, J. Rogelj and R. Knutti (2012). *Emission Pathways to Reach 2° Target. Model Results and Analysis*. Zurich: Infras.

Klinsky, S., H. Dowlatabadi, and T. McDaniels (2012). Comparing public rationales for justice trade-offs in mitigation and adaptation climate policy dilemmas. *Global Environmental Change* 22(4), 862–76.

Landis, F., and T. Bernauer (2012). Transfer payments in global climate policy. *Nature Climate Change* 2 (August), 628–33.

Leiserowitz, A. (2006). Climate change risk perception and policy preferences: the role of affect, imagery, and values. *Climatic Change* 77(1–2), 45–72.

Meinshausen, M., N. Meinshausen, W. Hare, S.C.B. Raper, K. Frieler, R. Knutti, D.J. Frame and M.R. Allen (2009). Greenhouse-gas emission targets for limiting global warming to 2°C. *Nature* 458(909), 1158–62.

Milinski, M., R.D. Sommerfeld, H.-J. Krambeck, F.A. Reed and J. Marotzke (2008). The collective-risk social dilemma and the prevention of simulated dangerous climate change. *Proceedings of the National Academy of Sciences* 105(7), 2291–94.

Morton, R.B., and K.C. Williams (2010). *Experimental Political Science and the Study of Causality*. New York: Cambridge University Press.

Potters, J., and F. van Winden (2000). Professionals and students in a lobbying experiment: professional rules of conduct and subject surrogacy. *Journal of Economic Behavior and Organization* 43(4), 499–522.

Ringius, L., A. Torvanger, and A. Underdal (2002). Burden sharing and fairness principles in international climate policy. *International Environmental Agreements* 2(1), 1–22.

Rogelj, J., W. Hare, J. Lowe, D.P. van Vuuren, K. Riahi, B. Matthews, T. Hanaoka, K. Jiang and M. Meinshausen (2011). Emission pathways consistent with a 2°C global temperature limit. *Nature Climate Change* 1 (November), 413–18.

Shue, H. (1999). Global environment and international inequality. *International Affairs* 75(3), 531–45.

Tavoni, A., A. Dannenberg, G. Kallis and A. Löschel (2011). Inequality, communication, and the avoidance of disastrous climate change in a public goods game. *Proceedings of the National Academy of Sciences* 108(29), 11825–29.

Tingley, D., and M. Tomz (2011). Conditional cooperation, international organizations, and climate change. Working Paper.

Tol, R.S.J. and R. Verheyen (2004). State responsibility and compensation for climate change damages – a legal and economic assessment. *Energy Policy* 32(9), 1109–30.

UNEP (2012). The Emissions Gap Report 2012. Nairobi: UNEP.

World Bank (2010). Public attitudes toward climate change: findings from a multi-country poll.

5 Negotiating to avoid "gradual" versus "dangerous" climate change

An experimental test of two prisoners' dilemmas

Scott Barrett and Astrid Dannenberg

Introduction

According to the Framework Convention on Climate Change, global collective action is needed to stabilize "greenhouse gas concentrations in the atmosphere at a level that would prevent *dangerous* [our emphasis] anthropogenic interference with the climate system." The Framework Convention thus implies that, on the far side of some critical concentration level, climate change will be "dangerous," while on the near side of the threshold, climate change will be "safe" (though perhaps still undesirable). Rather than be linear and smooth, the Framework Convention warns that climate change may be "abrupt and catastrophic."

What is the threshold for dangerous climate change? Climate negotiators first agreed on a value in the Copenhagen Accord, which recognizes "the scientific view that the increase in global temperature should be below 2 degrees Celsius." A year later, in Cancun, countries reaffirmed support for this goal, but added that the target might need to be strengthened to 1.5°C. The threshold, it seems, is uncertain. A close reading of the scientific literature confirms this. There exists a range of values for the change in mean global temperature needed to "tip" critical geophysical systems (Lenton *et al.* 2008).

The temperature threshold is not the only uncertainty. The level of greenhouse gas concentrations needed to avoid any particular change in mean global temperature is also unknown (Roe and Baker 2007). Rockström *et al.* (2009: 473) combine both uncertainties to recommend a single target in terms of atmospheric CO_2 concentrations – 350 parts per million by volume (ppmv). The value was chosen to preserve the polar ice sheets, and is derived from paleoclimatic evidence suggesting "a critical threshold between 350 and 550 ppmv." Rockström *et al.* essentially take a precautionary stance.

Even this may understate the uncertainties. Countries do not control atmospheric concentrations directly, they control emissions; and the relationship between emissions and concentrations is also uncertain. The stability of the carbon cycle itself cannot be taken for granted (Archer 2010).

All of this matters tremendously because recent research shows that uncertainty about the threshold for "dangerous" climate change can have a profound effect on international cooperation. Theory predicts that countries can coordinate to avoid a "catastrophic" threshold so long as the threshold is known (and certain other conditions are satisfied), but that collective action collapses if the threshold is uncertain (Barrett 2013). Experimental evidence confirms this behavior (Barrett and Dannenberg 2012). When the threshold is certain, players coordinate to stave off "catastrophe." When the threshold is uncertain, they fail to cooperate so as to stay on the good side of the threshold.[1]

The literature on international environmental agreements has assumed that the underlying climate change game is a prisoners' dilemma. Until recently, however, this literature has ignored the possibility of thresholds. The key insight of the research reported above is that the climate change game may be a prisoners' dilemma for a different reason than assumed previously. The reason may not be that climate change is "gradual." The reason may be that climate change is "abrupt and catastrophic" but with an uncertain threshold.

Does the distinction matter? Analytical game theory predicts that behavior should be identical in both of these situations. Using our particular analytical model, free-riding should cause countries to forego any abatement whether climate change is "gradual" and certain or "catastrophic" and uncertain. However, the consequences of free-riding are worse when countries face the prospect of a looming "catastrophe." Moreover, numerous experiments have demonstrated that people tend to contribute more than predicted by analytical game theory, though less than is needed to supply an efficient amount of a public good (Ledyard 1995). It is as if pure self-interest pulls players in one direction (towards free-riding), and group interest pulls them in another (towards full cooperation). In short, people are conflicted; they are, after all, caught in a dilemma. This suggests that behavior may differ when countries try to cooperate to mitigate "gradual" climate change as opposed to "abrupt and catastrophic" climate change with an uncertain threshold, since in the latter case they have much more to lose from the failure to cooperate.

In this chapter we provide an experimental test of this hypothesis. Experiments give valuable insights into people's behavior when facing a collective action problem. Unlike analytical studies, they do not assume any particular preferences (for example, as regards selfishness or a willingness to take risks). Instead, they reveal how real people, having their own preferences, behave when facing a collective action problem. Our results confirm that the uncertain prospect of "catastrophe" increases abatement as compared to the prisoners' dilemma for "gradual" climate change. However, this result is merely a silver lining in an otherwise dark cloud, for our results also confirm that collective action fails to prevent "catastrophe." Given the scientific evidence for thresholds, negotiators were right to emphasize early on the need to avoid "dangerous" climate change. By doing so, our research suggests, global abatement probably increased. But due to scientific uncertainty about the location of the threshold – uncertainty that is substantially irreducible – knowledge of the existence of a threshold only helps in limiting "gradual" climate change. It won't help us to avert "catastrophe."

A simple analytical model

Our underlying game-theoretic model assumes a one-shot setting with N symmetric countries, each able to reduce emissions by up to q^{A}_{max} units using technology A and by up to q^{B}_{max} units using technology B. The per unit costs of reducing emissions by these two technologies are constant but different, with $c^A < c^B$. Technology A may be thought of as representing low-cost "ordinary abatement" and B as a high-cost technology for removing carbon dioxide from the atmosphere (Keith 2009). To understand why we include the latter technology, note that concentrations today are about 400 ppmv CO_2. If the aim were to limit concentrations to 350 ppmv, as proposed by Rockström *et al.* (2009) and others, we not only need to cut emissions substantially, we need to remove CO_2 from the atmosphere. Avoiding "catastrophe" may require dramatic action.

Let Q denote the reduction in emissions by all countries collectively using both technologies. Every unit of emission reduction gives each country a benefit in the amount b, the marginal benefit of avoiding "gradual" climate change. Assuming $c^B > bN > c^A > b$ gives the classical prisoners' dilemma in which self-interest and collective interest diverge. For these parameter values, self-interest impels each country to abate 0, whereas collectively all countries are better off if each abates q^{A}_{max} units using technology A and 0 units using technology B. Air capture is not worth doing in a world facing only "gradual" climate change.

Since climate thresholds can be related to cumulative emissions (Allen *et al.* 2009; Zickfeld *et al.* 2009), threshold avoidance can be expressed in terms of abatement relative to "business as usual." Denote the threshold by $\bar{Q}$, a parameter. Abatement short of this value guarantees that the climate will tip "catastrophically," whereas abatement equal to or greater than this value preserves climate stability. Assume $N\left(q^{A}_{max}+q^{B}_{max}\right)>\bar{Q}>Nq^{A}_{max}$. That is, avoidance of the threshold is technically feasible and requires using technology B in addition to A. Abatement short of $\bar{Q}$ results in "catastrophic" loss of value X. We restrict parameter values so that when countries cooperate fully they can do no better than to abate $\bar{Q}$ precisely, with technology A being fully deployed everywhere and technology B being used as a "top up" to make sure $Q = \bar{Q}$.

Acting independently, each country will maximize its own payoff, taking as given the abatement choices of other countries. We restrict parameter values so that, facing a certain threshold, there are two symmetric Nash equilibria in pure strategies.[2] In one, every country abates 0 and the threshold is exceeded. In the other, every country abates q^{A}_{max} using technology A and $\bar{Q}/N-q^{A}_{max}$ using technology B, ensuring that the threshold is avoided, just.[3] By our restrictions, the latter equilibrium is universally preferred.[4] The game thus involves players coordinating to support this mutually preferred equilibrium.

With threshold uncertainty, $\bar{Q}$ is assumed to be distributed uniformly such that the probability of avoiding "catastrophe" is 0 for $Q<\bar{Q}_{min}$, $\left(Q-\bar{Q}_{min}\right)/\left(\bar{Q}_{max}-\bar{Q}_{min}\right)$ for $Q\in\left[\bar{Q}_{min},\bar{Q}_{max}\right]$, and 1 for $Q>\bar{Q}_{max}$. We assume $N\left(q^{A}_{max}+q^{B}_{max}\right)\geq\bar{Q}_{max}>\bar{Q}_{min}\geq Nq^{A}_{max}$ and restrict parameters so that when countries cooperate fully they abate $\bar{Q}_{max}$ collectively, eliminating threshold uncertainty, and when countries choose their abatement levels non-cooperatively, they do nothing to limit their emissions, making it inevitable that the threshold will be crossed. For purposes of comparison, we assume that the expected value of the threshold is the same in the uncertainty case as in the certainty case.

Our analytical results and the setting of parameters in the experimental treatments are summarized in Table 5.1. The full cooperative abatement level is different for all three treatments. It is higher under *Certain Threshold* than under *No Threshold* because of the assumption that more abatement is needed to avoid "catastrophe" than is worth doing to limit "gradual" climate change. It is higher under *Uncertain Threshold* than under *Certain Threshold* because in this model (even assuming risk-neutral preferences) countries want to eliminate any chance of "catastrophe" (by assumption, the expected value for the threshold is the same under *Certain Threshold* and *Uncertain Threshold*). That is, it is in the collective interests of countries to act as if according to a precautionary principle.

By contrast, the non-cooperative abatement level is predicted to be the same in the *No Threshold* and *Uncertain Threshold* treatments (zero abatement) but different for *Certain Threshold.* As explained before, for *Certain Threshold* there are two symmetric Nash equilibria, only one of which is efficient. With threshold certainty, abatement of greenhouse gases is a coordination game and, so long as the players can communicate, there is strong reason to believe that countries will coordinate around the more efficient equilibrium.[5]

Our aim is to test these qualitative and quantitative predictions in the lab. In the next section we explain how our experiment was designed to allow us to do this.

Experimental design

At the start of every game, each subject was given "working capital" of €11, distributed between Accounts A (€1) and B (€10). Contributions to the public good consisted of poker chips (abatement) purchased from these accounts. Chips purchased from Account A cost €0.10 each ($c^A = 0.1$), and there were 10 chips ($q^A_{max} = 10$). Chips paid for out of Account B cost €1.00 each ($c^B = 1$), and again there were 10 chips ($q^B_{max} = 10$). Every subject was also given an endowment fund of €20, allocated to Account C. This fund could not be used to purchase chips; it was included only to ensure that no player could be left out of pocket. When the game was over, each subject received a payoff equal to the amount of money left in his or her three accounts, after making the following adjustment: each subject was given €0.05 for every poker chip contributed by the group regardless of who had contributed that chip and from which account ($b = 0.05$). This treatment gives the classical prisoners' dilemma and is called *No Threshold.* Two more treatments included an additional adjustment: each subject's payoff was reduced by €15 ($X = 15$) unless $\bar{Q}$ or more chips were contributed. In the *Certain Threshold* treatment, $\bar{Q}$ was set equal to 150 and in the *Uncertain Threshold* treatment, $\bar{Q}$ was assumed to be distributed uniformly between 100 and 200. All parameter values are consistent with the expressions shown in Table 5.1.

Table 5.1 Model summary and parameterization of the experimental treatments

Treatment	*Game*	*Threshold*	*Full cooperation*	*Non-cooperation*	*Avoid "catastrophe"?*
No Threshold	Prisoners' dilemma for "gradual" climate change	—	Nq^A_{max}	0	—
			100	0	
*Certain Threshold**	Coordination	$\bar{Q}$	$\bar{Q} > Nq^A_{max}$	0, $\bar{Q}$	Yes**
		150	150	0, 150	
*Uncertain Threshold**	Prisoners' dilemma for "dangerous" climate change	$[\bar{Q}_{min}, \bar{Q}_{max}]$	$\bar{Q}_{max} > \bar{Q} > Nq^A_{max}$	0	No
		[100, 200]	200	0	

*These treatments are taken from Barrett and Dannenberg (2012). **Assumes coordination on the efficient Nash equilibrium.

The experimental sessions were conducted in a computer laboratory at the University of Magdeburg, Germany, using students recruited from the general student population (recruiting software Orsee; see Greiner 2004). In total, 300 students participated in the experiment, 100 per treatment: 10 groups × 10 students per group. In each session, subjects were seated randomly at linked computers (game software Ztree; see Fischbacher 2007). A set of written instructions including several numerical examples and control questions was handed out. The instructions involved a neutral frame for the experiment in order to avoid any potential biases the subjects may have regarding climate change. The control questions tested subjects' understanding of the game to ensure that they were aware of the available strategies and the implications of making different choices.

At the beginning of each session, subjects were assigned randomly to 10-person groups and played five practice rounds, with the membership of groups changing after each round. After a final reshuffling of members, each group played the game for real. To ensure anonymity, the members of each group were identified by the letters A through J. The game was played in stages; subjects first proposed a contribution target for the group and pledged an amount they each intended to contribute individually. It was common knowledge that these announcements were non-binding but would be communicated to the group. After being informed of everyone's proposals and pledges, subjects chose their actual contributions in the second stage. The decisions in both stages were made simultaneously and independently. After the game, subjects were informed about everyone's decisions and asked to complete a short questionnaire, giving a picture of their reasoning, emotions, and motivation during the game. In the *Threshold Uncertainty* treatment, "Nature" chose the threshold in a third stage: a volunteer was invited to activate a computerized "spinning wheel" to determine the value for the threshold. This novel way of demonstrating a uniform distribution placed the minimum and maximum value of the threshold range (100 and 200) at the "ends" of the wheel at 12 o'clock.[6] Every subject was able to observe the wheel being spun and see where the arrow came to rest. At the end of each session, students were paid their earnings in cash.

Compared with the earlier literature, our experiment involved a threshold public goods game with no rebate (contributions above the threshold are not returned) and no refund (contributions are not returned if they fall short of the threshold) where the provision threshold is set to zero (*No Threshold*), or 150 (*Certain Threshold*), or is a random variable distributed uniformly between 100 and 200 (*Uncertain Threshold*).[7] Table 5.2 shows total contributions and individual payoffs corresponding to the three treatments.

As shown in the table, compared to *No Threshold*, *Uncertain Threshold* increases the gap between the full cooperative and non-cooperative outcomes in terms of both contributions and payoffs. Previous papers have not made this same comparison, but they have tested for the effect of increases in the gap between the full cooperative and non-cooperative payoffs in linear public goods games by increasing the marginal per capita return from the public good. There is robust evidence that an increase in the marginal return increases contributions (see Davis and Holt 1993; Ledyard 1995; and the literature cited therein). Therefore, although the theory predicts free-riding behavior in both treatments, we may expect larger contributions in *Uncertain Threshold* than in *No Threshold*.

Results

Table 5.3 shows the summary statistics of the experimental data averaged across groups for each treatment.

Look first at the proposals. In the *No Threshold* treatment, the mean proposal for the group target was 116. This is higher than the full cooperative level (100). However, the full cooperative level was proposed more often (by 63 percent of subjects) than any other value. The mean proposal for *Uncertain Threshold* was

Table 5.2 Full cooperative and non-cooperative outcomes

Treatment	*Total contributions*			*Individual payoffs*		
	Full cooperation	*Non-cooperation*	*Difference*	*Full cooperation*	*Non-cooperation*	*Difference*
No Threshold	100	0	100	€35	€31	€4
Certain Threshold	150	150*	0	€32.5	€32.5*	€0
Uncertain Threshold	200	0	200	€30	€16	€14

*Assumes coordination on the efficient Nash equilibrium.

Table 5.3 Summary statistics

	Proposal		*Pledge*		*Contribution*		*Group contribution*
Treatment	*Mean (Std dev)*	*Mode (%)*	*Mean (Std dev)*	*Mode (%)*	*Mean (Std dev)*	*Mode (%)*	*Min / max*
No Threshold	115.8 (14.53)	100 (63%)	10.64 (1.18)	10 (65%)	4.9 (1.90)	0 (42%)	22 / 75
Certain Threshold	151.9 (1.57)	150 (83%)	14.7 (0.51)	15 (74%)	15.1 (0.77)	15 (56%)	136 / 159
Uncertain Threshold	166.3 (9.85)	200 (29%)	15.8 (1.69)	20 (32%)	7.7 (1.67)	10 (36%)	55 / 107

Mean and modal values for proposals, pledges, and contributions; standard deviations calculated with the group average taken as the unit of observation; percentages are shares of individuals per treatment. The rightmost column shows minimum and maximum group contributions for each treatment.

166. This is lower than the full cooperative level (200). Once again, however, the full cooperative level was proposed more frequently than any other value (but by only 29 percent of subjects in this case). Finally, the mean proposal for *Certain Threshold* was 152. This is almost precisely equal to the full cooperative level (150), which was also by far the most frequent proposal (83 percent of subjects, a remarkable degree of concordance). Taking groups as the unit of observation, a series of Mann-Whitney–Wilcoxon (MWW) tests shows that the differences in proposals between all three treatments are statistically significant ($n = 20$, $p = .00$ each; see Table 5.4).

Now consider the pledges. The mean pledge in the *No Threshold* treatment was 11. This is just a little over the full cooperative level, 10, which was the most frequently made pledge (announced by 65 percent of subjects). For the other prisoners' dilemma, *Uncertain Threshold*, the mean pledge was 16. This is below the most frequent pledge, which once again equals the full cooperative level (20, announced by 32 percent of subjects). Finally, in *Certain Threshold*, the mean pledge was equal to the most frequent pledge (15, announced by 74 percent of subjects), a value equal to the full cooperative level. The differences in pledges are significant between the *No Threshold* and the other two treatments (MWW test, $n = 20$, $p = .00$ each, see Table 5.4) and are weakly significant between *Certain Threshold* and *Uncertain Threshold* ($n = 20$, $p = .06$).

Lastly, look at the actual contributions. For *No Threshold*, the mean group contribution was 49, but the distribution of contributions varied widely, with most subjects contributing either zero (42 percent) or 10 (36 percent). Of course, the theory predicts that contributions should equal zero, but in this experiment chips contributed from Account A are very cheap. For *Uncertain Threshold*, the mean group contribution was 77. As in the *No Threshold* treatment, most subjects chose a contribution of either 10 (36 percent) or zero (30 percent). Behavior in the two prisoners' dilemma games was thus very similar. What differed was the level of provision. Unlike the players in *No Threshold*, a few players in *Uncertain Threshold* threw in some of their expensive chips (25 percent), though only 2 out of 100 contributed all their expensive chips (see Figure 5.1 below). Finally, the mean group contribution for *Certain Threshold* was 151. This is just a hair over the predicted 150 (t-test, $n=10$, $p=.72$), and in this case the most frequent individual contribution level equals the full cooperative level (15, chosen by 56 percent of subjects). The differences in contributions are significant between all three treatments (MWW test, $n=20$, $p=.00$ each; see Table 5.4). The contributions in the two prisoners' dilemma games (*No Threshold* and *Uncertain Threshold*) are not only lower but also more erratic than those in the coordination game (*Certain Threshold*) (Levene test, $n=20$, $p<.05$ each; see Tables 5.3 and 5.4). However, contributions in both *No Threshold* and *Uncertain Threshold* are significantly greater than the predicted zero (one-sided t-test, $n=10$, $p=.00$ each).

For *Certain Threshold* and *Uncertain Threshold*, the probability of "catastrophe" differs dramatically (MWW test, $n=20$, $p=.00$). In *Certain Threshold* 8 out of 10 groups avoided "catastrophe." By contrast, "catastrophe" occurred in 9 out of 10 cases for the *Uncertain Threshold* treatment, with the outlying group reducing the probability of "catastrophe" by just 7 percent.

Contributions in both of the prisoners' dilemma games are significantly lower than the proposals and pledges (Wilcoxon Signed-Rank test, $n=10$, $p<.01$ each; see Figure 5.1). By contrast, contributions in *Certain Threshold* are nearly equal to the amounts proposed and pledged.

Table 5.4 Significance of treatment differences

	No Threshold			*Certain Threshold*		
	Proposal	*Pledge*	*Contribution*	*Proposal*	*Pledge*	*Contribution*
Certain Threshold	.0002 (.0001)	.0002 (.1642)	.0002 (.0044)			
Uncertain Threshold	.0002 (.1144)	.0002 (.2655)	.0041 (.5692)	.0002 (.0052)	.0638 (.0170)	.0002 (.0137)

p-values from a Mann-Whitney Wilcoxon rank-sum test of treatment differences in mean values; in parentheses p-values from a Levene test of treatment differences in variances.

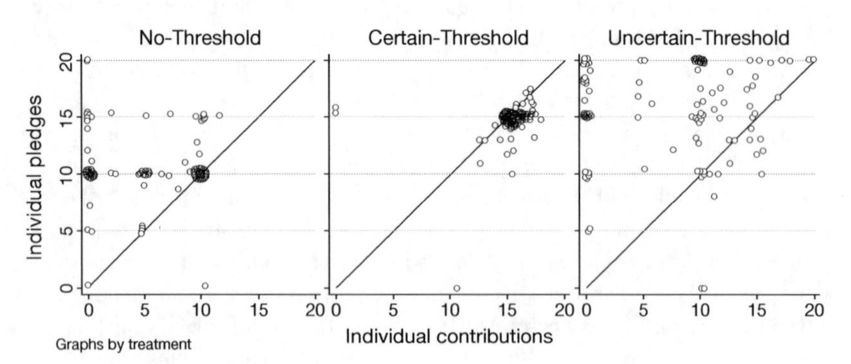

Figure 5.1 Correlation between pledges and contributions

Note: A small noise (3%) has been inserted to make all data points visible.

Figure 5.1 shows the relationship between individual pledges (vertical axis) and individual contributions (horizontal axis). The correlation is positive and highly significant for *Certain Threshold* (Spearman's correlation test, $n = 100$, $rho = .38$, $p = .00$), while it is insignificant for both *No Threshold* ($n = 100$, $rho = -.03$, $p = .80$) and *Uncertain Threshold* ($n = 100$, $rho = .10$, $p = .34$). For *Certain Threshold,* almost all players (98 percent) contributed at least as much as they pledged, while only a few players did so in *No Threshold* (33 percent) and *Uncertain Threshold* (18 percent).

Table 5.5 presents the responses to the ex-post questionnaire. Whenever questions are about the game rather than about general attitudes, the responses of the participants in the different treatments vary considerably, particularly between the prisoners' dilemma and coordination games. For example, while fairness and trust are important driving forces for the contribution decisions in *Certain Threshold,* the coordination game, they are less relevant in *No Threshold* and *Uncertain Threshold,* the two prisoners' dilemmas. The proposals, and in particular the pledges, are perceived as being much more useful in *Certain Threshold* than in *No Threshold* and *Uncertain Threshold.* In coordination games, communication is particularly important.

The theory assumes that players are risk-neutral. Is this a reasonable assumption? As can be seen in Table 5.5 (see question 12), subjects' risk aversion does not differ significantly between treatments; the percentage of risk-averse subjects in each treatment is between 56 percent and 62 percent. Moreover, analysis of behavior within each treatment shows that there is no significant correlation between individual risk aversion and individual contributions or between the number of risk-averse members in a group and group contributions (Spearman's correlation test, $p > .10$ each). Thus we cannot reject the hypothesis that risk aversion and behavior in the games are independent. People tend to play the same way in these games, irrespective of their preferences about risk.

Table 5.6 presents subjects' responses to the open questions about their motivation for making proposals, pledges, and contributions. The responses were classified according to key words and assigned to certain response categories. A large majority of the students playing *Certain Threshold* were motivated to state their proposal so as to maximize the joint group payoff (82 percent). By contrast, in the two prisoners' dilemma games, the motivation for making proposals was spread more evenly across three different responses: wanting to maximize joint payoffs, being realistic, and stimulating contributions by others. A large majority of the students playing in *Certain Threshold* used the pledge to signal truthfully their intended contribution and to create trust within the group (71 percent). By contrast, most subjects playing in *No Threshold* and *Uncertain Threshold* used the pledge to stimulate contributions by the other players (48 percent and 66 percent, respectively). As for the contribution decision, responses indicate that subjects in *Certain Threshold* were motivated either to contribute their fair share of the burden (56 percent) or to compensate for potentially missing contributions (33 percent). Most subjects playing in the *No Threshold* and *Uncertain Threshold* treatments said that they chose their contributions because they wanted to

Table 5.5 Responses to the ex-post questionnaire (percentages of subjects per treatment)

Question	*Response*	*No Threshold*	*Certain Threshold*	*Uncertain Threshold*
1) Were you generally satisfied with the game's outcome?	Very much	28	63	10
	Somewhat	48	18	31
	Little	16	5	26
	Not at all	8	14	33
2) Knowing how the game was played, with the benefit of hindsight, do you wish you had made a different contribution?	Very much	4	2	11
	Somewhat	15	19	17
	Little	17	27	22
	Not at all	64	52	50
3) Did fairness play a role for your contribution decision?	Very much	29	61	24
	Somewhat	23	16	10
	Little	17	11	21
	Not at all	31	12	45
4) Did trust play a role for your contribution decision?	Very much	23	58	18
	Somewhat	28	22	12
	Little	21	9	23
	Not at all	28	11	47
5) Do you agree with the statement that the exchange of proposals was helpful?	Very much	11	49	6
	Somewhat	37	27	28
	Little	33	13	34
	Not at all	19	11	32
6) Do you agree with the statement that the exchange of pledges was helpful?	Very much	13	68	10
	Somewhat	41	24	30
	Little	28	5	27
	Not at all	18	3	33
7) Generally speaking, do you trust other people?	Very much	24	25	21
	Somewhat	52	60	60
	Little	23	13	17
	Not at all	1	2	2
8) Generally speaking, do you agree with the statement that, if a person fails to keep his or her word, they deserve another chance?	Very much	40	24	41
	Somewhat	50	54	45
	Little	9	18	14
	Not at all	1	4	0
9) Generally speaking, do you try to keep your word?	Always	41	56	36
	Often	56	41	60
	Sometimes	2	1	4
	Rarely	1	1	0
	Never	0	1	0
10) Did you trust the other players to make the contributions they pledged?	Very much	8	47	10
	Somewhat	37	43	23
	Little	32	8	26
	Not at all	23	2	41

Question	Response	No Threshold	Certain Threshold	Uncertain Threshold
11) Knowing how the game was played, with the benefit of hindsight, do you feel that some of the other players betrayed your trust in them?	Very much	7	10	16
	Somewhat	24	12	21
	Little	34	37	23
	Not at all	35	41	40
12) Please imagine the following situation in another unrelated experiment: You have an initial endowment of €40. There is a 50% possibility that you will lose your €40. However, you can avoid this loss by paying €20 up front. Would you rather pay this amount and get €20 for certain or would you rather accept the risk of losing the €40 with probability 50%?	€40 uncertain	21	15	25
	Indifferent	23	27	13
	€20 certain	56	58	62

Table 5.6 Responses to the ex-post questionnaire (open questions)

Question	Response	No Threshold	Certain Threshold	Uncertain Threshold
1) What was the most important reason for your proposal for the group contribution?	Joint payoff maximization	33	82	22
	Fairness	5	3	1
	Safety	2	8	0
	Stimulation of others' contributions	34	2	31
	Realistic target	19	0	39
	Other reason	7	5	7
2) What was the most important reason for your pledge for your own intended contribution?	Signaling of intended contribution/ creation of trust	32	71	24
	Stimulation of others' contributions	48	17	66
	Safety	4	5	4
	Other reason	16	7	6
3) What was the most important reason for your contribution?	Fair share to reach target/own pledge	25	56	12
	Compensation of potentially missing contributions/safety	1	33	0
	Own payoff maximization	38	10	24
	Resignation/distrust	20	0	30
	Cheap chips/compromise between group and own interest	13	0	33
	Other reason	3	1	1

Percentages of subjects per treatment. These questions were posed as open questions; subjects' responses were classified by keyword search.

maximize their own payoff (38 percent and 24 percent, respectively), because they distrusted the other players (20 percent and 30 percent), or because the chips were cheap (13 percent and 33 percent). Thus there is a remarkable difference in the motivation and reasoning between the coordination and prisoners' dilemma games, indicating that the context of the games shapes people's beliefs and perceptions of appropriate behavior.

Conclusions

Theory predicts that behavior should be the same in the two prisoners' dilemma games: *No Threshold*, which corresponds to "gradual" climate change, and *Uncertain Threshold*, which corresponds to "dangerous" climate change with an uncertain threshold for "catastrophe." Our experimental results show that the motivations for the players were very similar in these games. However, the contributions varied. In both cases, the contributions exceeded the predicted amount, with the students playing the *Uncertain Threshold* prisoners' dilemma contributing more than the students playing the *No Threshold* prisoners' dilemma. This suggests that the framing of the negotiations around the need to avoid "dangerous" climate change has been advantageous. Countries may reduce their emissions a bit more when facing this prospect than when they are ignorant about the prospect of "catastrophe." But our results also suggest that the warning about "dangerous" climate change will not suffice to cause countries to reduce their emissions by enough to avoid crossing the threshold.

The result that contributions are higher under *Uncertain Threshold* than under *No Threshold* confirms findings in other experimental settings, primarily linear public goods games. It is well established that people do not play a prisoners' dilemma precisely as predicted by analytical game theory; cooperation tends to be partial rather than completely absent. People are torn when playing a prisoners' dilemma. Their motives are mixed. Under *Uncertain Threshold*, playing only to please one's self-interest comes at a painful collective cost. And yet our results should offer only little comfort. The players are not able to avoid the threshold. This is in complete contrast to how students play the game in which the threshold is certain. In this case, cooperation is enforced by "Mother Nature," which provides a sharp punishment for deviations from the full cooperative outcome. In the two prisoners' dilemma games, deviations from full cooperation are individually profitable for the players; to deter free-riding, players must provide the punishment and enforcement themselves.

The policy implication is clear. The central challenge for a climate agreement is enforcement, and the Kyoto Protocol lacks an effective enforcement mechanism. Kyoto did not stop the United States from failing to participate. Nor did it create incentives for Canada to comply or to remain a party. Of course, many states have reduced their emissions a little (Kyoto only aimed to reduce the emissions of Annex I countries by about 5 percent). However, this behavior is consistent with the players in our experiment contributing some of their cheap chips. The emission reductions needed to avoid the tipping points identified in the scientific literature require greater sacrifices. To avoid these levels, countries will have to hand in their expensive chips. Our research thus offers the following advice to negotiators: it is less important that countries agree on the collective target needed to avoid dangerous climate change than that they negotiate effective "strategic" mechanisms for enforcement (Barrett 2003). If "Mother Nature" doesn't provide enforcement, strategic mechanisms are needed to create the same conditions that, as our experiment has revealed, exist under a certain threshold.

Acknowledgements

We are grateful to James Rising for programming our "spinning wheel," and to the MaXLab team at Magdeburg University for allowing us to use their lab. This work was supported by the Swedish Research Council for the Environment, Agricultural Sciences and Spatial Planning through the program Human Cooperation to Manage Natural Resources.

Notes

1. This same theory also predicts that uncertainty about the impact of crossing a threshold should have no effect on behavior (Barrett 2013) – another behavior confirmed by our experiment (Barrett and Dannenberg 2012). It is uncertainty in the threshold that matters.
2. Of course, there are also many asymmetric Nash equilibria, but in our set up, contributions that are approximately symmetric are particularly focal.
3. Starting from a situation in which every country abates $\bar{Q}/N$, should any one country reduce its abatement unilaterally by one unit, it will save c^B but lose $b + X$. Play $\bar{Q}/N$ is thus a Nash equilibrium so long as $X \geq c^B - b$.
4. Avoiding "catastrophe" is mutually preferred so long as $b\bar{Q} - c^A q^A_{\max} - c^B\left(\bar{Q}/N - q^A_{\max}\right) \geq -X$.
5. If there is one thing climate negotiators can do, it is communicate, which is why we include this possibility in our experiment. The experimental literature on communication has shown that restricted and anonymous communication, such as the proposals and pledges in our experiment, improves coordination but works much less reliably for cooperation (for reviews, see Balliet 2010; Bicchieri and Lev-On 2007; Chaudhuri 2011; Crawford 1998; Croson and Marks 2000). A previous threshold experiment by Milinski *et al.* (2008) found that the (certain) threshold was often crossed. However, this experiment did not allow communication. Tavoni *et al.* (2011) modified this experiment to show that communication significantly improves coordination.
6. For details, see the Supplementary Information for Barrett and Dannenberg (2012).
7. On threshold public goods experiments, see Bagnoli and McKee (1991); Croson and Marks (2000); on rebate rules in threshold public goods experiments, see Marks and Croson (1998).

References

Allen, M.R., D.J. Frame, C. Huntingford, C.D. Jones, J.A. Lowe, M. Meinshausen, and N. Meinshausen (2009). Warming caused by cumulative carbon emissions towards the trillionth tonne. *Nature* 458, 1163–66.

Archer, D. (2010). *The Global Carbon Cycle.* Princeton: Princeton University Press.

Bagnoli, M. and M. McKee (1991). Voluntary contribution games: efficient private provision of public goods. *Economic Inquiry* 29(2), 351–66.

Balliet, D. (2010). Communication and cooperation in social dilemmas: a meta-analytic review. *Journal of Conflict Resolution* 54(1), 39–57.

Barrett, S. (2003). *Environment and Statecraft: the strategy of environmental treaty-making.* Oxford: Oxford University Press.

——(2013). Climate treaties and approaching catastrophes. *Journal of Environmental Economics and Management* 66(2), 235–50.

Barrett, S. and A. Dannenberg (2012). Climate negotiations under scientific uncertainty. *Proceedings of the National Academy of Sciences* 109(43), 17372–76.

Bicchieri, C. and A. Lev-On (2007). Computer-mediated communication and cooperation in social dilemmas: an experimental analysis. *Politics, Philosophy & Economics* 6(2), 139–68.

Chaudhuri, A. (2011). Sustaining cooperation in laboratory public goods experiments: a selective survey of the literature. *Experimental Economics* 14(1), 47–83.

Crawford, V. (1998). A survey of experiments on communication via cheap talk. *Journal of Economic Theory* 78, 286–98.

Croson, R.T.A. and M.B. Marks (2000). Step returns in threshold public goods: a meta- and experimental analysis. *Experimental Economics* 2(3), 239–59.

Davis, D. and C. Holt (1993). *Experimental Economics.* Princeton: Princeton University Press.

Fischbacher U. (2007). Z-Tree: Zurich toolbox for ready-made economic experiments. *Experimental Economics* 10(2), 171–78.

Greiner B. (2004). An online recruitment system for economic experiments. In K.Kremer and V. Macho (eds.), *Forschung und wissenschaftliches Rechnen 2003.* GWDG Bericht 63, Göttingen, Ges. für Wiss. Datenverarbeitung, pp. 79–93.

Keith, D.W. (2009). Why capture CO_2 from the atmosphere? *Science* 325(5948), 1654–55.

Ledyard, J. (1995). Public goods: a survey of experimental research. In J.H. Kagel and A.E. Roth (eds.), *Handbook of Experimental Economics.* New Jersey: Princeton University Press, pp. 111–94.

Lenton, T.M., H. Held, E. Kriegler, J.W. Hal, W. Lucht, S. Rahmstorf and H.J. Schellnhuber (2008). Tipping elements in the earth's climate system. *Proceedings of the National Academy of Sciences* 105(6), 1786–93.

Marks, M. and R. T. A. Croson (1998). Alternative rebate rules in the provision of a threshold public good: an experimental investigation. *Journal of Public Economics* 67(2), 195–220.

Milinski M., R.D. Sommerfeld, H.J. Krambeck, F.A. Reed and J. Marotzke (2008). The collective-risk social dilemma and the prevention of simulated dangerous climate change. *Proceedings of the National Academy of Sciences* 105(7), 2291–94.

Rockström, J., W. Steffen, K. Noone, Å. Persson, S. Chapin III, E.F. Lambin, T.M. Lenton, M. Scheffer, C. Folke, H.J. Schellnhuber, B. Nykvist, C.A. de Wit, T. Hughes, S. van der Leeuw, H. Rodhe, S. Sörlin, P.K. Snyder, R. Costanza, U. Svedin, M. Falkenmark, L. Karlberg, R. W. Corell, V.J. Fabry, J. Hansen, B. Walker, D. Liverman, K. Richardson, P. Crutzen and J.A. Foley (2009). A safe operating space for humanity. *Nature* 461, 472–75.

Roe, G. H and M. B. Baker (2007). Why is climate sensitivity so unpredictable? *Science* 318(5850), 629–32.

Tavoni, A., A. Dannenberg, G. Kallis and A. Löschel (2011). Inequality, communication, and the avoidance of disastrous climate change in a public goods game. *Proceedings of the National Academies* 108(29), 11825–29.

Zickfeld, K., M. Eby, H.D. Matthews and A.J. Weaver (2009). Setting cumulative emissions targets to reduce the risk of dangerous climate change. *Proceedings of the National Academy of Sciences* 106(38), 16129–34.

6 US climate policy and the shale gas revolution

Guri Bang and Tora Skodvin

Introduction

Under the presidency of Barack Obama, climate change has been elevated to the top of the US political agenda several times. Climate legislation was introduced and debated in the 111th Congress (2009–2010), but never gained traction in the Senate. In 2010, following a Supreme Court decision in 2007, the President charged the US Environmental Protection Agency (EPA) with crafting new GHG regulations under the Clean Air Act. EPA regulations were met with protests on Capitol Hill, and a host of unsuccessful attempts to kill such regulations. In June 2013, President Obama launched a Climate Change Action Plan to strengthen US climate policy. Meanwhile, since 2009, recoverable natural gas reserves have increased tremendously, thanks to new technology combining horizontal drilling and hydraulic fracturing. The resulting "shale gas revolution" has dramatically changed the US energy market, making cheap natural gas a realistic alternative to coal for electricity generation, and offering a fuel alternative that is cleaner than diesel – a development that, in the short run, has the potential to reduce US GHG emissions independent of regulation.

This chapter explores two main questions. First, to what extent might the new abundance of natural gas in the US domestic energy market alter the climate policy debate, and help legislators overcome barriers to federal GHG regulations that have prevailed during the past several decades? Second, what implications does abundant natural gas have for US participation in an international climate agreement?

To explore these questions we analyzed 12 key hearings on energy policy that took place in the Senate Committee on Energy and Natural Resources during the 112th Congress (2011–2012). Most observers see the vast resources of both shale gas and shale oil as an opportunity for the United States to alleviate concerns about energy security. The extent to which the objective of energy security is linked to a climate policy agenda, however, varies among key actors. The US shale gas revolution is currently in its earliest stages, and it is difficult to predict with any level of certainty its long-term effects on US energy policy in general or climate policy in particular. Our analysis should therefore be seen primarily as an exploration of how this major development in the US energy market may (or may not) alter energy and climate policies. While congressional hearings are a good source of information about the positions held by various groups, hearings are not comparable to congressional floor debates, and are thus an indirect way of identifying legislators' positions.

Our analysis is organized into four parts. The first summarizes US climate policy debates since the early 1990s, with a particular focus on barriers to federal climate legislation. The second considers the shale gas revolution that began in 2009 and identifies three energy-policy alternatives that are often mentioned in political and scholarly debates regarding the potential role of shale gas in a low-carbon energy future. The third part assesses the frequency with which the three policy alternatives were supported in 12 key congressional hearings held in the course of the 112th Congress. The chapter concludes by considering potential implications of the shale gas revolution for US participation in an international climate agreement.

The persistence of barriers to federal GHG regulations

Climate change has been on the US policy agenda since the early 1990s, when international negotiations for the United Nations Framework Convention on Climate Change (UNFCCC) began. In 1993, a Democrat-controlled Congress rejected President Bill Clinton's proposal for a carbon tax (the so-called Btu tax). Ten years later, a Republican-controlled Senate rejected the Climate Stewardship Act, proposed by Senators John McCain (R-AZ) and Joseph Lieberman (D-CT) (US Senate 2003). In 2005, when Senators McCain and Lieberman proposed an amendment to the Energy Policy Act that would have capped GHG emissions, it was rejected by the Republican-controlled Senate (US Senate 2005). In 2008, Senators Lieberman and John Warner (R-VA) proposed the Climate Security Act, which was rejected by the Democrat-controlled Senate (US Senate 2008). Finally, in 2009, a Democrat-controlled House of Representatives approved the American Clean Energy and Security Act, but the bill was never put to a vote in the Democrat-controlled Senate.

Two key factors seem to determine US legislators' positions on climate change: party affiliation and the natural resource endowments of their home state. For 20 years, despite shifts in political majorities, these two determinants have led to persistent congressional opposition to policies designed to address climate change. Analyses suggest that Democrats are generally more supportive of environmental regulations than Republicans (Shipan and Lowry 2001). Climate policies seem to follow this trend. Stavins suggests that "the unprecedented degree of political polarization that has paralyzed both houses of Congress" is a key obstacle to the realization of President Obama's climate policy ambitions for his second term (Stavins 2013). The most recent climate votes in the Senate (2003, 2005, and 2008) also reflect ideological divergence: 88.6 percent of Republican senators voted no to climate legislation, while 82 percent of Democratic senators voted yes.[1, 2] The 2009 American Clean Energy and Security Act passed "with support from 83 percent of the Democrats, but only 4 percent of the Republicans" (Stavins 2013).

The marked difference between Democrats and Republicans in environmental policies may be related to an ideological divide concerning the role of the state: traditionally, Republicans tend to be more skeptical of government intervention than Democrats. Environmental policies tend to expand the reach of governmental regulations and often limit the freedom of (polluting) businesses and industries, both of which make such policies less attractive to Republicans (Dunlap *et al.* 2001). Furthermore, Republicans tend to be more skeptical of the claim that human-induced climate change is actually occurring (McCright and Dunlap 2003, 2011).

Despite the strength of the ideological dimension, the natural resource base of legislators' home states creates a cross-cutting cleavage: legislators from fossil fuel dependent states display stronger opposition to GHG emissions regulations than legislators whose home-state economies depend less on fossil fuel resources (Fisher 2006). Five states are major petroleum producers, 25 states are small or large coal producers. Equally important, until 2009, 22 states relied on coal for more than 50 percent of their electricity – and of those, 10 relied on coal for more than 70 percent of their electricity (EIA 2012b). GHG regulations tend to drive up energy prices, particularly for the most fossil fuel dependent regions (i.e., those whose state economies are closely aligned with fossil-based energy resources). The level of fossil fuel dependence illustrated by these numbers suggests that a large share of legislators' constituents will be adversely affected by GHG regulations, which induces these legislators to oppose such regulations. The most recent climate policy votes in the Senate reflect the same pattern: on average, the home states of the senators who voted against climate legislation depended on coal for 58.2 percent of their electricity. In contrast, the home states of senators who voted yes on GHG regulations on average relied on coal for 36 percent of their electricity (EIA 2012b).[3]

If the shale gas revolution is to help legislators overcome barriers to federal limitations on GHGs, it must address the ideological dimension by creating opportunities for GHG policies that will be less interventionist, and the material dimension by creating opportunities for GHG policies that will be less costly for the most fossil-fuel-dependent states.

The significance of shale gas in US climate change strategy

In less than a decade, the US energy situation has changed dramatically from scarcity to abundance: new technology combining horizontal drilling and hydraulic fracturing has made it both technologically and economically viable to extract and process the natural gas trapped in shale formations. Between 2006 and 2010, US shale gas production increased from 1 trillion cubic feet to 4.8 trillion cubic feet (EIA 2011a). While estimates of the recoverable shale gas resource base are uncertain, it is likely large enough to make the United States self-sufficient for decades. In short, shale gas is a "game changer" for the US natural gas market (EIA 2011b).

Since climate change first surfaced on the US policy agenda, political leaders have attempted to link GHG regulations to other key policy priorities to attract support. In particular, climate change has often been linked to energy security – a connection that makes sense, in that "many of the actions that make for a more secure energy system also reduce the warming emissions that come from energy supplies" (Victor 2012: 4). Increasing energy efficiency and diversifying energy resources and technologies, for example, address both energy security and climate change. While energy security is a priority for most legislators, its link to GHG regulations and investments in "clean energy" is contested. Specifically, whereas high-carbon fuels such as coal and heavy oil may have a place in achieving energy security, they have little or no role in a low-carbon energy economy. Similarly, if intermittent renewable energy resources (such as wind and solar power) are to play a role in energy security, major transformations of electricity supply systems will be required (Victor 2012).

The new abundance of shale gas represents a tremendous opportunity in terms of jobs, economic growth, and increased energy security. The debate in the US Congress now concerns two key issues: (1) How, and to what extent, can government regulation help reap the opportunities associated with shale gas? (2) What role should climate change have in shale gas governance? Three main policy alternatives are often referred to in political and scholarly debates. They can be distinguished in terms of the extent to which energy security or climate change concerns take priority, and with respect to the level of government intervention that follows with the policy.

The first policy alternative holds that the new abundance of shale gas represents an opportunity to simultaneously strengthen energy security and reduce GHG emissions without government interference in the energy market. Since 2009, the availability of cheap natural gas has led to a rapid shift from coal to gas in electricity generation, which has also resulted in significant reductions in energy-related CO_2 emissions.[4] In the first quarter of 2012, energy-related CO_2 emissions were at their lowest level in 20 years. Energy Information Administration (EIA) projections suggest that energy-related CO_2 emissions will continue to drop for several years (by 2020, to 9 percent below 2005 levels), after which they will gradually begin to increase. But even by 2040, EIA projects that energy-related CO_2 emissions will be more than 5 percent lower than 2005 levels (EIA 2012a). Although such projections reflect existing federal policies (such as tighter fuel economy standards and energy efficiency standards), the shift from coal to gas will play a major role in emissions reductions and can occur without government interference.

In the second policy alternative, the abundance of shale gas represents a tool in a US low-carbon strategy, but only if regulatory policies are implemented to lock in the displacement of coal by natural gas, so as to achieve sufficient long-term reductions in GHG emissions to prevent "dangerous" climate change.[5] While it is likely that the price of gas will remain low, natural gas prices are notoriously volatile. Therefore, incentives must be provided for utilities to continue to generate electricity from natural gas rather than coal, even if natural gas prices increase. Ensuring a permanent transition from coal to natural gas will require regulatory measures to increase the cost of coal in comparison to natural gas (Plautz 2012; Celebi *et al.* 2012).

In the third policy alternative, shale gas is a potential bridge to a low-carbon future, but only in the context of regulatory policies that will stimulate continued investment in renewables and other low-carbon energy sources. To bring GHGs down to a level that can prevent dangerous climate change requires new, clean-energy technologies. Since low natural gas prices may out-compete renewable and other low-carbon energy sources, sustained government support is essential for the development and deployment of renewable and low-carbon technologies, including carbon capture and sequestration (Moniz *et al.* 2011).

Three lines of argument are often used to support these policy alternatives:

- *Argument 1:* The new abundance of shale gas represents an addition to the US energy-resource base whose exploitation should follow market mechanisms, with as little government intervention as possible.
- *Argument 2:* Regulatory policies to limit GHG emissions are necessary to reduce the competitiveness of coal-fired power plants vis-à-vis natural gas and ensure a permanent transition to natural gas.
- *Argument 3:* The new abundance of shale gas represents a challenge to the development of economically viable clean-energy technologies. Continued governmental support is required, in the form of subsidies, tax incentives, and/or funding for research, development, and deployment.

Whereas Argument 1 is logically incompatible with Arguments 2 and 3, Arguments 2 and 3 could very well be pursued simultaneously.

Shale gas agendas and policy alternatives

To explore key actors' perspectives on the role of climate change in federal policies governing shale gas extraction and exploitation, we analyzed 12 hearings held before the Senate Energy and Natural Resource Committee in the 112th Congress (2011–2012) on the subjects of natural gas, clean energy, and the outlook for the energy market.[6] More specifically, the shale gas revolution and natural gas usage in the United States were discussed in relation to challenges in the energy market, investments in renewables, and, in many cases, climate change. Twelve Democratic senators and eight Republican senators participated in the hearings, and 56 witnesses from government agencies, academic institutions, think tanks, and business organizations gave testimony.

There is a long tradition in scholarly literature for using congressional hearings as a data source (e.g., McCright and Dunlap 2003; Arnold 1990). Congressional hearings are the standard means of obtaining information and opinions from interest groups (US GPO 2013) and are regarded by legislators as a trusted source of information. Thus they have a prominent place in US policy making. Normally, a range of witnesses are invited to testify, representing government agencies, academia, think tanks, and key economic sectors, including industry and finance. Witnesses have the opportunity not only to provide Congress with information related to the legislation under discussion but also to give their opinion on alternative policy proposals. Since hearings feature the direct exchange of information and opinions between witnesses and members of Congress, and all participants' statements are recorded and made publicly available, hearings represent a good opportunity to map congressional interests and preferences. The availability of transcripts allows the analysis and comparison of arguments.

Under Senate rules, the majority party and the committee chairman dominate the proceedings with respect to which issues are discussed and which witnesses are called to testify. The Democrats held a 51 percent majority in the 112th Congress and a 12–10 majority on the Energy and Natural Resources Committee. Hence we expected the hearings to reflect the Democratic policy agenda, at least to some extent.

Our aim was to identify the frequency of support for one or more of the three arguments identified in the previous section, as well as which witnesses and senators supported which argument. To determine the frequency of support for each argument, we selected phrases or words that participants in debates often use when expressing alignment with a particular argument, then entered those terms into the search function in Adobe Acrobat. Thus, to identify Argument 1 we searched for the following: "gas price," "market-driven," "mandate," "jobs," and "balanced energy policy." For Argument 2 we searched for "regulation," "generation," "power plant," and "emissions." Finally, for Argument 3 we searched for "subsidies," "renewable," "winners and losers," and "low carbon." All of the search terms produced hits, although not necessarily across all the 12 hearings we explored. Analyzing hits in the context of the text gave us an overview of the positions of hearing participants. It is important to emphasize that this was an exploratory and qualitatively oriented analysis. Nevertheless, patterns were readily identifiable.

Table 6.1 gives an overview of witnesses' affiliations and the incidence of support for the three main arguments. In the 12 hearings we examined, 56 witnesses participated, giving a total of 68 statements. Of the 56, 15 represented various federal agencies. As would be expected, most of the witnesses from federal agencies supported Arguments 2 and 3. All witnesses from academia, think tanks, and the renewable energy industry did as well, along with 8 out of 9 witnesses from the financial sector.

Witnesses who supported Arguments 2 and 3 often claimed that unchecked development of unconventional gas resources would most likely lead to lower prices for natural gas and unconventional oil — and that, as a result, the market would show a preference for natural gas to produce electricity, and for natural gas or unconventional oil to produce transportation fuels. Without governmental support to achieve commercialization and price parity, renewable energy alternatives such as wind- or solar-powered electricity, and electric-powered vehicles, could be squeezed out of the market. Supporters of Arguments 2 and 3 also argued that a truly diversified US energy portfolio – including renewable energy alternatives – would further improve the country's energy security, while cutting carbon emissions substantially in the long run.

Also as might be expected, all witnesses representing the fossil fuel industry and the manufacturing sector supported Argument 1. Along with most of the

Table 6.1 Witnesses' affiliations and incidence of support

	Federal agencies	*Renewable energy or alternative vehicles industries*	*Fossil-fuel industry*	*Manufacturing*	*Utilities*	*Finance sector/ investors*	*Energy-sector consultants*	*Think tanks*	*Academia*	*N*
Support for A1	2		3	5	3	1	3			17
Support for A2	4				1	2	1	1	3	12
Support for A3	13	5			2	6		2	11	39
N	19	5	3	5	6	9	4	3	14	68

Notes: Fossil-fuel industry includes natural gas and natural-gas-powered vehicles; because some witnesses expressed support for more than one argument, the number of instances of support is higher than the number of witnesses.

energy sector consultants, they expressed more concern about energy security and low energy prices than about ensuring long-term cuts in carbon emissions. Their arguments emphasized that the new domestic energy reserves can ensure a stable supply and low prices for many years ahead, especially if the Keystone XL pipeline is completed, which will connect Canadian oilfields with the US market. A secure supply and low prices would also boost the manufacturing sector and create new jobs in that sector. Adherents of this position argued that the best thing for the government to do is to stand back and let the market forces work, and not implement new regulations that will prematurely or too rapidly increase the demand for natural gas. Manufacturing industry witnesses, for instance, warned that new EPA regulations limiting GHG emissions from existing power plants could cause too rapid a shift away from coal, thereby disrupting the balance between demand and supply in the natural gas market and inflating natural gas prices. Among the potential results could be a nationwide retraction of planned manufacturing investment and a concomitant loss of jobs. Several witnesses from the fossil fuel industry and the manufacturing sector, along with energy sector consultants, pointed out that previous government efforts to intervene in the natural gas market, in the 1970s and 1990s, had caused a great deal of harm. Thus they prefer to allow competition in the energy sector to create a price-driven transition from coal to natural gas in the utility sector, which would produce energy at the lowest possible prices. Clearly, avoiding new regulations would benefit the material interests of the witnesses who supported Argument 1.

Finally, Table 6.1 shows that witnesses representing the utility sector were split with respect to their support for the three arguments. Three utility representatives spoke in support of Argument 1, whereas two witnesses from the utility sector supported Argument 3. We interpret this split as a reflection of the portfolios of the utilities in question – that is, whether portfolios were dominated by investments in fossil fuels or renewable energy.

Table 6.2 gives an overview of senators' support for the three main arguments in the 12 hearings we examined. Twelve Democratic and eight Republican senators participated[7]; Democrats gave 66 statements and Republicans 30. Table 6.2 indicates that ideological differences between Republicans and Democrats are strong. More than 96 percent of the statements from Republican senators supported Argument 1. Senator Lamar Alexander (R-TN), who had been invited specifically to participate in a hearing on advanced vehicle technologies, was the only Republican who voiced support for government regulation of clean technology development (Argument 3), but he did so only once, in that particular context.

Given the relatively small percentage of the entire Senate that was represented on the committee, it is not possible to draw a clear inference regarding the material dimension on the basis of the data we collected. Although more than 74 percent of statements by Democrats were in support of Arguments 2 or 3, Table 6.2 suggests that the material dimension may be more important for Democrats than for Republicans. More than 22 percent of the statements made by Democrats were in support of Argument 1; nevertheless, such statements came primarily from two senators: Senator Joe Manchin (D-WV), who represents a coal state, and Senator

Table 6.2 Senators' party affiliations and incidence of support

	Democrats		*Republicans*	
	%	*N*	%	*N*
Support for A1	22.7	15	96.7	29
Support for A2	3.0	2	0	0
Support for A3	74.2	49	3.3	1
N	100	66	100	30

Mary Landrieu (D-LA), who represents an oil state. Among Republicans on the Committee, of which six out of ten represent fossil fuel dependent states, only 3.3 percent of statements were in support of Argument 3.

Looking more closely at the arguments, we find that the main difference between Democrats and Republicans is their appetite for government intervention in the energy market. Almost all Democrats supported the notion that renewable energy is key to a clean, diversified, and secure energy future, and that renewable energy technologies need government support to survive competition with cheap natural gas and coal. In contrast, almost all Republicans argued that previous experience shows that long-term subsidies for specific energy technologies do not work, and that a better alternative is to support research and development, which will push through market-competitive technologies of all types, including renewable energy technologies, without government interference.

Tables 6.1 and 6.2 both indicate that support for Argument 2 was much lower than support for either of the other two arguments, especially among legislators. The reason that only two senators supported Argument 2 – Senator Al Franken (D-MN) and Senator Maria Cantwell (D-WA) – was probably the controversial nature of the argument. In a worst-case scenario, regulations designed to penalize the use of coal could coincide with an unexpected rise in natural gas prices, trapping utilities in a high-cost situation. At a time when the US economy is still struggling to recover from the financial crisis, support for regulations that could potentially lead to an increase in electricity prices seems to be perceived as politically risky, especially given the traditionally volatile nature of natural gas prices. In fact, many hearing participants – especially Republican senators, one Democratic senator (Manchin), and witnesses from the manufacturing and fossil fuel industries – voiced strong opposition to the Clean Air Act regulations currently being developed by EPA, which are designed to make the economics of coal-fired power plants unfavorable when compared with those of natural gas plants.

Nearly all participants emphasized the importance of shale gas to US energy security, but the policy alternatives they suggested varied. Participants who supported Argument 1 typically stressed two important aspects of shale gas governance: first, assuming that the huge reserves of unconventional natural gas can be extracted, the need for both oil and natural gas imports to the United States will drop substantially during the coming decades. Although natural gas prices are known to be volatile, large reserves are likely to keep them low; and as low natural gas prices continue, natural gas liquids will, to some extent, replace diesel as a transportation fuel. Natural gas liquids, in combination with domestically produced unconventional oil and biofuels, will further improve energy security. New, stricter emissions standards from vehicles implemented by the Obama administration since 2009 will provide further incentives to decrease fossil fuel use. Taken together, these factors will help reduce oil imports and increase reliance on and use of domestically produced energy. Further government regulations may have a disruptive effect, however, and should be avoided.

Second, supporters of Argument 1 stressed that the abundance of shale gas, and the resulting low prices for natural gas, will strengthen the economy by helping to keep electricity costs down. The economics of power generation from all other energy sources – coal, nuclear, renewables – will also be affected by the drop in natural gas prices, which will likely lead to a substantial increase in natural gas fired power generation. Not only is the outlook for natural gas availability very bright, but the cost of building a new, natural gas fired power plant is lower than the cost of building plants powered by nuclear or renewable energy sources. To ensure expedited licensing and permits for natural gas extraction and thereby guarantee consumers low electricity prices, government policies should be limited. Thus supporters of Argument 1 were more concerned with energy security and ensuring low energy prices than with reducing GHG emissions.

In contrast, participants who supported Argument 3 typically emphasized that unless new initiatives are implemented to slow the growth of fossil fuel use, energy-related CO_2 emissions will not be reduced enough to avoid dangerous effects on Earth's climate. Increasing the use of natural gas would lower emissions levels when compared to coal combustion, but it would still emit too much CO_2 to be sustainable. Hence, these participants emphasized that policy regulations are needed to ensure, first, the increased development and deployment of renewable, fossil-free energy technologies; and second, that over the long term, renewable energy will supply a substantial portion of the base load of US energy needs, in addition to natural gas. In sum, climate change was integral to these participants' perceptions of energy security.

Participants who supported Argument 2 emphasized that a truly secure energy future must take climate change into account, and that government regulations are therefore necessary to internalize the environmental cost of coal combustion – in other words, regulations are needed that will make it more expensive to burn coal than to use natural gas and other energy sources with lower CO_2 emissions. These participants supported EPA regulations for new and existing power plants proposed by the Obama administration, and a clean-energy standard that would provide incentives for the use of low-carbon energy sources.

To summarize, our qualitative exploratory analysis of the 12 hearings indicates that the shale gas revolution is not likely to fundamentally change positions on federal climate policy among legislators, experts, or stakeholders. Ideological differences related to the role of government in regulating the energy market seem to prevail, and are continuing to color arguments senators are using to set the agenda for shale gas governance. Similarly, alternative solutions presented by witnesses reflected long-held views, material interests, and beliefs regarding the benefits of government regulation, whether in the form of EPA rules targeting the economics of coal plants, or subsidies and incentives for the development of renewable and clean-energy technologies.

Consequences for US participation in an international climate agreement

The United States is a key actor in international efforts to control GHG emissions. It is the second-largest emitter of GHGs in the world (surpassed only by China) and has the highest GHG emissions per capita. The United States is also a key actor in the sense that its participation in an international agreement may spur participation by other key actors, notably China (Underdal *et al.* 2012). Thus an international agreement to regulate GHG emissions that does not include the United States is inefficient. Although the United States is a party to the UNFCCC, it withdrew from the Kyoto Protocol in 2001. As the international community has turned its attention to negotiating a new climate agreement in 2015, it is pertinent to ask whether the United States will be more willing than it has been in the past to take on legally binding commitments. Specifically, will the shale gas boom make it easier for the United States to participate in the next climate agreement?

US participation in international agreements is strongly linked to US domestic politics. Although the US Constitution allows the president to negotiate international agreements on behalf of the country, he can ratify such agreements only with the "advice and consent" of the Senate – and only if a two-thirds Senate majority (67) concurs with the provisions of the agreement. The Constitution also states, however, that if an agreement is accepted by a two-thirds majority of the Senate and has been ratified by the president, the provisions of the international agreement constitute the "supreme law of the land" (Article VI) and are subject to the same strict implementation regime as other federal legislation (see e.g., Skodvin and Andresen 2009; Hovi *et al.* 2012). In most cases, it is necessary to adopt new federal legislation – "enabling legislation" – that specifies how the provisions of an international agreement are to be implemented (Hovi *et al.* 2012). US ratification of international agreements can thus be seen as a two-stage process in which implementation of the Senate's endorsement of the provisions of an agreement depends on the will and ability of Congress to adopt enabling legislation. As a consequence, it is often difficult for the United States to ratify international agreements that have provisions not already reflected in federal legislation (Bang 2011).

Our analysis suggests that opposition to federal climate legislation remains strong in the United States. The shale gas revolution seems unlikely to alter the fundamental political cleavages that have dominated US climate policies for the past two decades, and is thus also unlikely to alter the positions of key domestic actors that have blocked climate legislation in the past. In the absence of federal legislation, the willingness and ability of the US government to accept legally binding international targets for GHG emissions is likely to remain very low (Bang 2011; Bang *et al.* 2012).

It could be argued that the shale gas revolution will help reduce US GHG emissions even without regulations, and that this development in itself may enable the United States to join an international climate agreement. It is important to recognize, however, that the changes observed so far in the US energy market are mainly price- and market-driven. Even though natural gas abundance and low prices may temporarily reduce the use of coal in electricity production – and hence energy-related GHG emissions – a reversion to coal use is possible, and even likely, in the long term, in the absence of policies to prevent it. Moreover, from a long-term perspective, the shale gas revolution will do little to reduce US fossil fuel dependence unless it is bolstered by government policies that support and encourage the development of low-carbon and renewable energy technology. The data we have analyzed here offers no indication that the political feasibility of such policies has improved. On the contrary, our analysis indicates that the fundamental and enduring political barriers to GHG regulations remain intact, and that the likelihood of US participation in an ambitious international climate agreement in the near future remains very low. The shale gas revolution, therefore, does not seem to represent a "sweet spot" with respect to the three challenges that have dominated US energy and climate policy debates for several decades: energy security, low energy prices, and environmental protection.

Acknowledgements

We thank Arild Underdal and one anonymous reviewer for very useful comments on a previous draft, and Sandy Chizinsky for excellent language editing.

Appendix

List of hearings from the Senate Energy and Natural Resources Committee, 112th Congress (2011–2012), used in the analysis.

1 S. Hrg. 112-35 – ENERGY AND OIL MARKET OUTLOOK February 3, 2011.
2 S. Hrg. 112-52 – GLOBAL INVESTMENT TRENDS IN CLEAN ENERGY March 17, 2011.
3 S. Hrg. 112-21 – CLEAN ENERGY DEPLOYMENT ADMINISTRATION May 3, 2011.
4 S. Hrg. 112-38 – ADVANCED VEHICLE TECHNOLOGIES May 19, 2011.
5 S. Hrg. 112-146 – NATURAL GAS July 19, 2011.
6 S. Hrg. 112-215 – LIQUEFIED NATURAL GAS November 8, 2011.
7 S. Hrg. 112-188 – QUADRENNIAL ENERGY REVIEW ACT November 15, 2011.

8 S. Hrg. 112-378 – US GLOBAL ENERGY OUTLOOK FOR 2012 January 31, 2012.
9 S. Hrg. 112-435 – GASOLINE PRICES March 29, 2012.
10 S. Hrg. 112-466 – CLEAN ENERGY May 17, 2012.
11 S. Hrg. 112-536 – AMERICAN ENERGY INNOVATION REPORT May 22, 2012.
12 S. Hrg. 112-587 – USAGE OF NATURAL GAS July 24, 2012.

Source: US GPO (2013).

Notes

1 These were all votes on whether to stop filibuster attempts by invoking cloture. We interpret a no vote to invoke cloture as an indirect no vote regarding the passage of the legislative proposal.
2 These are aggregate numbers for all three votes and were calculated on the basis of Senate voting records (Senate 2003, 2005, 2008).
3 These figures are calculated on the basis of electricity-generation data for each year voting took place (2003, 2005, and 2008).
4 However, the production of shale gas involves methane leakages, which are potentially more serious for the climate than the higher CO_2 emissions caused by coal combustion (see e.g., Howarth *et al.* 2012). Low gas prices may also make it less profitable to capture and use natural gas released in the oil drilling process, hence increasing flaring (see Crooks and Makan 2013).
5 The ultimate objective of the UNFCCC is to avoid "dangerous anthropogenic interference with the climate system" (Article 2). This formulation has been interpreted to imply that average global warming should not exceed 2°C (see e.g., the 2009 Copenhagen Accord).
6 See the appendix for a listing of individual hearings.
7 Senators Alexander (R-TN) and Merkley (D-OR) were invited to participate in a hearing on advanced vehicle technologies. They were not members of the Committee, but cannot be regarded as ordinary witnesses. We have therefore classified and counted them as legislators in Table 6.2.

References

Arnold, D.R. (1990). *The Logic of Congressional Action.* New York: Yale University Press.

Bang, G. (2011). Signed but not ratified: limits to US participation in international environmental agreements. *Review of Policy Research* 28(1), 65–81.

Bang, G., J. Hovi and D. Sprinz (2012). US presidents and the failure to ratify multilateral environmental agreements. *Climate Policy* 12(6), 755–63.

Celebi, M., F. Graves and C. Russell (2012). Potential coal plant retirements: 2012 update. Executive Summary, *Brattle Group Report,* October. Retrieved 27 February 2013 from www.brattle.com/_documents/UploadLibrary/Upload1081.pdf

Crooks, E. and A. Makan (2013). Flares take shine off shale boom. *Financial Times.* 27 January 2013. Retrieved 26 April 2013 from www.ft.com/intl/cms/s/0/ac8b0726-6894-11e2-9a3f-00144feab49a.html#axzz2QniSepAw

Dunlap, R.E., C. Xiao and A.M. McCright (2001). Politics and environment in America: partisan and ideological cleavages in public support for environmentalism. *Environmental Politics* 10(4), 23–48.

EIA (Energy Information Administration) (2011a). Analysis and projections: review of emerging resources: US shale gas and shale oil plays. Retrieved 24 February 2013 from www.eia.gov/analysis/studies/usshalegas/

——(2011b). Review of emerging resources: US shale gas and shale oil plays. Retrieved 24 February 2013 from www.eia.gov/analysis/studies/usshalegas/pdf/usshaleplays.pdf

——(2012a). Annual energy outlook 2013: early release overview. Retrieved 24 February 2013 from www.eia.gov/forecasts/aeo/er/pdf/0383er(2013).pdf

——(2012b). Net generation by state by type of producer by energy source: annual data 1990–2011 (EIA-906, EIA-920, EIA-923). Release date 2 October 2012. Retrieved 12 November 2012 from www.eia.gov/electricity/data/state/

Fisher, D. (2006) Bringing the material back in: understanding the US position on climate change. *Sociological Forum* 21(3), 467–94.

Hovi, J., D. Sprinz and G. Bang (2012). Why the United States did not become a party to the Kyoto Protocol: German, Norwegian and US Perspectives. *European Journal of International Relations* 18, 129–50.

Howarth, R.W., R. Santoro and A. Ingraffea (2012). Venting and leaking of methane from shale gas development: response to Cathles *et al. Climatic Change* 113, 537–49.

McCright, A.M. and R.E. Dunlap (2003). Defeating Kyoto: the conservative movement's impact on US climate change policy. *Social Problems* 50(3), 348–73.

——(2011). The politicization of climate change and polarization in the American public's views of global warming, 2001–2010. *Sociological Quarterly* 52 (2011), 155–94.

Moniz, E., H.D. Jacoby, A.J.M. Meggs, *et al.* (2011). The future of natural gas. *MIT Energy Initiative Report: Executive Summary.* Retrieved 7 March 2013 from http://mitei.mit.edu/system/files/NaturalGas_ExecutiveSummary.pdf

Plautz, J. (2012). Is Obama really waging a war on coal? *Bloomberg News*, October 17. Retrieved 7 March 2013 from www.bloomberg.com/news/2012-10-17/is-obama-really-waging-a-war-on-coal-.html

Shipan, C.R., and W.R. Lowry (2001). Environmental policy and party divergence in Congress. *Political Research Quarterly* 54(2), 245–63.

Skodvin, T. and S. Andresen (2009). An agenda for change in US climate policy? Presidential ambitions and congressional powers. *International Environmental Agreements* 9, 263–80.

Stavins, R.N. (2013). Is Obama's climate change policy doomed to fail? Maybe not. *PBS Newshour: The Business Desk*, March 1. Retrieved 16 March 2013 from www.pbs.org/newshour/businessdesk/2013/03/why-climate-change-skeptics-ar-1.html

Underdal, A., J. Hovi, S. Kallbekken and T. Skodvin (2012). Can conditional commitments break the climate change negotiations deadlock? *International Political Science Review* 33(4), 475–93.

US GPO (2013). Congressional hearings. Retrieved 2 February 2013 from www.gpo.gov/fdsys/browse/collection.action?collectionCode=CHRG&browsePath=112%2FSENATE%2FCommittee+on+Energy+and+Natural+Resources&isCollapsed=false&leafLevelBrowse=false&isDocumentResults=true&ycord=0

US Senate (2003). US Senate roll call votes 108th Congress – 1st Session. Vote no. 420, 30 October 2003. Retrieved 24 April 2013 from www.senate.gov/legislative/LIS/roll_call_lists/roll_call_vote_cfm.cfm?congress=108&session=1&vote=00420

——(2005). US Senate roll call votes 109th Congress – 1st Session. Vote no. 148, 22 June 2005. Retrieved 24 April 2013 from www.senate.gov/legislative/LIS/roll_call_lists/roll_call_vote_cfm.cfm?congress=109&session=1&vote=00148

——(2008). US Senate roll call votes 110th Congress – 2nd Session. Vote no. 145, 6 June 2008. Retrieved 24 April 2013 from www.senate.gov/legislative/LIS/roll_call_lists/roll_call_vote_cfm.cfm?congress=110&session=2&vote=00145

Victor, D. (2012). White paper on energy security and global warming from the World Economic Forum's Global Agenda on Energy Security. Retrieved 24 February 2013 from www3.weforum.org/docs/WEF_GAC_WhitePaperEnergySecurityGlobalWarming_2012.pdf

Part II

Resolution

Paths toward a new agreement

7 International environmental agreements with endogenous minimum participation and the role of inequality

David M. McEvoy, Todd L. Cherry and John K. Stranlund

Introduction

International action to address global climate change presents a social dilemma in which the interests of individual countries do not correspond to the interests of the global community. In the absence of appropriate institutions, voluntary actions to address climate change by countries will fall short of the actions needed to achieve desired outcomes. The analysis and design of institutions to coordinate voluntary actions are important for understanding how countries may overcome this social dilemma. Related research has mostly focused on the performance of exogenous institutions; however, in the case of international climate agreements there is no governing body that can impose and enforce a set of rules on a group of countries. Because nations are sovereign, any institutional arrangement created to facilitate sufficient action to address climate change has to be developed endogenously.

In this chapter we experimentally explore a form of endogenous institution formation in which agents voluntarily form agreements to provide a public good. Prior to making the decision whether to join the agreement, all players vote on the minimum number of members required for an agreement to form. An agreement forms if enough players voluntarily join so that the minimum membership requirement is met or surpassed, and only then are members required to make contributions to the public good. Many international treaties and almost all international climate agreements contain minimum participation requirements like the one explored in this chapter. For example, the Kyoto Protocol to the United Nations Framework Convention on Climate Change required ratification by at least 55 parties accounting for at least 55 percent of total 1990 greenhouse gas emissions. Similarly, the Montreal Protocol on Substances that Deplete the Ozone Layer required at least 11 countries representing at least two-thirds of 1986 ozone consumption to ratify it before it entered into force in 1989. Many non-environmental examples also exist. For example, the Treaty on the Non-Proliferation of Nuclear Weapons required the ratification of the five nuclear nations (at that time) plus 40 additional nations.

Although minimum participation rules are ubiquitous in international negotiations, very little is known about their effectiveness in facilitating cooperation among resource users. Economic experiments provide a platform for testing such mechanisms in a simple, isolated environment in which policy components can be evaluated individually. This study adds to the recent, but growing, literature on experimental tests concerning international environmental negotiations (e.g., Kosfeld *et al.* 2009; Burger and Kolstad 2010; Dannenberg *et al.* 2010; McEvoy 2010; McEvoy *et al.* 2011; Barrett and Dannenberg 2012; Dannenberg 2012; McGinty *et al.* 2012; Cherry and McEvoy 2013). More generally, this study also adds to the literature on endogenous institution formation to confront social dilemmas (see, for examples, Walker *et al.* 2000; Gurerk *et al.* 2006; Tyran and Feld 2006; Kroll *et al.* 2007; Sutter *et al.* 2010). Our work has strong similarities to that of Kosfeld *et al.* (2009). They also analyze the formation of a voluntary coalition designed to increase contributions to a public good. In their game players first decide whether to join a coalition, and then the members vote on whether to contribute to the public good. Hence, in their analysis members of a coalition decide what the coalition should accomplish *after* they make their participation decision. In contrast, the players in our study understand *ex ante* what they are required to do in a coalition (agreement). This is closer to the actual international environmental agreement formation process, in which countries typically decide the commitments of agreement members and what triggers an agreement to enter into force *before* they decide whether to join.

Our experiments have two stages. In the first stage players vote on the minimum number of members required for an agreement to form, and in the second stage they decide whether to join the agreement and contribute to the public good. Our first experimental treatment explores agreements when players have identical payoffs, and efficiency requires that all players join a cooperative agreement to provide a public good. In this treatment, agreements formed most of the time and most of these were efficient. In these cases we observe significant efficiency gains compared to previous studies that use similar incentives but without endogenous minimum membership requirements (Kosfeld *et al.*, 2009; Dannenberg *et al.* 2010; McEvoy *et al.* 2011). We also explore treatments in which the efficient agreement size is less than all the members of the group. In these treatments the marginal benefit of additional cooperators turns to zero after a subset of individuals join the agreement. This is relevant if less-than-global participation in a climate agreement is efficient. We implement a kinked benefit function of additional participation because of its simplicity and for the fact that it allows for crisp theoretical predictions. That is, members and non-members will co-exist in the efficient outcome. In these partial participation treatments, the minimum participation mechanism was significantly less effective. Groups tended to vote for minimum participation constraints that were higher than the efficient level, and many profitable agreements were deliberately blocked from forming. When heterogeneity was introduced in the payoff functions, the effects were exacerbated. These results are not consistent with a presumption that players are only concerned with material payoffs, but they are consistent with players being averse to inequality.

Experiments

The experimental design entails two stages, a voting stage and an agreement formation stage. If an agreement forms its members are required to contribute a single unit of a public good, while non-members do not contribute their unit of the good. We implement four experimental treatments that differ according to the returns to individual contributions to the public good. In two treatments, the marginal return to contributions is constant so that the efficient agreement size requires full participation. In the other two treatments, the marginal return to public good contributions is constant up to an aggregate contribution and then is zero so that the efficient agreement requires only a subset of the players. Each pair of treatments includes one in which players have homogeneous payoffs and another in which payoffs are heterogeneous. We parameterized the treatments so that the equilibrium predictions given standard preferences were the same under both homogeneous and heterogeneous payoffs. We include treatments with heterogeneous payoffs to examine the influence of heterogeneity on agreement formation.

Suppose that n players decide whether to contribute a single unit to a public good. The basic payoff function of a player that contributes to the public good is:

$$\pi_i = \begin{cases} A + bs - c, & \text{for } s \leq \bar{s}, \\ A + b\bar{s} - c, & \text{for } s \geq \bar{s}; \end{cases} \tag{1}$$

where A is a positive constant, s is the number of individuals contributing to the public good, b is the marginal return to individual contributions to the good up to $\bar{s} \leq n$ contributions, and c is the individual cost of a contribution. We chose parameter values of $n = 6$, $A = 10$, $c = 10$ for all four treatments. In one of the treatments in which all subjects had identical payoff functions we chose $b = 4.5$ and $\bar{s} = 6$. Since $nb - c > 0$ for all contributions, the efficient agreement in this case is six individuals (we call this treatment *full participation*). In the other treatment with identical payoffs, $b = 4.5$ and $\bar{s} = 3$. Note from (1) that $b = 0$ for contributions greater than 3, so the efficient agreement size in this case is three individuals (we call this treatment *partial participation*). In one of the treatments in which players had heterogeneous payoffs, $\bar{s} = 6$, and three individuals had $b_h = 5$ (high earners) while the other three had $b_l = 4$ (low earners). Full participation is efficient in this treatment. In the other heterogeneous payoff treatment, $\bar{s} = 3$ and the efficient agreement in this case contains three members of any combination of high and low earners. The payoff parameters are such that all three-player agreements in all treatments are profitable in the sense that individuals with standard preferences would prefer to be a member of a three-player agreement than have no agreement form. However, since the cost of an individual contribution exceeds the marginal return in all cases, no player would contribute to the public good in a standard, non-cooperative Nash equilibrium.

All sessions were run at Appalachian State University using software specifically designed for this experiment. In each session, three groups of six subjects were in the laboratory. These groups of six remained constant throughout the experiment (i.e., partners design) which lasted 20 periods. We ran one session for each of the four treatments and therefore we have 72 subjects who generated 1,440 individual-level observations. These observations include their votes for the membership requirement in the first stage and their decision to join an agreement in the second stage.

In the first stage of the experiments, subjects voted on the minimum membership requirement for an agreement to form in the second stage. They were given 60 seconds to vote by selecting a number, in the range of 1 to 6, from a listbox. Subjects could change their votes during the voting period, but the number selected in the subject's listbox at the end of the 60 seconds was recorded as their vote. A important feature of our voting protocol was that a subject's selection from the listbox was viewed by the other five group members. Although only the subject's final selection was recorded as their vote, this feature allowed for a type of structured group communication in which subjects exchanged information about their intended votes. This communication feature better represents the process of determining minimum participation requirements for international agreements compared to simultaneous voting without the ability to communicate. A modal response (plurality) voting rule was used. The membership requirement that received the most votes was implemented in the second stage of the game. Ties were settled by a random draw.

In the agreement formation stage, each subject decides whether or not to join the agreement. Again, players are given 60 seconds to make their decision. Once a subject had made their choice they could not change it. The decisions in the agreement formation stage were made sequentially. During the stage, all subjects were given real-time information about the decisions made by the other five members in the group. Specifically, they were informed about the number of other subjects that had joined the agreement, the number of others that had decided not to join, and the number of others that had not yet made a decision. Therefore, each player knew whether their participation decision was critical for the agreement to form. If someone failed to make a decision within the allotted time, they were made not to join the agreement by default. The software was designed with a number of features to ensure that when making their choice subjects had perfect information regarding the choices made by the other group members. For example, if more than one subject made a decision within the same second, only the first decision was recorded. In those situations the subjects whose decision did not register received a message informing them that their action was not recorded and instructed them that the group's information had changed. Those subjects could then reevaluate their position and were given the opportunity to make another decision. In addition, if a group member made a decision within the last five seconds of the round when some subjects remained undecided, an additional five seconds were added to the time remaining. This feature provided undecided subjects enough time to assimilate changes before making their decision.

If enough players joined the agreement to satisfy the minimum membership requirement, then the agreement "formed" (entered into force) and those that joined contributed to the public good. Those who did not join did not contribute. If too few subjects joined the agreement to satisfy the membership requirement, then no agreement formed and no one contributed to the public good. This stage of the experiment was very similar to traditional threshold public goods experiments that utilize money-back guarantees (Dawes *et al.* 1986; Erev and Rapoport 1990; Bagnoli and McKee 1991; Cadsby and Maynes 1999).

If subjects have standard preferences in the sense that they seek to maximize their expected payoffs, theoretical results due to Carraro *et al.* (2009) suggest that subjects will vote to make the efficient size agreement the minimum membership requirement in the first stage, and this agreement will form in the second. This prediction holds whether efficiency requires all players or only a subset, and, given our chosen parameters, under homogeneous and heterogeneous payoffs. Given a voting outcome from the first stage, the sequential nature of decision making in the agreement formation stage mitigates the coordination problems that plague many simultaneous decision games. For example, suppose the result of the voting in the first stage was a three-player agreement. Because a three-player agreement is profitable, it is expected to form. However, because players are better off free-riding on profitable agreements rather than being members, there is an incentive for three players not to join. In our experiments, the sub-game perfect equilibrium involves the first three players opting out of the agreement and the final three joining the agreement. Indeed, no player is expected to join unless they are critical; meaning that if they do not join the agreement fails to form.[1]

However, if some players care about their payoffs relative to others, and are specifically averse to inequality, other equilibria are possible (Kosfeld *et al.* 2009). This is especially true when the efficient agreement size is smaller than the grand coalition; that is, when inequality itself is efficient. In the experiments in which full participation is efficient, we expect that this agreement will form even if some subjects are inequality averse because there is no inequality when the grand coalition forms. If inequality aversion is important, we expect its effects to show up in the experiments for which three-player agreements are efficient. In these cases, it is possible that inefficiently large agreements may form, because inequality averse individuals may vote to implement membership requirements that exceed the efficient coalition size. Moreover, inequality averse individuals may block an efficient agreement from forming. An individual can block an agreement in the following way. Given others' earlier decisions to join or not join an agreement, let a player be critical if the agreement fails to form if they decide not to join the agreement. A critical player blocks an agreement from forming if they refuse to join.

In sum, in the experiments in which three-player coalitions are efficient, we may observe a significant number of larger coalitions forming. In addition, if the three-person membership requirement is ever implemented, we may observe that it is blocked a significant number of times.

Results

Our experimental data suggest the following broad conclusions. In the treatments for which full participation with an agreement is efficient (the *full participation* treatments), individuals voted overwhelmingly to implement the six-player membership requirement. Moreover, agreements of size six formed in a significant majority of trials, leading to high efficiency as measured by the ratio of group earnings to maximum group earnings. Performance was significantly worse in the treatments for which a three-player agreement was efficient (the *partial participation* treatments). Agreements formed in only about half of the trials, and only half of these were efficient. Consequently, average efficiency was significantly lower in these treatments than in the *full participation* treatments. These results, and the individual voting decisions and decisions to join agreements that we now look at in detail, are not consistent with the presumption of standard self-interested preferences. They are, however, consistent with the presence of inequality-averse players.

We first examine the data on voting for the membership requirement in the first stage of the experiments. Table 7.1 provides votes and referenda outcomes by membership requirement and treatment. The first row in each cell contains the number of votes and percentage of total votes (out of 360 for each treatment) for that minimum membership requirement. The second row in each cell contains the number of times and percentage of trails in the treatment (out of 60) that membership requirement was implemented. Kolmogorov-Smirnov (K-S) tests of the null hypothesis that the samples of votes under homogenous and heterogeneous payoff treatments are drawn from the same distribution are rejected but only at the 10 percent level of significance ($p = 0.097$ for both tests under *partial* and *full participation*). K-S tests of the null hypotheses that the samples of adopted membership requirements under homogenous and heterogeneous payoff treatments are drawn from the same distribution cannot be rejected ($p = 0.809$ under *partial participation*; 0.925 under *full participation*). Since the patterns of votes and membership requirements for the homogeneous and heterogeneous treatments are so similar, we discuss only the combined results for the *partial* and *full participation* treatments.

Under the *full participation* treatments, the six-player membership requirement received 61.1 percent (440 of 720) of total votes, which is considerably more than the 18.8 percent received by the second most preferred option of a five-player membership requirement. The remaining four options received even fewer votes. This voting behavior resulted in the selection of the efficient agreement size in 71.7 percent (86 of 120) of referenda. This behavior is generally consistent with our predictions for the *full participation* treatments.

Voting under the *partial participation* treatments was more complicated. Even though the three-player membership requirement received the greatest percentage of votes, this is significantly less than the corresponding percentage when full participation was required (43.1 percent vs. 61.1 percent, $p = 0.000$).[2] Moreover, the subjects under the *partial participation* treatments showed a considerable

Table 7.1 Stage One Results – Individual votes and referenda outcomes by minimum membership requirement and treatment

	Minimum membership requirement					
	1	*2*	*3*	*4*	*5*	*6*
Full Participation						
Homogeneous	6 (1.7%)	7 (1.9%)	17 (4.7%)	26 (7.2%)	72 (20.0%)	232 (64.4%)
	1 (1.7%)	1 (1.7%)	2 (3.3%)	4 (6.7%)	6 (10.0%)	46 (76.7%)
Heterogeneous	19 (5.3%)	18 (5.0%)	21 (5.8%)	31 (8.6%)	63 (17.5%)	208 (57.8%)
	1 (1.7%)	3 (5.0%)	4 (6.7%)	4 (6.7%)	8 (13.3%)	40 (66.7%)
Combined	25 (3.5%)	25 (3.5%)	38 (5.3%)	57 (7.9%)	135 (18.8%)	440 (61.1%)
	2 (1.7%)	4 (3.3%)	6 (5.0%)	8 (6.7%)	14 (11.7%)	86 (71.7%)
Partial Participation						
Homogeneous	25 (6.9%)	39 (10.8%)	160 (44.4%)	22 (6.1%)	33 (9.2%)	81 (22.5%)
	2 (3.3%)	6 (10.0%)	33 (55.0%)	4 (6.7%)	3 (5.0%)	12 (20.0%)
Heterogeneous	16 (4.4%)	40 (11.1%)	150 (41.7%)	21 (5.8%)	19 (5.3%)	114 (31.7%)
	2 (3.3%)	7 (11.7%)	26 (43.3%)	3 (5.0%)	3 (5.0%)	19 (31.7%)
Combined	41 (5.7%)	79 (11.0%)	310 (43.1%)	43 (6.0%)	52 (7.2%)	195 (27.1%)
	4 (3.3%)	13 (10.9%)	59 (49.2%)	7 (5.8%)	6 (5.0%)	31 (25.8%)

Top of each cell: Number of votes for each minimum membership requirement (percent of total votes by treatment). Note there are 360 individual votes per treatment.
Bottom of each cell: Number of times each minimum membership requirement was implemented (percent of total trials by treatment). Note there are 60 group-level observations per treatment.

tendency to vote to implement higher membership requirements, in particular the six-player membership requirement (27.1 percent of the votes). Membership requirements mirror the votes, with the three-player requirement being implemented in 49.2 percent of the trials and the six-player requirement implemented in 25.8 percent of the trials. While groups under the *partial participation* treatments implemented the efficient coalition size as the membership requirement considerably more often than the six-player requirement, we will see shortly that this difference did not result in the efficient agreements forming significantly more often than six-player agreements.

When players are materially self-interested, the model predicts that groups will always implement efficient agreement sizes as minimum membership requirements. This hypothesis is clearly violated in our treatments requiring only three-player agreements. However, the observed behavior is consistent with inequality-averse players. While the three-player agreement is efficient, this agreement size requires that a subset of players does not join the agreement and free-ride off contributions of the members. By requiring that everyone join an agreement for it to form, groups can eliminate the free-riding. The support for three-player and six-player membership requirements suggests a fundamental tension between efficiency and eliminating inequality in our experiments.

Table 7.2 contains results concerning agreement formation by minimum membership requirement and treatment. For each membership requirement/ treatment combination we provide the number of times an agreement formed under the membership requirement, this number as a percentage of total trials, and agreement formations as a percentage of times the membership requirement was adopted. In the final column in Table 7.2 we present the number and percentage of trials an agreement of any size formed. A z-test comparing the proportion of agreement formation between homogeneous and heterogeneous treatments suggests the two are equivalent ($p = 0.583$ under *partial participation*; $p = 0.307$ under *full participation*) and therefore we examine the combined data.

Under the *full participation* treatments, note that agreements of any size formed in 102 of 120 (85 percent) trials. Eighty-two of these agreements (80.4 percent) were efficient. Thus agreements formed quite frequently in these treatments and the greatest majority of these were efficient. Other agreements formed far less frequently. When groups implemented the six-player membership requirement, the agreement formed 95.4 percent of the time. This suggests that the main reason that smaller than efficient agreements formed in a minority of the trials is because groups sometimes failed to implement the six-player membership requirement.

In contrast, agreements formed far less frequently under the *partial participation* treatments and only about half of these were efficient. Agreements formed in 47.5 percent of trials under these treatments, which is significantly lower than the 85.0 percent agreement formation rate under the *full participation* treatments ($p =$ 0.000). Of the 57 agreements that formed only 28 of these were efficient. The rate at which efficient agreements formed under the *partial participation* treatments (23.3 percent) is far lower than the rate at which the efficient agreement formed under the *full participation* treatment (68.3 percent, $p = 0.000$). One reason for the

Table 7.2 Stage Two Results – Agreement formation by minimum membership requirement and treatment

	Minimum membership requirement						
	1	*2*	*3*	*4*	*5*	*6*	*Total*
Full Participation							
Homogeneous	1	0	2	1	4	45	53
	1.7%	0.0%	3.3%	1.7%	6.7%	75.0%	88.3%
	100.0%	0.0%	100.0%	25.0%	66.7%	97.8%	
Heterogeneous	1	0	2	2	7	37	49
	1.7%	0.0%	3.3%	3.3%	11.7%	61.7%	81.7%
	100.0%	0.0%	50.0%	50.0%	87.5%	92.5%	
Combined	2	0	4	3	11	82	102
	1.7%	0.0%	3.3%	2.5%	9.2%	68.3%	85.0%
	100.0%	0.0%	66.7%	37.5%	78.6%	95.4%	
Partial Participation							
Homogeneous	1	1	15	1	0	12	30
	1.7%	1.7%	25.0%	1.7%	0.0%	20.0%	50.0%
	50.0%	16.7%	45.5%	25.0%	0.0%	100.0%	
Heterogeneous	0	3	13	0	1	10	27
	0.0%	5.0%	21.7%	0.0%	1.7%	16.7%	45.5%
	0.0%	42.9%	50.0%	0.0%	33.3%	52.6%	
Combined	1	4	28	1	1	22	57
	0.8%	3.3%	23.3%	0.8%	0.8%	18.3%	47.5%
	25.0%	30.8%	47.5%	14.3%	16.7%	71.0%	

Top of each cell: Number of times agreements formed. **Middle of each cell**: Percentage agreement formation by number of trials per treatment. **Bottom of each cell**: Percentage agreement formation by adopted membership requirement.

low rate of efficient agreement formation is the low rate at which the three-player membership requirement was implemented (49.2 percent of trials from Table 7.1). Recall that the six-player membership requirement was implemented in a significant number of trials. In fact, the grand coalition formed in 18.3 percent of all trials in the *partial participation* treatments, while the efficient agreement formed in 23.3 percent of trials.

Another reason the efficient agreement failed to form in the *partial participation* treatments is that it formed only 47.5 percent of the time when the three-player membership requirement was adopted. The three-player membership requirement was implemented in 59 out of 120 trials (49.2 percent from Table 7.1) in the combined homogeneous and heterogeneous payoff treatments under *partial participation*. Agreements failed to form in 31 of these trials because a player deliberately blocked a profitable agreement from forming. All of these blocks would be inconsistent with an agreement formation model with only individuals with standard preferences. They are, however, consistent with the presence of inequality-averse subjects.

Table 7.3 Public good provision and efficiency

Treatment	*Average Public Good Provision*	*Efficiency*
Full participation		
Homogeneous	5.05	90.25%
	(0.26)	(2.74)
Heterogeneous	4.55	84.78%
	(0.30)	(3.15)
Combined	4.8	87.5%
	(0.20)	(2.09)
Partial participation		
Homogeneous	2.07	70.12%
	(0.31)	(2.57)
Heterogeneous	1.83	67.34%
	(0.30)	(2.37)
Combined	1.95	68.7%
	(0.21)	(1.74)

Standard errors are in parentheses. Each treatment consists of 60 group-level observations.

We complete our data analysis with results on average public good provision and average efficiency in Table 7.3. Efficiency for each group in each period is calculated as the ratio of aggregate payoffs to maximum payoffs. As expected, public good provision was lower in the *partial participation* treatments than in the *full participation* treatments. More importantly, the inability of subjects under the *partial participation* treatments to form efficient agreements consistently produced significantly lower efficiency as compared to the *full participation* treatments. For the homogeneous and heterogeneous payoff treatments combined, subjects earned 87.5 percent of maximum earnings under the *full participation* treatments, while subjects earned significantly less, 68.7 percent, under the *partial participation* treatments ($p = 0.000$). On average, public good provision and efficiency were less under the heterogeneous treatment than under the homogeneous treatments, but these differences are not significant (for public good provision, $p = 0.213$ and $p= 0.585$; for efficiency, $p = 0.194$ and $p = 0.428$ for *full participation* and *partial participation* respectively).

Conclusion

We have explored an endogenous agreement formation game in which players determine a minimum participation requirement before deciding whether they will join. Almost all international environmental agreements contain minimum participation constraints, and this chapter tests the effectiveness of the institution in a simple laboratory experiment. If players only care about maximizing their expected financial payoffs, the prediction is that players will implement efficient agreement sizes, regardless of whether efficiency requires full or partial participation. However, if some players are averse to inequality they can move groups to inefficient outcomes, particularly if an efficient agreement requires less than full participation. Inequality-averse individuals can vote to cause larger-than-efficient agreements to form, and they may block the formation of efficient agreements.

Our experimental results are generally consistent with players having an aversion to inequality. In treatments for which the efficient agreement was the grand coalition, agreements formed 85 percent of the time and most of these were efficient. While this result is predicted by a model of players having standard preferences, adopting the grand coalition is also consistent with inequality aversion since this agreement size minimizes inequality. But in treatments for which the efficient agreement required only a subset of the group, agreements formed in just over half the trials, but only half of these were efficient. In fact, the grand coalition formed at about the same rate as the efficient agreement size. We also find that efficient agreements that allowed for free-riding were blocked about half the time. Although individuals with standard payoff-maximizing preferences would never adopt larger-than-efficient agreements or block efficient agreements from forming, these actions are consistent with inequality-averse individuals.

Our results also demonstrate the value of the endogenous implementation of a minimum membership requirement in promoting agreement formation and public good provision when the efficient agreement size is the grand coalition. When efficiency required full participation in our experiments, agreements were 87.5 percent efficient on average. This efficiency level is high compared to other experiments on agreement formation with similar incentives as ours but without minimum membership requirements. For example, Kosfeld *et al.* (2009), Dannenberg *et al.* (2010) and McEvoy *et al.* (2011) report average efficiency measures of 60.5, 22 and 57.9 percent respectively.

The research also provides insight into the role preferences for equity play in the design of effective institutions that govern international environmental agreements. One of the striking results from this study is the high frequency of trials in which groups adopted the grand coalition even though efficiency required some degree of free-riding. This finding may help explain the choice of participation requirements in many existing voluntary institutions, in particular the fact that many international environmental agreements require full or very high levels of participation (see Barrett 2003 for a detailed list of IEAs and minimum membership requirements). In light of our results it is possible that some of these existing agreements have participation thresholds that are inefficiently high in order to limit the extent of free-rider payoffs.

Many economists now appreciate the role that equity and fairness play in the design of effective institutions to resolve social dilemma situations. The typical result in the literature is that inequality aversion can help foster cooperation between group members in public goods and common-pool resource games and lead to more efficient outcomes. Our study contributes to this growing theoretical and experimental literature on inequality. Within the agreement formation game that we consider, we show that inequality aversion can actually move groups away from efficient outcomes. The complete picture shows that inequality aversion can either foster or frustrate cooperation among group members, and which prevails likely depends on whether resolving the social dilemma mitigates or exacerbates inequality measures.

Notes

1 The experiments derive from a theoretical analysis which is not included in this chapter, but is available from the authors upon request.
2 We report unconditional summary statistics in our tables. However, we recognize that our observations are not entirely independent because the same subject makes repeated decisions as part of a stable group. To address this issue our hypotheses tests use linear regression techniques to control for period, subject (for individual-level data) and group (for group-level data) fixed effects. We report these *p*-values throughout.

References

Bagnoli, M., and M. McKee (1991). Voluntary contribution games: efficient private provision of public goods. *Economic Inquiry* 29(2), 351–66.

Barrett, S. (2003). *Environment and Statecraft: the strategy of environmental treaty making*. Oxford: Oxford University Press.

Barrett, S. and A. Dannenberg (2012). Climate negotiations under scientific uncertainty. *Proceedings of the National Academy of Sciences* 109(43), 17372–76.

Burger, N.E. and C.D. Kolstad (2010). International environmental agreements: theory meets experimental evidence. Available at: http://fiesta.bren.ucsb.edu/~kolstad/HmPg/papers/IEA percent20Experiments.9.2010.pdf (accessed: August 2013).

Cadsby, C.B. and E. Maynes (1999). Voluntary provision of threshold public goods with continuous contributions: experimental evidence. *Journal of Public Economics* 71(1), 53–73.

Carraro, C., C. Marchiori and S. Oreffice (2009). Endogenous minimum participation in international environmental treaties. *Environmental and Resource Economics* 42(3), 411–25.

Cherry, T.L. and D.M. McEvoy (2013). Enforcing compliance with environmental agreements in the absence of strong institutions: an experimental analysis. *Environmental and Resource Economics* 54(1), 63–77.

Dannenberg, A. (2012). Coalition formation and voting in public goods games. *Strategic Behavior and the Environment* 2(1), 83–105.

Dannenberg, A., A. Lange and B. Sturm (2010). On the formation of coalitions to provide public goods – experimental evidence from the lab. *NBER Working Paper No. 15967.*

Dawes, R.M., J.M. Orbell, R.T. Simmons and A.J.C. Van De Kragt (1986). Organizing groups for collective action. *The American Political Science Review* 80(4), 1171–85.

Erev, I., and A. Rapoport (1990) Provision of step-level public goods: the sequential contribution mechanism. *Journal of Conflict Resolution* 34(3), 401–25.

Gurerk, O., B. Irlenbusch and B. Rockenbach (2006). The competitive advantage of sanctioning institutions. *Science* 312(5770), 108–11.

Kosfeld, M., A. Okada,and A. Riedl (2009). Institution formation in public goods games. *American Economic Review* 99(4), 1335–55.

Kroll, S., T.L. Cherry and J.F. Shogren (2007). Voting, punishment, and public goods. *Economic Inquiry* 45(3), 557–70.

McEvoy, D.M. (2010). Not it: opting out of voluntary coalitions that provide a public good. *Public Choice* 142(1), 9–23.

McEvoy, D.M., J. Murphy, J. Spraggon and J. Stranlund (2011). The problem of maintaining compliance within stable coalitions: experimental evidence. *Oxford Economic Papers* 63(3), 475–98.

McGinty, M., G. Milam and A. Gelves (2012). Coalition stability in public goods provision: testing an optimal allocation rule. *Environmental and Resource Economics* 52(3), 327–45.

Sutter, M., S. Haigner and M.G. Kocher (2010). Choosing the carrot or the stick? Endogenous institutional choice in social dilemma situations. *Review of Economic Studies* 77(4), 1540–66.

Tyran, J.-R. and L.P. Feld (2006). Achieving compliance when legal sanctions are non-deterrent. *Scandinavian Journal of Economics* 108(1), 135–56.

Walker, J.M., R. Gardner, A. Herr, and E. Ostrom (2000). Collective choice in the commons: experimental results on proposed allocation rules and votes. *The Economic Journal* 110 (460), 212–34.

8 Climate policy coordination through institutional design

An experimental examination

Matthew E. Oliver, Jamison Pike, Shanshan Huang and Jason F. Shogren

Introduction

The effects of institutions on economic outcomes are important when considering common-pool resources. Institutions of all varieties have become ubiquitous in the evolution of human society in relation to most if not all natural resources. Crawford and Ostrom (1995) view institutions as "enduring regularities of human action in situations structured by rules, norms, and shared strategies." A majority vote, the institutional "rule" examined in this chapter, would seem to pre-date recorded history. Ostrom (1986) defines a "rule" as a "prescription commonly known and used by a set of participants to order repetitive, interdependent relationships." As we will see, even an institution as basic as a majority voting rule has the potential to coordinate the otherwise rationally self-interested and myopic behavior of individual players toward a more socially beneficial outcome.

There are certain cooperative public endeavors that challenge the standard nomenclature of economics. For example, as an international treaty, the Kyoto Protocol focuses on emission control strategies to mitigate or reduce the probability of a global disaster from climate change (see, for example, Kane and Shogren 2000). Mitigation in this sense is a global public good. That which is mitigated, a climate-related disaster (or the probability thereof), could be considered a global public *bad*. Emissions reduction, then, is both a global public good and mitigation of a global public bad. As a second example, consider the eradication of a disease, the global effort to eradicate smallpox in the twentieth century. Barrett (2003, 2006) defines this achievement as a global public good, but similar to the goal of the Kyoto Protocol, the eradication of smallpox can be thought of as the elimination of a public bad.

The psychological, sociological, and philosophical implications of "good" versus "bad" are beyond the scope of this chapter. We make this distinction in terminology, however, because the prevailing literature on public goods experiments has focused on just that – public *goods* provision. Behavioral economics provides precedence for the validity of such a distinction in a provision game via the concept of loss aversion.[1] The key assumption is that "losses loom larger than gains." We do not seek to test for loss aversion per se, but instead use it as a point of departure from previous public goods experiments. In this chapter we describe an experiment in which individuals are asked to contribute toward the threshold elimination of a public bad. We frame the problem such that the payoff to individuals if the threshold is met is the avoidance of greater loss. In so doing, our experiment assumes loss aversion *ex ante*, such that elimination of a public bad (in this case a monetary loss to all individuals) is a desirable end for collective action.

To the non-economist, whether the mitigation of climate change should be termed a "public good" as opposed to "elimination of a public bad" is a matter of semantics. It has characteristics of both. Mitigation of climate change is a public good insofar as everyone in society is better off if climate damages are avoided. Economic welfare benefits include reduced damages from less frequent climate related natural disasters such as 2012's "super-storm" Sandy, failing crop yields, more intense storms, forest fires, droughts, flooding, heat waves, melting of the west Antarctic ice sheet, less variation in rain patterns and agricultural productivity, and other private gains from activities like tourism and global commerce (see the detailed discussion in the Stern Report, Stern 2007). Climate change is a public bad given the risk to affect negatively everyone in a society. This holds in poverty-stricken regions where adaptation strategies are too costly or non-existent. The greatest damages are likely to occur in the countries poorly equipped to deal with them. For example, Srinivasan *et al.* (2008) find that predicted climate change impacts are disproportionately severe for low-income countries, and have been driven by emissions from high- and middle-income countries. As noted by these authors, "through disproportionate emissions of greenhouse gases alone, the rich group may have imposed climate damages on the poor group greater than the latter's current foreign debt." In developing countries where vast numbers live on marginal lands and have no financial or technological means for adapting to changing climate conditions, the task of mitigating the damages on these populations is left to the global community, as are its costs. Industrialized nations have been unable to enforce climate treaties yet continue to be the source of the bulk of global emissions. Meanwhile, many developing nations in regions like Africa, South Asia and the Pacific, and Latin America contribute minimally to global emissions, but are expected to bear the brunt of global damages.

Although the damages associated with global climate change are serious, the provision requirements for their mitigation have in many cases outpaced institutional ability to coordinate global responses (Walker *et al.* 2009). The coordination challenge is to create institutions that build trust and commitment across countries differing in wealth and population. One proposed solution is a *polycentric* order that clusters global players into homogenous groups. The concept of polycentricity is related to the economic theory of endogenous coalition formation, in which a group of players may decide to act together as a unit (Hart and Kurz 1983). The key is that the formation of a coalition does not preclude the decision-making abilities of its individuals. Similarly, Ostrom (1999: 57) defines a polycentric order as "one where many elements are capable of making mutual adjustments for ordering their relationships with one another within a general system of rules where each element acts with independence of other elements." Individuals can group together and independently accede to collectively agreed-upon institutions – much like countries in the European Union (EU) or the US states. Modern institutional economics has sought to distinguish collective entities from the individuals who comprise them. The concept of methodological individualism emphasizes that individuals differ from each other in preferences, abilities, goals, and ideas. The implication is that collective entities (such as a sub-global bloc of countries) should not be seen to behave as if they were individual agents. A theory of social phenomena should begin with the behaviors of the individual members whose actions result in the collective outcome (Furubotn and Richter 1997: 2). Polycentricity allows individuals to operate within more tightly-knit collectivities, which eases the coordination burden. Trust, commitment, and reciprocity are more effectively organized between individuals within smaller-scale governance units linked through information networks that encourage experimentation and rational evaluation of alternative strategies, and establish means to compare results across locations (Ostrom 2009).[2]

Are polycentric structures alone sufficient to coordinate global action? In this chapter we present evidence of greater coordination when an intra-mural voting system is used within a polycentric structure. Andersson and Ostrom (2008) stress that

> all government institutions are imperfect responses to the challenge of collective action problems. Because these imperfections may exist at any level of governance, analysts should consider the extent to which complementary back-up institutions exist at higher or lower levels of governance that can help offset some of the imperfections at any one level.

To this end, we designed an experiment to examine contributions toward the threshold elimination of a public bad in a polycentric structure with and without a second hierarchical back-up institution – a voting mechanism within each cluster of players. Our experiment attempts to address the distinction between the individual and the collective by way of this polycentric design, but invokes the secondary institution as a means to bolster coordination. The voting rule is implemented at the group level, allowing individual participants of the experiment to engage in collective action *at the group level only*, the effects of which are felt at the global level.

Our experimental results suggest that the institution of intra-mural voting helps coordinate individual contributions within this stylized setting. The implications for coordinating global action to combat climate change are rather straightforward. We do not mean to imply that sub-global blocs of nations can vote on whether to reduce carbon emissions as a group, and that such an arrangement would solve the global coordination problem associated with climate change. Rather, this experiment demonstrates that the organization of individual countries into coalitions or any other polycentric order may not be sufficient to support global climate coordination. Strong institutional arrangements should be embedded within polycentric group formation if the global community is to be successful at coordinating a response to climate related threats – arrangements that enhance communication, transparency, enforceability, and accountability between individual players.

Related experimental economics literature

How can rich and poor nations more effectively coordinate their efforts to deal with climate change? Global institutions must have the authority to enforce binding agreements. Only then can the necessary coordination be achieved. One way to advise the creation of an effective institutional apparatus is through economic experimentation. Researchers use experiments to examine incentive and contextual questions that arise when allocating resources, both in private markets and in public institutions (see for example, Bohm 2003). Most experiments follow a process that combines the trial-and-error experimentation of the artisan and the deductive method of reasoning of the mathematician. By combining pattern recognition with theoretical insight, the researcher uses experimental methods to help clarify how incentives and contexts affect the way in which people allocate resources, given different sharing and cost rules as defined by the nature of the public good.

A sizable literature has emerged using experimental methods to study the economics of climate change, much of which is rooted in the context of public good provision. As noted by Milinski *et al.* (2006), "Maintaining the Earth's climate within habitable boundaries is probably the greatest 'public goods game' played by humans." These authors describe an experiment in which participants contribute to a common pool that is not divided among them, but is instead intended for investment in programs to encourage a reduction in fossil fuel use. This result requires altruism on the part of each individual, because no player benefits directly from contributing to the common pool. They show that "given the right set of circumstances," altruistic behavior can emerge, and players will contribute to the climate protection program. Players' contributions also increase when they can contribute publicly, as they gain "social reputation" that is in turn rewarded by other players.

The idea of a polycentric structure in experiments in public goods provision has also gained new ground. Tackling global challenges such as climate change often involve the formation of institutions among "coalitions" of players, which is similar to the exogenously imposed polycentric structure used in the experiment described in this chapter. Experimental research on endogenous coalition formation has shown that contributions toward the public good increase when individuals are allowed to form coalitions (Burger and Kolstad 2009). Other researchers have found that fewer players will form a coalition when the internalization of mutual benefits of members is prescribed by the institution. Contrary to theory, coalitions that attempt to reduce free-riding by requiring lower contributions from their members do not attract additional members. Substantial efficiency gains occur, however, when individual coalition members can propose a minimum contribution level, with the smallest common denominator being binding (Dannenberg *et al.* 2010).

Emissions reduction targets such as those laid out by the Kyoto Protocol can be thought of as a threshold public goods game. Cadsby and Maynes (1999) describe a seminal experiment that examines the effects of allowing individuals to contribute any desired proportion of their equal endowments toward a threshold public good. In this study, each member of a group is asked to make a private contribution toward the production of the public good. If enough contributions are made such that the stated threshold is reached, the good is provided. If insufficient contributions are made, players lose their contributions and the good is not provided. Everyone is better off if the good is provided than if it is not, but those who do not contribute are better off than those who do, regardless of the outcome. The threshold could be an emissions target (see for example, Brick and Visser 2010). Or, like the experiment in this chapter, it could be a "point-of-no-return," where all players irreversibly suffer a welfare loss if the threshold is not met (as in Fischbacher *et al.* 2011).

The threshold could be some reduction in emissions that, if not reached, would result in a catastrophic warming trend that would destroy life on Earth as we know it. But as Weitzman (2009) demonstrated, uncertainty plays a powerful role in our assessment of the risk of catastrophic (or even more manageable) damages associated with climate change. Uncertainty in outcomes has been shown to reduce contributions toward a public good (Burger and Kolstad 2009). Threshold uncertainty, both in terms of the threshold's value and its probability distribution, reduces coordination to the detriment of the public good's provision (Dannenberg *et al.* 2011). Climate negotiations can turn into a coordination game when the fear of crossing a dangerous (but uncertain) threshold is strong. Such fears can help ensure collective action to avoid a dangerous threshold. Coordination breaks down when uncertainty about the threshold's true value turns the game back into a prisoners' dilemma (Barrett and Dannenberg 2012). In a related paper, Barrett (2013) assumes a discontinuity at the threshold in the climate catastrophe mitigation benefit function, demonstrating that "if the threshold is known with *certainty* [author's emphasis], and the loss from catastrophe vastly exceeds the costs of avoiding it, then the collective action problem changes fundamentally." Under such a scenario, non-binding treaties such as the Copenhagen and Cancun agreements are sufficient to coordinate behavior. The key implication is that as long as uncertainty about catastrophic damages looms over the negotiation process, self-enforcing international treaties are likely to be "grossly inadequate."

Another important characteristic of the climate change problem that has been studied in public goods games is that different players (or at a macro scale, different sectors or even different countries) may differ in their abilities to contribute to climate change mitigation. Heterogeneity may be in terms of abatement costs (as, for example, in Brick and Visser 2010, 2012). In some cases, though, heterogeneity may be in terms of players' endowments. Cherry *et al.* (2007) designed an experiment to explore the impact of heterogeneous endowments and earned endowments on observed contributions in a linear public good game. Their results suggest that contribution levels were significantly lower when groups had heterogeneous rather than homogeneous endowments, independent of the origin of endowment. Like the experiment presented in this chapter, the Cherry *et al.* (2007) experiment tests homogeneity *within* a group, with heterogeneity *between* groups.

Kroll *et al.* (2007) follow this with an experiment that explores whether the institution of a voting rule with and without endogenous punishment increases contributions to a public good. This experiment's results suggest that voting by itself does not increase cooperation, but if voters can punish violators, contributions increase significantly. While costly punishment increases contributions at the price of lower efficiency, overall efficiency for a voting-with-punishment rule still exceeds the level observed for a voting-without-punishment rule. The voting rule implemented in this chapter differs from that of Kroll *et al.* (2007) in that the vote is not enforced through punishment, but through majority rule obligation. Brick and Visser (2012) study the effect of democratically-chosen emissions reduction policies on reaching emissions targets, and find that voting does not improve coordination when subjects have heterogeneous abatement costs.

The contribution of experimental economics to the climate change debate is ongoing. Researchers are studying the effects of various characteristics and treatments in public good provision and climate change, including inequality and fairness (Tavoni *et al.* 2010, 2011; Lamb *et al.* 2011), mitigation versus adaptation (Hasson *et al.* 2009), communication (Brick *et al.* 2012), enforcement (McEvoy *et al.* 2011; Cherry and McEvoy 2013), dynamic externalities (Saijo *et al.* 2009), and globalization (Buchan *et al.* 2009). Given the complexity of the climate change problem, there is no shortage of possible permutations on the public goods provision experimental framework that can help expand our knowledge about the efficient formation of effective institutions.

Experimental design and procedure

The experiment laid out in this section is a threshold public good coordination game, although threshold public good provision is not necessarily the primary focus. Public goods provision games have been studied, but we depart from the canonical literature by making the objective of our game threshold provision of the elimination of a public bad. A fundamental theme of this chapter is the effect of an institution on the collective achievement of a socially desirable objective when individuals have incentive to free-ride. To this end, we divide each experimental session into a control set with no voting rule, and a treatment set in which the voting rule is implemented. We test the null hypothesis: intra-mural majority voting within a polycentric framework has no effect on global public good coordination. If we reject the null, it supports the Andersson–Ostrom position that an intra-mural mechanism supports coordination.

In an experimental setting, we administer a repeated coordination game with a static payoff structure based on continuous contributions toward the threshold elimination of a public bad, i.e. climate-induced damages. The primary objective of our experiment is to test whether institution of the majority voting rule at the group level is sufficient to induce a cooperative outcome at the global level. One of the unique features of our experiment is that we introduce two levels of heterogeneity: one at the level of the individual, and another at the level of a group of individuals.[3] We achieve this by dividing individuals into groups of different sizes with different per-capita endowments. This heterogeneous polycentric order is the baseline institutional structure. In each period, players are asked to contribute toward the threshold. In the control game, there is no vote, and individuals are free in each period to determine their own contributions privately. We then introduce a majority voting treatment at the group level to test for the effects of an additional institutional mechanism. With the implementation of the voting treatment, a group majority can vote in favor of contributing, effectively forcing all members to contribute. In either game, if insufficient contributions are made, all players lose what they contributed and the public bad occurs, resulting in the loss of a portion of their remaining endowment. Similar to Cadsby and Maynes' (1999) threshold public goods experiment, everyone is better off if the threshold is met and the public bad is avoided, but those who do not contribute are better off than those who do, regardless of outcome. Our results show that implementation of the voting rule increases the frequency with which the threshold is met.

We now turn to a detailed description of the experiment, and discuss some of the theoretical aspects of the game itself. For each experimental session, a total of 32 participants are required, constituting two sets of 16. The first set serves as the control, the second as the treatment. Each participant plays a repeated static game (see Figure 8.1) for a minimum of 10 periods. In each period, aggregate contributions must be at least 25 percent of the aggregate endowment for damages to be avoided. Starting at the tenth period, a random stopping device determines whether the game continues to another period. Each participant receives a minimum payout for participating in the experiment, plus an additional cash sum based on his/her accumulated payoffs.[4]

Control

The 16 subjects in the control set are each randomly assigned to one of four groups, indexed $i = A, B, C, D$. The groups are heterogeneous with respect to both population and endowment of "tokens" per person (per period), ω_{ij}, where $j = 1, \ldots, n$ is an index of individuals within group i. Once each participant is assigned to a group, they play a repeated game in which contributions are made at the individual level toward an elimination threshold for a public bad. No individual is required to contribute any amount of his/her endowment, i.e., all contributions are voluntary. The aggregate endowment per period is 1000 tokens. Groups are constructed in the following way:

- Group A has population 5, and an endowment per person of 100 tokens.
- Group B has population 5, and an endowment per person of 25 tokens.
- Group C has population 3, and an endowment per person of 100 tokens.
- Group D has population 3, and an endowment per person of 25 tokens.

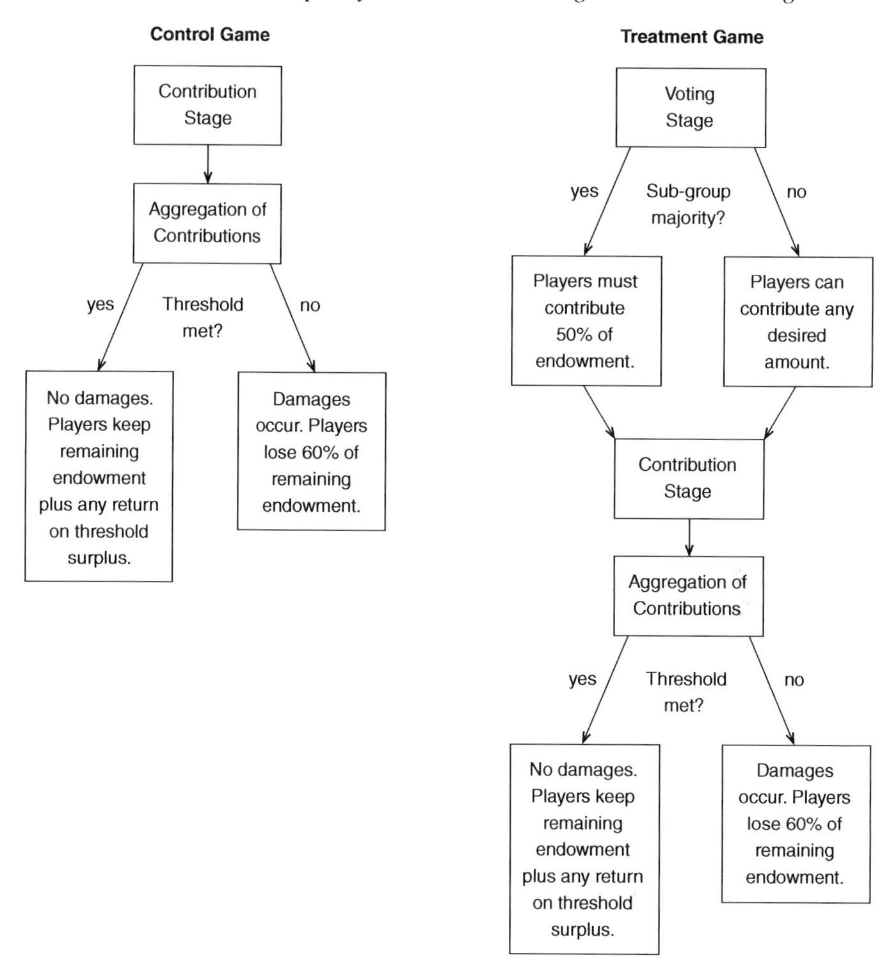

Figure 8.1 Schematic representation of control and treatment games

The elimination threshold is known to all participants before making their individual contributions, and is equal to one-fourth of the aggregate endowment, or 250 tokens. In each period, all individuals make a voluntary contribution, m_{ij}, which may equal any amount between zero and his/her full endowment.[5] The contributions are then aggregated. If the aggregate contribution meets the threshold exactly, each player gets a payoff for that period that is the endowment minus the amount contributed (see Appendix, Equation A1.1, for a formal mathematical representation). So, for example, if a given player in group A contributes 25 tokens in a period and aggregate contributions sum to 250, that player's payoff for that period is 75. If the aggregate contribution surpasses the threshold, each player (regardless of group) gets an equal return on the surplus. This return is meant to represent an investment in an alternative public good.[6] The payoff to each individual in the event of a surplus is the remainder of their endowment, plus a fraction, α, of the provision surplus (see Equation A1.2).[7] If the aggregate contribution falls short of the threshold, all individuals incur additional damage, D_{ij}, that is equal to 60 percent of the remainder of their endowments (see Equation A1.3).[8]

Equilibria in the control game

Previous experimental work on threshold public goods provision by Cadsby and Maynes (1999) points to two "symmetric" equilibria:

> In the experimental framework, N individuals are provided with identical endowments . . . Of the many possible pure strategy Nash equilibria, two are symmetric in that individuals, all of whom face the same opportunities and constraints, follow identical strategies. The first symmetric pure strategy equilibrium is one in which [contributions] $c = 0$ for all participants ... Contributing anything other than zero is suboptimal if player i believes that nobody else will make a contribution. The second symmetric pure strategy equilibrium is characterized by $c = \frac{T}{N}$ for all participants [where T is the threshold]. Such an outcome just achieves the threshold . . . This equilibrium exists if and only if $R \geq \frac{T}{N}$ [where R is the reward from achieving the threshold]. Contributing $\frac{T}{N}$ is a best response for player i (weakly best if $R = \frac{T}{N}$) if he believes that each other player will also contribute $\frac{T}{N}$, so that others contribute a total of $(N-1)\left(\frac{T}{N}\right)$, making his $\frac{T}{N}$ contribution just enough to achieve the threshold. The symmetric nature of these two equilibria may make them focal points, around which a group of non-communicating individuals might be expected to coalesce.

As a corresponding example, Barrett's (2006) analysis of smallpox eradication echoes this result by stating that "given the contribution levels of other countries, each country will either contribute nothing or an amount just sufficient to meet the threshold." He goes on to say that "there would also exist many other Nash equilibria to this game in which many if not all countries made positive contributions." Since solving analytically for the infinite asymmetric equilibria is not insightful, here we focus only on the two symmetric equilibria of our game – the first being that all individuals contribute zero and the second that all individuals contribute an equal proportion of their endowments.

In theory, individual rationality leads all players to have the dominant strategy to contribute zero. From the perspective of a single player j, one of two outcomes is possible, the first being that all *other* players' contributions sum up to meet the threshold; the second being that they do not. Consider the first case: a contribution of zero maximizes player j's payoff. If it is instead the case that all other players' contributions do not meet the threshold, it is still individually rational to have contributed nothing. Given that this is a simultaneous move game, player j does not know whether his/her individual contribution will sufficiently cover the gap required for the sum of contributions to reach the threshold. The unique Nash equilibrium for this game is that all players should contribute zero $m_{ij} = 0, \forall i, \forall j$. This equilibrium will be referred to as the "free-riding" equilibrium.

One key feature to the Cadsby and Maynes experiment is that all individuals have the same endowment. But rich and poor countries are heterogeneous in their ability to provide contributions toward the eradication threshold. The polycentric structure addresses this problem. Each group is not a player unto itself, but comprises several individual players. Since these individuals are heterogeneous in their endowments, depending on which group they belong to, the second symmetric pure strategy equilibrium referred to in Cadsby and Maynes does not exist in our model. Instead, the Pareto-optimal equilibrium is for all individuals to contribute toward the threshold an amount *proportional* to their endowment. Given that the threshold is set at 25 percent of the aggregate endowment, the optimal strategy equilibrium occurs where each individual contributes exactly 25 percent of his/her personal endowment. This equilibrium will be referred to as the "fair share" equilibrium. It is the Pareto-optimal pure strategy equilibrium and is symmetric, in the sense that each individual has the same strategy – although the actual amount contributed varies.

Polycentric voting treatment

We seek to show whether introducing an institutional mechanism at the group level is sufficient to induce coordination at the global level. We have chosen a binding majority vote similar to that imposed in Kroll *et al.* (2007). Before individuals make their contributions, groups vote on whether each member will be obligated to contribute half of their endowment (as we explain below). This voting rule has the potential to overcome free-riding within groups, but not between groups. The set up for the treatment set is the same as the control set, with each of the 16 subjects randomly assigned to one of four groups, indexed $i = W, X, Y, Z$. Group W is equivalent in size and endowment to group A, as X is to B, and so on. The difference between the control and treatment set is that a voting stage is added to the static game for the treatment set. This is similar to the "participation stage" introduced in Kosfeld *et al.* (2009). In the voting stage, each member in group i anonymously votes on whether the entire group should participate. If a group votes to participate, each member is then bound to contribute half of his/her endowment to the provision threshold. Note that the required contributions (in the event of a majority vote to participate) are set such that the provision threshold could be met by group W alone, or by groups X, Y, and Z in combination. This arrangement allows us to test whether free-riding exists at the group level. If a group votes not to participate, no member is required to contribute, but neither is any member prohibited from contributing. The detailed structure of the treatment set's repeated static game is as follows:

- *Voting stage* – each member of group $i = W, X, Y, Z$ anonymously votes whether the group should participate or not.

- *Contribution stage* – if a majority in group i has voted in favor of participation, each member is then required to contribute half of his/her endowment. If a majority has voted not to participate, there are no contribution requirements or restrictions (except that the full endowment is the maximum any individual can contribute in a given period).

The threshold and payoff structure are the same as for the control group. If the aggregate contribution meets the abatement threshold, each individual's payoff is the remainder of his/her endowment (Equation A1.1). Surplus is "invested in a low-return public good." The payoff to each individual, π_{ij}, is the remainder of their endowment, plus a fraction, α, of any provision in excess of the threshold (Equation A1.2). If the aggregate contribution falls short of the threshold, all individuals then incur additional damage, D_{ij}, that is equal to 60 percent of the remainder of their endowments (Equation A1.3). The individual contribution is non-refundable.

Equilibria in the treatment game

Again, our key question: is the institution of a binding majority vote at the group level enough to induce collective action, threshold provision of the elimination of a global public bad? Given that enough individuals recognize the social welfare gains associated with a coordinated provision effort, theoretically they would be able to force provision through the voting mechanism, but there is no guarantee of this outcome.

Perhaps counter-intuitively, the social optimal strategy equilibrium in the treatment group's game is for all individuals to vote against participation and then to contribute one-fourth of their endowment, as in the control group's game. The reason is that if all groups vote to participate, they contribute too much – 50 percent of their total endowment when the threshold is only 25 percent. Because individuals are not refunded for a contribution surplus, there is an incentive to meet the threshold exactly. But voting against participation and privately choosing contributions such that they meet the threshold exactly is an unstable equilibrium.[9] It is also likely to be unattainable. Because all groups have voted not to participate, no individual has any reason to assume that the threshold will be met, leading to the dominant strategy to contribute zero. The second-best outcome would be a participation majority in all groups, which would lead to half of the aggregate endowment being contributed, and the returns from the surplus being distributed among all participants.

The objective of the game, though, is not necessarily to achieve either the Pareto-optimal or second-best outcomes, but to meet the provision threshold. There is still potential for free-riding between groups. The damages associated with not meeting the threshold are such that group *W* has the incentive to vote "yes" and meet the threshold unilaterally. Likewise, even if group *W* votes not to participate, groups *X*, *Y*, and *Z* still have the ability and incentive to coordinate. By each voting "yes," they can meet the threshold without the participation of *W*.

Testable Hypotheses

There are two testable null hypotheses for this experiment, which we will denote as H_0^1 and H_0^2. The fundamental and natural null hypothesis for this experiment is H_0^1. If implementation of the institutional mechanism at the group level enhances the ability of the treatment set to coordinate and reach the global provision threshold on a more consistent basis, this hypothesis can be rejected. If, on the other hand, the treatment set shows no significant increase in frequency of provision, this suggests that the group level voting mechanism has not succeeded as an institution capable of coordinating aggregate collective action.

H_0^1: Application of the group majority voting treatment has no effect on the frequency with which the treatment set reaches the provision threshold, i.e., the control and treatment sets show no significant difference in their abilities to coordinate provision.

H_0^2: All treatment groups will vote to participate, and $\sum_i \sum_j m_{ij} = 500$.

The second null hypothesis represents the second-best outcome in the treatment group game, and follows from the incentive for each of the treatment groups to vote for participation. As stated before, group *W* has the incentive to participate and provide the threshold amount unilaterally, while groups *X*, *Y*, and *Z* have the incentive to coordinate their actions and provide the threshold even without the participation of *W*. Since all groups have the incentive to participate, the contribution of group *W* should equal the combined contributions of *X*, *Y*, and *Z*. If it does not, free-riding exists between groups, and this hypothesis should be rejected. For example, say group *Z* votes against participation when *W*, *X*, and *Y* vote to participate. The threshold will be met by the participating groups, and group *Z* individuals would contribute less than 50 percent of their endowments yet would avoid damages just the same.

Experimental results

We now present the results of three experimental sessions. We reject the null hypothesis that intra-mural voting has no effect on the frequency with which the voting set reaches the provision threshold. Over three sessions, the control set reached the threshold a total of 12 of 32 periods, or 37.5 percent. The treatment set reached the threshold a total of 20 of 30 periods, or 66.7 percent. The voting treatment nearly doubled the percentage of periods in which the threshold was met. The marked increase in the ability of the treatment set to collectively reach the threshold suggests that the voting rule is sufficient to induce provision on a more consistent basis. Table 8.1 summarizes the unconditional data on contributions. The mean aggregate contribution for the control set is below the threshold of 250, while that of the treatment set is above it. The range of aggregate contributions was greater for the treatment set. This is related to the information conveyed in the voting stage. A group-level majority vote to participate obligates all group members to contribute half of their endowments, but a majority vote *not* to participate provides even more incentive for an individual to contribute zero. The final column reports the mean aggregate final payout per session.[10] To the extent that this can be thought of as a measure of welfare, the treatment is an improvement. At the group level, more insight into individual behavior becomes clear. Recall that the cooperative equilibrium in the control set is for all individuals to contribute one-fourth of their endowment. The final column of Table 8.1 suggests that, on average, this was nearly the case. The relative inability of the control set to reach the threshold indicates that individuals were unable to coordinate their efforts consistently. “On average” is insufficient to get everyone over the threshold.

Other summary statistics in Table 8.1 stand out, particularly with respect to the treatment set. Group *X*, representing a large group of poor countries, voted “yes” only once in 30 periods. Individuals in group *X* found it more advantageous to take the risk, free-ride, and hope the other groups “picked up the tab.” This strategy paid off, since the treatment set reached the threshold two-thirds of the time, even without the participation of *X*. At the other end of the spectrum, group *Y*, representing a small group of rich countries, voted “yes” 17 of 30 periods. The implication is that even though group *Y* is unable to reach the threshold unilaterally by voting “yes,” (the total contribution from *Y* in this case is 150), they essentially “shoulder the load” for the other three groups. The other groups were able to vote “no,” contribute a smaller amount, and reap the benefits. Groups *W*, *X*, and *Z* were able to partially free-ride due to the apparent social responsibility of group *Y*. What is not shown in Table 8.1 is that in treatment session 3, it was group *W* that voted “yes” more consistently. The total contribution from *W* in the case of a “yes” vote is 250, making it possible for the other three groups to free-ride to the full extent possible. In no period did all four treatment set groups vote “yes.” No more than two groups ever voted “yes” in the same period. Based on this result, it is our conclusion that the second null hypothesis, H_0^2, is rejected. Free-riding exists, at both the individual and group levels.

We confirm our experimental results with two conditional regressions. First, we estimate the effect of voting on the probability of meeting the threshold using a logit regression (see Equation A1.4). Second, we explore how voting affects contributions with a panel regression of the total group contribution (see Equation A1.5). Table 8.2 presents the results of these regressions. Consider the logit results. The majority voting rule resulted in better than a fourfold increase in the odds of meeting the threshold in a given period.[11] For the panel regression, the coefficient estimate for the voting treatment dummy is not statistically significant. These two results suggest that voting did not increase contributions, but improved coordination as defined by the probability the threshold would be met. The relative inability of the polycentric-only structure to hit the threshold consistently seems driven by a lack of coordination, not a lack of contributions. The voting rule effectively improved overall welfare. We also see from the coefficients on the group dummies that the group with the highest population-endowment characteristic, Group *A* for the control set and Group *W* for the treatment set,

Table 8.1 Experimental results – summary statistics

Set	*Sub-Group*	*Population*	*Endowment per capita*	*# Periods Voted "yes"*	*Sub-group contributions*			*Mean contribution per person*
					Mean	*Min*	*Max*	
Control	A	5	100	n/a	140.2	60	225	28
	B	5	25	n/a	21.9	1	45	4.3
	C	3	100	n/a	60	0	170	20
	D	3	25	n/a	9.3	0	25	3.1
Treatment	W	5	100	7	106.9	0	250	21.4
	X	3	25	1	21	0	62.5	4.2
	Y	5	100	17	109.7	0	155	36.6
	Z	3	25	9	21.4	0	50	7.1
			Periods Played	*Threshold Reached*	*Aggregate Contributions*			*Average payout ($)*
					Mean	*Min*	*Max*	
Control			32	12	231.3	143	380	485.89
Treatment			30	20	259	24.25	422	597.17

Table 8.2 Regression results

Logit regression: Binary dependent variable where $y_{i,t} = 1$ indicates the threshold was met				
	coeff.	*st. err.*	*t*	*p*
Treatment Dummy	1.44	0.601	2.402	0.016
Likelihood Ratio X^2	15.46			

OLS panel regression: Continuous dependent variable- total sub-group contribution $C_{i,t}$				
	coeff.	*st. err.*	*t*	*p*
Treatment Dummy	6.22	5.55	1.12	0.264
Sub-group B/X	–102.62	7.72	–13.29	0.000
Sub-group C/Y	–40.06	7.72	–5.19	0.000
Sub-group D/Z	–108.99	7.72	–14.11	0.000
R-squared	0.55			

contributed the most toward the public good. This result appears to be consistent with reality: the United States is one of the largest contributors toward global public goods. For example, Sandler (2001: 33) notes in the World Bank, a "country-specific benefit is promoted by assigning members' votes in the Bank based on the size of its subscription . . . In 1999, the United States held over 16 percent of the votes on the Bank's policies owing to its generous subscription. [I]n return for carrying a greater burden of the World Bank financing, *a large subscriber gains greater autonomy* over the Bank's policy decisions and direction" (italics in original). We do not report the results for the period dummies because in both regressions none of their coefficient estimates are statistically significant, indicating that timing effects were absent from the game.

Discussion

Our experimental results suggest that low income individuals and groups have less influence on the aggregate ability to reach the threshold. Regardless of the perceived likelihood that the threshold will be reached, low income individuals and groups contribute little or nothing. High income individuals and groups have the option of varying their contributions over a greater range. If a high income individual contributes his/her entire endowment of 100, or two-fifths of the threshold, other players need only to contribute the remaining three-fifths. The odds of everyone else reaching the threshold improve greatly as the contribution of one high income individual or group increases. Our experimental results seem driven by the updated belief about the possibility of achieving the threshold based on previous periods' outcomes.

This last point could motivate future research. Our variation of the threshold provision experiment uses a repeated game in which the threshold must be reached *in each period*. But what if the threshold were cumulative? The central question of this experiment is whether an institution can induce cooperative action more consistently, as measured by the percentage of static periods in which the threshold is reached. The corresponding question for a dynamic game with a cumulative threshold is: can a similar institution induce provision of the cumulative threshold more quickly, i.e., in fewer periods?

Aside from extending this experiment into a different direction such as using a dynamic cumulative threshold game, there are modifications to the existing game that could be implemented. For example, in designing this experiment there was much deliberation over the proper rebate rule for amounts in excess of the threshold. We chose an "equal return" rule, in which each individual receives the same amount regardless of their contribution. Other work on rebates includes Spencer *et al.* (2009), which examines the empirical performance of voluntary contributions in a threshold public good experiment for alternative rebate rules. The objectives of this research were to identify empirically efficient rebate rules when the public good is provided and identify rebate rules that are demand revealing on the aggregate and individual levels. Among the rebate rules studied, a rebate proportional to contributions is the logical alternative for use in our experiment. Spencer *et al.* (2009) define the proportional rebate rule such that returns to a contributor are a share of the excess contributions in proportion to his/her contribution relative to the total contributions collected. No secondary investment is assumed. The rebate is similar to a dividend payment. Payments depend on how large a share of the aggregate contribution each individual has contributed. Contributors of a larger share get a higher rebate. With the proportional rebate rule, the surplus is allocated according to each person's contribution toward the total. Since no secondary investment opportunity is considered in this situation, only the contribution matters. A secondary investment, as in a low-return public good, may lead high income individuals to contribute less. This is because the high income individual finds it disadvantageous for the aggregate contribution to exceed the threshold. For example, if the high income individual contributes 50, and the threshold is exceeded by, say, 30, that individual could have contributed 20 and increased his/her payoff by 30. Instead, the individual receives only a fraction of this amount.[12]

We close this section with some final notes regarding the voting rule. First, the probability of reaching a majority "yes" vote is higher for a smaller group (as evidenced by the voting patterns in the experiment: each of the smaller groups voted "yes" in more periods than both the larger groups combined). If a global vote were implemented instead, the probability of reaching a majority would decrease significantly, and the threshold would be more difficult to achieve. This suggests, and could easily be tested in an experimental setting, that the group majority voting rule is superior to a global majority vote in inducing the desired global collective action. The implication is that a decentralized decision-making process would be more effective in a cooperative game between collective sub-entities such as nations contributing to a global provision threshold. Allowing each group to hold its own vote increases the probability that at least one of the rich groups will vote to participate, effectively easing the burden on the other groups.

Conclusion

In this chapter we have explored how institutional design affects collective action when two levels of heterogeneity impede coordination. We developed an economic experiment using a game of public good provision. We departed from the existing literature on public good provision experiments by modeling our experiment as one of threshold elimination of a public bad, and added to it a voting treatment at the group level. The results suggest that the group voting rule is effective. The control sets, which had no voting rule, achieved the threshold in 37.5 percent of periods played. The treatment sets used the voting rule, and achieved the threshold in 66.7 percent of periods played. Welfare, as measured by accumulated payoffs, was higher on average for the treatment sets. Encouraging as these results may be, more sessions are always better in that they provide for a more robust empirical analysis of the experimental results.

This experimental evidence is suggestive: an intra-mural mechanism enhances a polycentric structure by inducing more coordination in public good provision. Localized institutions play a key role in conditioning individual and collective behaviors (also see Kroll and Shogren 2008 on the power of two-level games). Viable extensions of this experiment would test the robustness of how institutional polycentricism bolstered by hierarchical support schemes can promote global objectives for the mitigation of climate related risk. For example, climate negotiations involve many different countries, each of which can be home to a variety of cultures and social attitudes. As demonstrated by Henrich *et al.* (2001), behavioral experimental results can differ across cultures. These authors ask whether the heterogeneity between individuals' behaviors is at least in part attributable to those individuals' cultural environments and socio-economic backgrounds. They go on to suggest that such questions cannot be answered by existing research "because virtually all subjects have been university students, and while there are cultural differences among student populations throughout the world, these differences are small compared to the range of all social and cultural environments." To study the effects of cultural heterogeneity on economic behavior, Henrich *et al.* conducted economic experiments based on several well-known paradigms (including public goods games) in a variety of non-industrialized societies. Among the key findings were that none of the groups studied displayed behavior consistent with the classic economic assumption of purely self-interested behavior, and that higher payoffs to cooperation led unambiguously to greater cooperation in experimental games. The implication is that future extensions of this experiment or others like it should test whether the results are robust when using a more heterogeneous pool of subjects than just university students. Other modifications to this experiment could be useful in developing a more complete understanding of how institutions might affect the threshold public good game. How might the results change if the damages associated with not meeting the threshold were *not* proportional to the players' remaining endowments? We discussed earlier that the burden of climate change is likely to fall disproportionately on low income countries. Perhaps a payoff structure in which low-endowment groups suffered greater proportional damages would reduce the incentive of high-endowment groups to contribute, because their own damages would be relatively less severe. In contrast, we can wonder how inter-group altruism might be a factor under such an arrangement. These are key behavioral questions that can potentially be answered through further experimentation.

Another interesting treatment would involve allowing individuals within groups to communicate with one another before voting. Climate negotiations involve intense communication over multiple rounds. By designing the game such that it is repeated over many rounds, individuals are able to learn the benefits and costs of more/less coordination. With the voting treatment, however, fewer individuals need to have internally recognized the benefits of coordination and/or cooperation for the threshold to be met – it requires as many individuals as it takes to have a majority vote to contribute. Second, the voting treatment itself is a form of commitment – if there are a sufficient number of "yes" votes within a group, the group is effectively committing to a specific contribution. Preceding the voting stage with a communication stage might further enhance coordination, because perceptive individuals may be able to convince other group members of the benefits of participation. Allowing groups to communicate with each other might help provide greater assurance of other groups' participation, reducing between-group free riding. Other variations on the polycentric public good game include different voting rules (see Kroll *et al.* 2007) and different rebate schemes for amounts in excess of the threshold (see Spencer *et al.* 2009). Another, more ambitious, extension would be to recast the experiment using a dynamic game with a cumulative threshold.

Appendix: model equations

In a given period, if the threshold is met exactly, the payoff structure is given by

$$\text{If } \sum_i \sum_j m_{ij} = \overline{M} \rightarrow \pi_{ij} = \omega_{ij} - m_{ij}, \qquad \textbf{(A1.1)}$$

where m_{ij} are individual contributions, $\overline{M}$ is the threshold, π_{ij} are individual payoffs, and ω_{ij} are individual endowments.

If aggregate contributions exceed the threshold, the payoffs are

$$\text{If } \sum_i \sum_j m_{ij} \geq \overline{M} \rightarrow \pi_{ij} = \omega_{ij} - m_{ij} + \alpha\left(\sum_i \sum_j m_{ij} - \overline{M}\right), \qquad \textbf{(A1.2)}$$

where α is the return on the threshold surplus.

If aggregate contributions fall short of the threshold, individual payoffs are

$$\text{If } \sum_i \sum_j m_{ij} < \overline{M} \rightarrow \pi_{ij} = \omega_{ij} - m_{ij} - D_{ij}, \qquad \textbf{(A1.3)}$$

where $D_{ij} = 0.6\ (\omega_{ij} - m_{ij})$ are damages.

For our empirical analysis of the experimental results, our logit regression model is

$$y_{i,t} = \frac{\exp(x_{i,t}\beta)}{1+\exp(x_{i,t}\beta)} \quad \textbf{(A1.4)}$$

where the subscripts index game i played in period t. The dependent variable is binary: $y_{i,t} = 1$ indicates the threshold was met; otherwise, $y_{i,t} = 0$. The vector of characteristics, $x_{i,t}$, includes a voting treatment dummy and period dummies to control for time effects (period 1 baseline). The key coefficient is that of the voting treatment dummy. Our panel regression model is

$$C_{i,t} = \beta_0 + \beta_1 x_{i,t} + \sum_{s=2}^{10} \beta_5 P_{i,t}^s + \sum_g \beta_g G_{i,t}^g, \quad \textbf{(A1.5)}$$

where $C_{i,t}$ are group contributions. The index g indicates groups *B*/*X*, *C*/*Y*, and D/Z. Again the voting treatment dummy, which is $x_{i,t}$ in regression equation (A1.5), yields the coefficient of interest. We again control for time effects with a period 1 baseline using period dummies, $P_{i,t}^s$ (s = 2, …, 10) but also for population and endowment effects using group dummies, $SG_{i,t}^g$ (*A*/*W* baseline), noting that groups *A* and *W*, *B* and *X*, etc., have the same sub-group dummy variable data because they are identical in population and endowment per person.

Notes

1 For a complete discussion of loss aversion, see Tversky and Kahneman (1991), who explain preference reversals with reference-dependent utility functions.
2 Also see Milinski *et al.* (2006), Rothstein (2005), and Poteete *et al.* (2010).
3 As opposed to the entire experimental set of participants.
4 The results depicted in this chapter are taken from pilot sessions of this experiment conducted at the University of Wyoming in the fall of 2010. Our subjects were University of Wyoming graduate and undergraduate students, and the minimum payout was $10.
5 All contributions, once made, are sunk.
6 Any global organization responsible for administering a climate change mitigation program would be likely to invest any such surplus in alternative public goods such as adaptation and awareness programs. We model our experiment to represent a scenario where any contribution in excess of the threshold is invested in a low-return public good.
7 Setting $\alpha = \frac{1}{30}$ ensures that even at the (unlikely) maximum possible surplus of 750, no player can receive more than his/her initial endowment, eliminating the incentive to gamble.
8 The damage proportion of 60 percent is meant to represent a catastrophic loss from a climate disaster. This parameter is ad hoc. Further research seems worthwhile if it could determine the effect of the severity of the damages on the experimental outcome. For this experiment, however, we note 60 percent is sufficiently large to provide incentive for each sub-group to vote "yes" in the treatment game, which we explain below.
9 By unstable we mean that, although this strategy is an equilibrium in that no player would have incentive to deviate from it, if even a single player cheats and contributes

less than 25 percent, there is incentive for all players to then contribute zero. As noted by Benaïm *et al.* (2009), "if a Nash equilibrium is unstable, we would expect actual players, for example, subjects in an experiment, not to play that equilibrium or even to be close to it."

10 After 10 periods, since the first two control sets each went 11 periods. Does not include the $10 base payment paid to all participants.

11 The coefficient estimate, 1.444, is in a log-linearized form, meaning the estimate represents the change in the log-odds of meeting the threshold. The exponential of the coefficient represents the increase in the odds of meeting the threshold given implementation of the voting rule. Thus we say that, relative to the control set, the voting rule increased the odds of meeting the threshold in a given period by a factor of $e^{1.444}$=4.237, with a 95 percent confidence interval of (2.31, 7.70).

12 Spencer *et al.* (2009) test several alternative rebate rules, including: (1) winner-take-all with proportional probability; (2) winner-take-all with uniform probability among contributors only; (3) winner-take-all with uniform probability among all group members; (4) random full-rebate with uniform probability; and (5) random full-rebate with proportional probability. All of the rebate rules led to public-good provision, but only the proportional rebate rule is found to achieve both aggregate and individual demand revelation.

References

Andersson, K. and E. Ostrom (2008). Analyzing decentralized resource regimes from a polycentric perspective. *Policy Sciences* 41(1), 71–93.

Barrett, S. (2003). Global disease eradication. *Journal of the European Economic Association* 1(2–3), 591–600.

——(2006). The smallpox eradication game. *Public Choice* 130(1–2), 179–207.

——(2013). Climate treaties and approaching catastrophes. *Journal of Environmental Economics and Management* (in press).

Barrett, S. and A. Dannenberg (2012). Climate negotiations under scientific uncertainty. *Proceedings of the National Academy of Sciences* 109(43), 17372–76.

Benaïm, M., J. Hofbauer and E. Hopkins (2009). Learning in games with unstable equilibria. *Journal of Economic Theory* 144(4), 1694–709.

Bohm, P. (2003). Experimental evaluations of policy instruments. In K.G. Mäler and J.R. Vincent (eds.), *Handbook of Environmental Economics* Vol. 1, Amsterdam: Elsevier, pp. 438–60.

Brick, K. and M. Visser (2010). Meeting a national emission reduction target in an experimental setting. *Climate Policy* 10(5), 543–59.

——(2012). Heterogeneity and voting: a framed public good experiment. *Economic Research Southern Africa (ERSA) Working Paper* 298.

Brick, K., Z. Van der Hoven and M. Visser (2012). Cooperation and climate change: can communication facilitate the provision of public goods in heterogeneous settings? Resources for the Future (RFF), Environment for Development Discussion Paper Series, November 2013 EfD DP pp. 12–14.

Buchan, N., G. Grimalda, R. Wilson, M. Brewer, E. Fatas and M. Foddy (2009). Globalization and human cooperation. *Proceedings of the National Academy of Sciences* 106(11), 4138–42.

Burger, N. and C. Kolstad (2009). Voluntary public goods provision, coalition formation, and uncertainty. *National Bureau of Economic Research (NBER) Working Paper* No. 15543.

Cadsby, C.B. and E. Maynes (1999). Voluntary provision of threshold public goods with continuous contributions: experimental evidence. *Journal of Public Economics* 71(1), 53–73.

Cherry, T. and D. McEvoy (2013). Enforcing compliance with environmental agreements in the absence of strong institutions: an experimental analysis. *Environmental and Resource Economics* 54(1), 63–77.

Cherry, T., S. Kroll and S. Shogren (2007). The impact of endowment heterogeneity and origin on public good contributions: evidence from the lab. *Journal of Economic Behavior & Organization* 57(3), 357–65.

Crawford, S. and E. Ostrom (1995). A grammar of institutions. *American Political Science Review* 89(3), 582–600.

Dannenberg, A., A. Lange and B. Sturm (2010). On the formation of coalitions to provide public goods – experimental evidence from the lab. *National Bureau of Economic Research (NBER) Working Paper* No. 15967.

Dannenberg, A., A. Löschel, G. Paolacci, C. Reif and A. Tavoni (2011). Coordination under threshold uncertainty in a public goods game. *Grantham Research Institute on Climate Change and the Environment Working Paper* No. 64, London School of Economics.

Fischbacher, U., W. Güth and M.V. Levati (2011). Crossing the point of no return: a public goods experiment. *Jena Economic Research Papers* No. 2011.059.

Furubotn, E. and R. Richter (1997). *Institutions and Economic Theory*. Ann Arbor: University of Michigan Press.

Hart, S., and M. Kurz (1983). Endogenous formation of coalitions. *Econometrica* 51(4), 1047–64.

Hasson, R., Å. Löfgren and M. Visser (2009). Climate change in a public goods game: investment decision in mitigation versus adaptation. *Ecological Economics* 70(2), 331–38.

Henrich, J., R. Boyd, S. Bowles, C. Camerer, E. Fehr, H. Gintis, and R. McElreath (2001). In search of homo economicus: behavioral experiments in 15 small-scale societies. *The American Economic Review* 91(2), 73–78.

Kane, S. and J. Shogren (2000). Linking adaptation and mitigation in climate change policy. *Climatic Change* 45(1), 75–102.

Kosfeld, M., A. Okada and A. Riedl (2009). Institution formation in public goods games. *The American Economic Review* 99(4), 1335–55.

Kroll, S., T. Cherry and J. Shogren (2007). Voting, punishment, and public goods. *Economic Inquiry* 45(3), 557–70.

Lamb, L. and P.P. Tsigaris (2011). Public good provision and fairness issues for climate change mitigation. *Journal of Business Ethics Education* 8(1), 139–55.

McEvoy, D., J. Murphy, J.K. Stranlund and J. Spraggon (2011). The problem of maintaining compliance within stable coalitions: experimental evidence. *Oxford Economic Papers* 63(3), 475–98.

Milinski, M., D. Semmann, H. Krambeck and J. Marotzke (2006). Stabilizing the Earth's climate is not a losing game: supporting evidence from public goods experiments. *Proceedings of the National Academy of Science* 103(11), 3994–98.

Ostrom, E. (1986). An agenda for the study of institutions. *Public Choice* 48(1), 3–25.

——(1999). Polycentricity – Part 1. In Michael McGinnis (ed.), *Polycentricity and Local Public Economies: readings from the Workshop in Political Theory and Policy Analysis*. Ann Arbor: University of Michigan Press, pp. 52–74.

——(2009). A polycentric approach for coping with climate change. *Political Research Working Paper* No. 5095, Indiana University.

Poteete, A., M. Janssen and E. Ostrom (2010). *Working Together: collective action, the commons, and multiple methods in practice*. Princeton, NJ: Princeton University Press.

Rothstein, B. (2005). *Social Traps and the Problem of Trust*. Cambridge, UK: Cambridge University Press.

Saijo, T., K. Sherstyuk, N. Tarui and M. Ravago (2009). Games with dynamic externalities and climate change experiments. Working paper.

Sandler, T. (2001). On financing global and international public goods. Working paper, University of Southern California.

Spencer, M., S. Swallow, J. Shogren and J. List (2009). Rebate rules in threshold public good provision. *Journal of Public Economics* 93(5–6), 798–806.

Srinivasan, U.T., S. Carey, E. Hallstein, P. Higgins, A. Kerr, L. Koteen, A. Smith, R. Watson, J. Harte and R. Norgaard (2008). The debt of nations and the distribution of ecological impacts from human activities. *Proceedings of the National Academy of Sciences* 105(5), 1768–73.

Stern, N. (2007). *The Economics of Climate Change: The Stern Review*. Cambridge, UK: Cambridge University Press.

Tavoni, A., A. Dannenberg and A. Löschel (2010). Coordinating to protect the global climate: experimental evidence on the role of inequality and commitment. *ZEW Discussion Papers* No. 10-049.

Tavoni, A., A. Dannenberg, G. Kallis and A. Löschel (2011) Inequality, communication, and the avoidance of disastrous climate change in a public goods game. *Proceedings of the National Academy of Science* 108(29), 11825–29.

Tversky, A, and D. Kahneman (1991). Loss aversion in riskless choice: a reference-dependent model. *Quarterly Journal of Economics* 106(4), 1039–61.

Walker, B., S. Barrett, S. Polasky, V. Galaz, C. Folke, G. Engström, F. Ackerman, K. Arrow, S. Carpenter, K. Chopra, G. Daily, P. Ehrlich, T. Hughes, N. Kautsky, S. Levin, K. Mäler, J. Shogren, J. Vincent, T. Xepapadeas and A. de Zeeuw (2009). Looming global-scale failures and missing institutions. *Science* 325 (5946), 1345–46.

Weitzman, M. (2009). On modeling and interpreting the economics of catastrophic climate change. *Review of Economics and Statistics* 91(1), 1–19.

9 Improving the design of international environmental agreements

Matthew McGinty

Introduction

This chapter presents some insights from the last decade of theoretical work in the International Environmental Agreement (IEA) literature to illustrate the role of transfers when nations are asymmetric. The goal is to improve the design of IEAs and subsequently to increase both participation and abatement levels. The theoretical literature does not ask who is responsible for past emissions, nor imply normative judgments about who should pay for abatement, or even abate their emissions. Rather, the primary concern is what abatement requirements will generate a large number of signatories and substantial increases in abatement, compared to the outcome without a treaty. This analysis adopts the standard assumption in economics that nations are rational and choose actions that maximize their well-being, without regard for other nations. IEAs need to be self-enforcing in the sense that there is no supra-national institution that can force sovereign nations to adhere to the IEA. Nations join the IEA when they are better off by doing so, and leave if it is in their interest to do so. This means that the abatement requirements need to be chosen with the voluntary participation constraint in mind.

The fundamental problem with the Kyoto Protocol is the lack of abatement requirements for all nations. Under Kyoto, only some of the richest nations (listed in Annex B) were required to reduce emissions for the period 2008–2012. These included the United States, 15 members of the European Union, Japan, Canada, Australia and New Zealand. The rest of the world was not subject to binding abatement requirements. The United States never ratified the treaty, and Canada and New Zealand have subsequently withdrawn from future participation. An agreement that does not include abatement by the five most populated nations in the world (China, India, United States, Indonesia and Brazil) cannot solve the problem.

The rest of this chapter is organized as follows. First, the relationship between asymmetry and transfers is discussed. Next, the three main types of models used in the literature are presented, along with some recent results. A comparison between the early approach of cooperative game theory rules and the more recent approach to abatement requirements is then presented. The conclusion details the main ideas from the current literature and suggests some directions for future research. All equations are relegated to the appendix, which also contains a simple example to illustrate some of the main points.

Asymmetry

In December 2012 United Nations Secretary-General Ban Ki-moon told the delegates in Doha: "The climate change phenomenon has been caused by the industrialization of the developed world." Indeed, it is certainly the case that the current levels of greenhouse gases are primarily the result of the industrialized nations becoming rich. The Secretary-General continued "It's only fair and reasonable that the developed world should bear most of the responsibility." In short, the Secretary-General is advocating the Kyoto position that the so-called "firewall" between developed and developing nations be maintained and that only the developed nations have binding abatement requirements. The argument is that cheap energy, and the resulting emissions, are part of the development process. Nations that are currently developing should not be placed at a disadvantage compared to the developed nations. However, this sentiment is counter-productive to generating a meaningful agreement. The problem is not just past emissions of greenhouse gases, but rather present and future emissions. China is now the world's largest emitter and the majority of current emissions come from the developing world. Maintaining the current firewall between developed and developing nations will result in the failure to obtain a meaningful agreement with respect to greenhouse gas abatement. All large nations need to be a part of the solution.

Greenhouse gases mix uniformly in the upper atmosphere; thus from a global perspective the terrestrial location of emissions does not matter. This means that an IEA can break the link between actual abatement and required abatement. Nations can be responsible for abatement beyond their domestic amount by paying other nations to abate for them. Nations that currently use a high proportion of coal, such as China (67 percent) and India (42 percent) can provide abatement much more cheaply than nations such as France (4 percent), Brazil (5 percent) and Sweden (4 percent) that use very little coal (International Energy Agency, share of total primary energy supply, 2009). Clearly, the global cost of greenhouse gas abatement will be far less if much of the abatement is done in the developing world. This is clearly recognized by policy makers such as Zou Ji, the deputy director of China's National Center for Climate Change Strategy. Ji said "If China could replace coal with oil as a primary source of energy, emissions would drop by one third. If we could replace coal with natural gas, they would drop by two thirds. But China's main resource is coal" (Revkin 2013).

The benefit that each nation receives from abatement is in general different. The share of global benefits from abatement can depend on many factors such as population, geography and climate. Some nations, such as Bangladesh and Tuvalu, are particularly vulnerable to sea level rise and storm surge. Other nations are particularly vulnerable to drought or other water-related issues. In the absence of an agreement this means that on the benefit side nations have different incentives to provide abatement.

With increasing marginal abatement (MAC), the cost of the last unit of abatement will be different due to both the benefit and the cost sides. Asymmetric benefits generate a non-cooperative solution with different abatement levels for each nation. Thus even if MAC curves are identical, the MAC of the last unit is different and, without an agreement, abatement costs will vary greatly across nations. This generates potential gains from trade, and therefore asymmetry itself can generate incentives to join a properly designed agreement.

However, this does not mean that the developing world needs to be responsible for this abatement. Abatement by developing nations is simply taking advantage of gains to trade that exist due to the vast differences in marginal abatement costs across nations. The empirical estimates from Reports #40 and #41 of the MIT Joint Program on the Science and Policy of Global Change (Ellerman and Decaux 1998; Ellerman *et al.* 1998) show the stark differences in the cost curves of marginal abatement. Their model estimated that China could reduce emissions by up to 40 percent before the cost of the last unit of abatement rose above $50. India has an almost identical cost curve of marginal abatement. By contrast, Japan crosses the $50 threshold at a reduction of around 6 percent of emissions, and the cost of the last unit for a 22 percent reduction is $300. This asymmetry can be exploited, and the developing nations are key to this effort. Requiring domestic abatement by high cost nations is, in effect, penalizing those nations that already use a clean (in terms of GHG emissions) mix of fuels, such as France, Sweden and Brazil. Under a properly designed agreement, these nations can meet their abatement requirements by purchasing pollution permits.

Transfers

The recent theoretical literature shows how a zero-sum system of transfers among IEA signatories can increase both participation and abatement levels (McGinty 2007). A nation can meet its abatement requirements under the agreement in two ways: either by domestic abatement, or by purchasing pollution permits which involve paying other nations to abate for them. The cost curves imply that the developing nations are the cheapest sources of GHG abatement. However, this does not mean that the developing nations need to pay for this abatement. With a zero-sum system of transfers among signatories, nations that have steep MAC curves can pay those with flat MAC curves to provide abatement. An agreement that contains transfers from the developed to the developing world can result in a least-cost solution and could also have the benefit of reducing global inequality.

These transfers could be direct bilateral payments specified in the agreement. However, this may be politically difficult as it will entail large transfers of wealth, particularly when low-growth nations are making transfers to rapidly-growing developing economies. One method of transfers that avoids bilateral direct comparisons is a system of tradable pollution permits. A nation that abates more than its requirements can sell a permit to a nation whose abatement is less than required. Under permit trading the market price of a permit equals the marginal abatement cost of the last unit. Permit trading achieves the least-cost solution for a given level of abatement as the permit price equates the marginal abatement cost across all signatory nations. The equilibrium permit price is increasing in the total amount of abatement, but is decreasing in the number of nations and when the MAC curves are asymmetric (see the appendix for more details on these points). A system of tradable pollution permits means that the sum of abatement requirements is realized, not that any individual nation has to abate a given amount.

Furthermore, as the recent increase in coal use in Germany has indicated, there are potentially dramatic carbon leakage effects from a failure to have an agreement with full participation. Changes in fossil fuel prices, as the result of an IEA or for other reasons, can have significant impacts on emissions. In 2012, for example, 25.6 percent of Germany's electricity was generated by burning lignite (also known as brown coal), compared to 22.7 percent in 2010. Hydraulic fracturing (commonly referred to as fracking) has resulted in a dramatic increase in natural gas production in the United States and subsequently a fall in the domestic price. Not surprisingly, natural gas has replaced coal as a fuel source for many US power plants. This, in turn, has reduced both the use and the domestic price of coal. United States coal exports to Europe were 26 percent higher in the first nine months of 2012 than they were the previous year (Birnbaum 2013).

With limited IEA participation, carbon leakage is a likely result. For example, consider the effects of an IEA only among the 34 OECD nations with stringent GHG reductions. Sharp reductions in coal use in these nations will lower the world price of coal and make it a more attractive fuel source to those outside the agreement. The agreement will then result in carbon leakage where those outside the agreement increase emissions. This incentive is not a new observation. On July 25th, 1997 Senator Charles Hagel addressed the United States Senate during debate about the Kyoto Protocol. Senator Hagel said: "The main effect of the assumed policy would be to redistribute output, employment and emissions from participating to non-participating countries." The Senate then approved by a vote of 95-0 the Byrd-Hagel resolution that stated that "The United States should not be a signatory to any protocol to, or other agreement regarding, the United Nations Framework Convention on Climate Change of 1992" (Kuik and Gerlagh 2003).

Two elements of the Kyoto Protocol recognize the potential for cost savings from outside abatement opportunities. These are the Clean Development Mechanism (CDM) and Joint Implementation (JI). The CDM allows developed nations to earn tradable certified emissions reduction (CER) credits by undertaking projects in the developing world. Abatement for these projects is defined as relative to "what would otherwise have occurred" (UNFCCC.org). By contrast, JI allows developed nations to undertake abatement projects in other developed nations. While both mechanisms recognize cost savings and realize mutually beneficial transactions, neither increases the total amount of abatement nor protects against carbon leakage. They are in some respects a piecemeal approach to what is essentially trading on a project by project basis. This results in both monitoring costs and measurement issues. There is no reason to believe that either the CDM or JI result in a cost-minimizing solution for any particular project, given their limited scope. While these mechanisms are a step in the right direction, they fail to realize all the gains that a fully-functioning permit market would obtain.

Theoretical models

Three types of theoretical model are presented which reflect the modeling tension between complexity and tractability. The simplest models are linear and denoted as Type I in what follows. These are typically a prisoner's dilemma, where to pollute is a dominant strategy, while the socially optimal outcome is to abate emissions. Type II models have constant marginal benefit but increasing marginal abatement cost. These models result in positive levels of abatement at the non-cooperative solution. However, abatement levels for those outside the IEA are still a dominant strategy. Thus there is no response by non-signatories to changes in the signatories' abatement levels, that is, no carbon leakage. Type III models assume declining marginal benefit and increasing marginal abatement cost. In this case, abatement by those outside the IEA depends on both the number of signatories and their aggregate abatement level. For Type III models the non-cooperative outcome can result in abatement levels that are close to those with full participation, despite the fact that the agreement results in carbon leakage. Each of the three types of model is discussed as they relate to the role of transfers and asymmetry in improving both participation and abatement levels. Further details regarding the equations and references underlying these models are found in the appendix.

Type I models have linear payoff functions. This comes from the assumptions of both constant marginal benefit and constant marginal cost from each unit of abatement. Two examples of Type I models are Kolstad (2007) and Barrett (2001). Barrett (2001) considers a linear IEA modeled as a stage-game in which there are two types of nations (i.e., rich with high benefit and poor with low benefit). The single stage game is a prisoner's dilemma with a dominant strategy of "pollute." Barrett then considers a situation with the following four stages. In the first stage nations make a decision to participate or not in the IEA. In stage two all high benefit signatories collectively choose pollute or abate, and a transfer from the high benefit to the low benefit nations should they join the IEA and also choose abate. In stage three each of the low benefit nations decides to join the IEA and receive the transfer, or remain outside. In stage four, all nations outside the agreement choose pollute or abate.

In the four-stage game with transfers, Barrett (2001) finds that a stable IEA with full participation by the low benefit nations, and 33 out of 50 high benefit nations, can be obtained when "cooperation is for sale." High benefit nations pay low benefit nations in the form of a transfer to participate in the IEA and abate emissions. A key element of this model is that the equilibrium is "lynchpin." This means that the agreement is not very robust in the sense that should any signatory change their participation decision and pollute, then all other signatories pollute and the entire agreement collapses. However, the model shows that asymmetry and the proper transfers can dramatically improve the outcome.

McGinty (2010) extends Barrett's (2001) model both by allowing for two other types of linear games, and by looking at a different type of timing. In addition to a prisoner's dilemma where pollute is a dominant strategy, McGinty (2010) also considers coordination and hawk–dove games where the IEA is modeled as an evolutionary game, rather than the stage game in Barrett (2001). In an evolutionary game nations choose actions based on the current payoff differential, and an evolutionary equilibrium is robust to changes in the participation decision of a small number of other nations.

In a coordination game, there is a payoff advantage to choosing to pollute when IEA participation is low, but beyond an endogenous threshold there is a payoff advantage to abate. Cooperation is easily obtained in a coordination game, both with identical nations (single population) or with asymmetry in the form of two types of nations (two populations). Asymmetry and transfers are not needed in a coordination game since an IEA with full participation is obtained as long as the number of signatories is greater than the threshold.

In a hawk–dove game there is a payoff advantage to choosing abate when participation is low and a payoff advantage to choosing pollute when participation is high. In this case, as with the vast majority of the IEA literature, the stable IEA consists of very few members and consequently achieves very little compared to the non-cooperative outcome. However, this is when asymmetry and transfers can result in the most dramatic improvements. In the two-population model with asymmetry, there is partial participation in both populations without transfers. With a small degree of asymmetry across populations the surprising result emerges that a credible transfer from the low payoff group to the higher payoff group can increase both participation and payoff for all nations. Thus exacerbating inequality may make all nations better off. The reason is that abatement by the high payoff nations is greater, thus they bring more to the agreement.

When there is a large degree of asymmetry across populations there are two evolutionary equilibria. In this case credible transfers can ensure that the evolutionary equilibrium with the higher payoff for all nations is obtained. The high payoff nations can credibly commit to make a transfer to the low payoff nations, thus reducing inequality. Transfers have a role in determining which of the two equilibria is obtained.

The seminal paper in the IEA literature is Barrett (1994), which assumes identical nations. Much of the analysis consists of finding stable IEAs for the Type III model with declining marginal benefit, increasing marginal abatement cost, and Stackelberg leadership for the signatories. This means that IEA signatories choose their abatement prior to non-signatories. Those outside the agreement choose abatement after observing the signatories' choices. Barrett (1994) shows that when the benefits are high and the costs are low there is little difference between the non-cooperative outcome and the global optimum. In this case the stable IEA can consist of all nations, since the agreement does not dictate a substantial change in abatement. However, when the slopes of the global marginal benefit and marginal abatement cost are the same, the stable IEA consists of only three nations out of one hundred. In this case there is a large difference between the non-cooperative outcome and global optimum. Barrett (1994) concludes that there is an inverse relationship between the gains to cooperation and the number of signatories in a stable IEA. With identical nations, IEAs can achieve the least when they are needed the most.

McGinty (2007) modifies Barrett (1994) by allowing asymmetry on both the benefit and cost sides in a Type III model. When the high benefit nations are also low cost, the non-cooperative outcome is closer to global optimum and the gains to cooperation are smaller than the symmetric result. When the high benefit nations are high cost, the gains to cooperation are larger than symmetry would indicate. Asymmetry generates mutually beneficial transfers between nations. A zero-sum system of transfers can be implemented in the form of tradable pollution permits. These mutual benefit transactions can do the most when the high benefit nations are also high cost. In this case a small number of IEA signatories (5 out of 20 nations) can result in 47 percent of the abatement difference, when only 5 percent would be obtained with symmetry. While asymmetry can lead to much better results, the fundamental tradeoff from Barrett (1994) remains. Stable IEAs are smaller when the gains to cooperation are larger.

Abatement requirements

Abatement requirements under the Kyoto Protocol were specified to be a minimum of 5.5 percent of 1990 emissions for the developed (Annex B) nations. The total amount of abatement did not reflect the solution to any maximization problem, but rather was the result of negotiations. The theoretical IEA literature takes a very different approach. These models specify benefit and cost functions from abatement, and then, for a given set of signatories, ask what amount of signatory abatement maximizes their net benefit. Signatory net benefit is the sum of signatory benefit from abatement minus the sum of signatory costs. It is assumed that signatories to an IEA behave cooperatively, in the sense that they choose an aggregate abatement level that maximizes the sum of signatories' net benefits. This means that signatories to an agreement internalize the benefits that accrue to other signatories. Signatories do not consider the benefits that accrue to non-signatories. It is also assumed that non-signatories choose abatement levels that maximize their own payoff, without any regard for other nation's well-being. Together, this typically means that global abatement is increasing in the number of signatories. Abatement is then largest when the agreement has full participation and all nations are signatories.

Abatement of GHGs is an example of what economists call a global public good. This means that when one nation undertakes costly abatement, all nations benefit. Each nation bears the full cost, but only receives a portion of the global benefit from their abatement. This leads to the free-rider problem, where nations can enjoy the benefits from other nation's abatement even if they don't pay for any abatement themselves.

Barrett (1997) and Botteon and Carraro (2001) investigate IEA stability using the traditional cooperative game theory allocation rules. The idea is that since the signatories are behaving cooperatively, rules from cooperative game theory are appropriate. Both papers check for stable coalitions (d'Aspremont *et al.* 1983) after imposing abatement requirements that are consistent with the Shapley Value (SV), while Botteon and Carraro (2001) also consider the Nash Bargaining Solution (NBS). Cooperative game theory implicitly assumes that binding coalitions can form without cost, consequently they are less concerned with stability. Both the SV and the NBS work well when the grand coalition is stable. However, for public goods such as greenhouse gas abatement, the grand coalition is not often stable.

Barrett (1997) considers a Type II model with constant marginal benefit and increasing marginal abatement cost. He shows that in the absence of side payments the stable IEA will consist of at most three nations. Under the SV the stable IEA will consist of either two or three nations; thus side payments generating the SV solution do not improve the outcome. Recent work by Fuentes-Albero and Rubio (2010) and Pavlova and de Zeeuw (2013) have arrived at similar conclusions. Fuentes-Albero and Rubio (2010) allow for two types of nations in a Type II model and show that the largest stable IEA contains three nations in the absence of transfers. They then show that transfers cannot improve the outcome when abatement costs are asymmetric. However, with asymmetric benefits a large degree of cooperation can be obtained. Pavlova and de Zeeuw (2013) extend these results by allowing for both types of asymmetry simultaneously and then determining not only the size of the stable IEA, but also the increase in abatement levels. They show that by simultaneously allowing for both types of asymmetry large coalitions can form even without transfers. In this case transfers do not improve abatement levels very much since the IEA consists of low benefit nations who do substantially increase abatement when joining. Their results illustrate the fundamental tradeoff identified by Barrett (1994), that stable IEAs with a large number of nations are only possible when the gains to cooperation are small.

Barrett (1997) also considers the Type III declining marginal benefit and increasing marginal cost model from Barrett (1994), but allows for two types of nations. For the Type III model without side payments, he finds that the three high cost nations are members of the stable IEA, but none of the four low cost nations. When side payments consistent with the SV are implemented there are two stable IEAs, both of which contain three members. The IEA with the three low cost members results in greater global payoff, but still falls far short of the optimum, despite the fact that there are only seven nations. In short, Barrett finds that the SV with two types of nations does not improve the number of IEA signatories, and may only moderately improve payoffs.

Botteon and Carraro (2001) consider a Type III model with a five-region world. Barrett (1994) assumes that signatories to an IEA behave as a Stackelberg leader, choosing abatement before non-signatories. Those outside the IEA observe signatory abatement, then individually choose abatement to maximize their individual payoff. By contrast, Botteon and Carraro (2001) assume that abatement levels by all nations are chosen simultaneously, and thus signatories do not have a leadership advantage. They calibrate costs and benefits for the five regions and then run simulations to find which IEAs are stable. Using both the SV and NBS they show that under the standard assumptions coalitions larger than three cannot form.

Both the SV and the NBS allocate payoffs to coalition members and thus provide unique abatement requirements. However, neither of these allocation rules consider the payoff that a nation would earn if it were to individually leave the IEA. That is, neither the SV nor the NBS are designed with internal stability in mind. The SV is based on how much a nation increases the worth of the coalition. Specifically, the SV is a weighted sum of the marginal payoff contribution of that nation to all possible coalition formation paths leading to the grand coalition. The NBS is based only on the worth of a given coalition and the payoffs at the non-cooperative outcome with no coalition. The NBS is useful if no coalition is internally stable and the relevant threat point is the non-cooperative outcome. However, if the remaining coalition is stable after one nation leaves, then the non-cooperative outcome (no coalition) is not the relevant threat point.

The new approach to coalition stability depends on the outside payoff, not on what a nation brings to the coalition. A new class of sharing rules from Carraro *et al.* (2006), McGinty (2007) and Weikard (2009) explicitly address internal stability by awarding each nation their outside payoff, plus a share of any remaining surplus. The surplus is defined as the aggregate signatory payoff minus the sum of the payoffs that each nation would earn if it were to individually leave the coalition. If the surplus is positive, then under these new rules the IEA is internally stable and no nation has an incentive to leave. Neither the SV nor the NBS ensures that each nation receives at least its outside payoff, even when the surplus is positive. Put differently, it is indeed the case that IEAs that are stable under the new class of rules are not stable under SV and NBS. See Finus (2003) for a comprehensive survey and taxonomy of the previous IEA approaches.

McGinty (2011) proposes a sharing rule where each nation receives its outside payoff plus an equal share of the remaining surplus. This allocation is shown to be risk-dominant in the sense that participation remains a best response up to the largest degree of uncertainty regarding any other signatory's participation decision. This rule also imposes one notion of equity in an asymmetric setting where payoffs are different across nations both before and after an IEA enters into force. The equal surplus share allocation is illustrated using the asymmetric Type III model. It is shown that stable coalitions under the equal surplus share are unstable when the SV and NBS allocations are imposed.

These allocation rules clearly differ from the design of the Kyoto Protocol. Under Kyoto each Annex I nation had a fixed abatement requirement. Once the twin thresholds of 55 nations accounting for at least 55 percent of the abatement requirements had been ratified, the treaty entered force and the abatement requirements became binding. The thresholds are recognition of the free-rider incentives and impart a degree of strategic complementarity. That is, if one nation joins and the thresholds are met, then other signatories will abate. However, the arbitrary abatement requirements under Kyoto did not ensure either a cost-minimizing solution or an optimal global abatement level.

Under the SV, the NBS and the new sharing rules, signatory abatement levels are chosen to maximize aggregate signatory payoff and result in a cost-minimizing solution. This means that the abatement requirements depend on how many and which nations are IEA signatories. In other words, a properly designed IEA should not specify fixed abatement requirements. Rather, abatement requirements should also depend on which other nations are also in the agreement. A properly designed agreement is then a complete contingency plan. It specifies an abatement requirement for all possible IEA membership situations. This complete contingency plan is essentially what game theorists call a strategy.

It might be cause for concern that participation decisions can more easily be changed than actual abatement levels. It is certainly much easier to join or leave an IEA than it is to change a nation's abatement. This concern is valid if actual abatement is the same as an abatement requirement. However, with transfers and permit trading, actual abatement does not need to change when the requirement changes for any individual nation. The agreement does not dictate that actual abatement changes with participation decisions, but rather just the abatement requirement. With trading, a nation can purchase permits following an increase in their requirement.

Conclusion

The most important lesson going forward is that any effective agreement to reduce greenhouse gas emissions must include all large nations, regardless of their current standard of living. Indeed, it is the developing nations that can provide the cheapest abatement options. This does not mean the burden needs to be borne by the developing world, but rather that asymmetry and transfers provide an opportunity that can be exploited to make all nations better off. A properly designed IEA can exploit these differences by breaking the link between actual abatement and abatement requirements under the treaty. A zero-sum system of transfers implemented in the form of tradable pollution permits will achieve the least-cost solution for a given level of global abatement.

Abatement requirements need to be chosen with an explicit recognition of what payoff each nation would receive if they were outside the agreement. With voluntary participation decisions it is not important what a nation brings to the agreement, but rather what they would earn if they were to leave. An agreement should specify an abatement requirement for each nation given all possible combinations of members. Actual abatement need not change for all possible membership decisions, only the abatement requirement. This generates an element of strategic complementarity. The agreement then specifies an increase in abatement requirements for each existing member when a new nation joins.

Asymmetry and transfers have been shown to be able to increase both participation and abatement level; however, this does not mean that an IEA with full participation will be stable. Future research is needed to determine if more realistic models can improve the theoretical results. There is ample empirical evidence that there are increasing marginal abatement cost curves, and that the rates of increase vary dramatically across nations. The source of this variation is the relative carbon intensity of the underlying fuel sources. However, the shape of the benefit function from abatement is less clear. The current literature assumes either constant or decreasing marginal benefits from abatement. Decreasing marginal benefit means that the first unit of abatement provides more benefit than subsequent units. This assumption is analogous to a downward sloping demand curve in Economics. However, this differs from a standard dose–response function for pollutants in the Biology literature. Typically there is an S-shaped pollution damage function. This means that the marginal benefit from abatement first increases before eventually decreasing. The theoretical implications for different benefit functions have not been fully explored.

Finally, the implications of carbon leakage and other beliefs have not been fully investigated. Internal stability recognizes that other nations will change abatement levels when one nation changes their participation decision. However, for the abatement decision all the current literature adopts Nash beliefs of zero response. That is, nations assume others will not change their abatement level, even though they recognize that other nations have an incentive to change. This is the essence of carbon leakage. Allowing for different beliefs regarding abatement levels is another promising area for future research.

Appendix

Types of models

There are three types of models most commonly found in the literature.

The symmetric Type I model is linear with constant marginal benefit and constant marginal cost.

$$\pi_i = \frac{bQ}{n} - cq_i$$

The payoff to nation i is π_i, Q is aggregate abatement, q_i is abatement by nation i and n is the number of nations. These models are mostly a prisoner's dilemma with a dominant strategy of zero abatement. Two other types of linear games known as coordination and hawk–dove are also possible.

The symmetric Type II model is linear-quadratic with constant marginal benefit and increasing marginal cost.

$$\pi_i = \frac{bQ}{n} - \frac{c(q_i)^2}{2}$$

Type II models also generate a dominant strategy, but one with positive abatement levels. The dominant strategy means that there is no carbon leakage by non-signatories.

The asymmetric Type III model is quadratic-quadratic with decreasing marginal benefit and increasing marginal cost.

$$\pi_i = b\alpha_i\left(aQ - \frac{Q^2}{2}\right) - \frac{c_i(q_i)^2}{2}$$

The benefit share for nation i is α_i and the marginal abatement cost slope is c_i. Type III models generate downward sloping best-response functions. This means that those outside the agreement reduce abatement when the signatories increase abatement.

Coalition stability

Nation i has no incentive to leave coalition S when their payoff as a signatory is at least as great as what they would earn if they were to individually leave the coalition.

$$\pi_i^s(S) \geq \pi_i^n(S \backslash i)$$

Summing this internal stability requirement across all coalition members results in the following.

$$v(S) = \sum_{i \in S} \pi_i^s(S) \geq \sum_{i \in S} \pi_i^n(S \backslash i)$$

A coalition can be internally stable if the worth $v(S)$ is not less than the sum of payoffs from individually leaving the coalition. That is, a coalition is stable if there is a non-negative surplus $\sigma(S)$ defined as:

$$\sigma(S) = v(S) - \sum_{i \in S} \pi_i^n(S \backslash i) \geq 0$$

Marginal abatement cost curves

Type II and III models have increasing marginal abatement cost.

$$MAC_i = c_i q_i.$$

The rate of increase of MAC for each individual nation is c_i. This results in the aggregate MAC curve for a coalition S of IEA signatories.

$$MAC(S) = \frac{Q^S}{\sum_{i \in S}\left(\frac{1}{c_i}\right)}$$

It is immediately clear that the aggregate signatory MAC curve is lower the greater the number of signatories, and when those signatories are asymmetric in their c_i.

Transfers

Transfers allow for abatement requirements under the IEA $q_i^r(S)$ to differ from $q_i^s(S)$ given the constraint:

$$Q^s(S) = \sum_{i \in S} q_i^s(S) = \sum_{i \in S} q_i^r(S)$$

A system of tradable pollution permits then implies the permit price $p(S)$ for a given set of signatories S.

$$p(S) = MAC_i^s(q_i) \forall i \in S$$

Transfers $T_i(S)$ among IEA signatories are a zero-sum system of side payments where $q_i^s(S) - q_i^r(S)$ is the number of permits supplied. That is, actual abatement minus the requirement.

$$T_i(S) = p(S)[q_i^s(S) - q_i^r(S)]$$

The new class of allocation rules are found in Carraro *et al.* (2006), McGinty (2007, 2011) and Weikard (2009). See these papers for more details regarding the relationship between transfers, abatement requirements and internal stability. In McGinty (2011) abatement requirements imply that each nation receives their outside payoff plus an equal share of the surplus. Under this equal surplus sharing rule, abatement requirements solve:

$$\pi_i^n(S \backslash i) + \frac{\sigma(S)}{|S|}$$

Numerical example

Consider a Type II or Type III model example with two nations, where $c_1 = 1$ and $c_2 = 3$. For simplicity let the benefit shares be equal, thus $\alpha_1 = \alpha_2 = \frac{1}{2}$. If both nations are IEA members then the signatory MAC curve is $MAC(S) = \frac{Q^s}{(\frac{4}{3})}$. Suppose the optimal level of abatement is 60 units and in the absence of permit trading each nation is responsible for 30 units of abatement. Then the total abatement cost for nation 1 would be $\frac{(30)^2}{2} = 450$. The total abatement cost for nation 2 is $\frac{3(30)^2}{2} = 1{,}350$. The total cost of 60 units of abatement is then 1,800. Since the benefit shares are equal for this example, nation 1 will have a payoff that is higher by 900, the difference in their total abatement costs.

Now consider permit trading. Using the signatory MAC curve, the permit price will be $p(S) = \frac{60}{(\frac{4}{3})} = 45$. Setting each nation's MAC curve as equal to 45 implies that actual abatement for nations 1 and 2 will be 45 and 15, respectively. The total abatement cost for nations 1 and 2 are 1,012.5 and 337.5 respectively, for a total of 1,350. Permit trading reduces the total abatement cost by 450. Since benefit shares are equal, one possible agreement would be equal cost burdens, making each nation responsible for 675, or half of the 1,350. Thus the transfer solves

$T_1(S) = p(S)[q_1^S(S) - q_1^r(S)] = 1{,}012.5 - 675$. Using the permit price of 45 and actual abatement for nation 1 of 45, this implies abatement requirements of 37.5 for nation 1 and 22.5 for nation 2. These abatement requirements imply that nation 1 sells 7.5 permits to nation 2 for 45 each. The transfer of 337.5 from nation 2 to nation 1 results in an equal share of the cost burden, and in this case equal payoffs.

References

Barrett, S. (1994). Self-enforcing international environmental agreements. *Oxford Economic Papers* 46, 878–94.

——(1997). Heterogeneous international environmental agreements. In C. Carraro (ed.), *International environmental negotiations. Strategic Policy Issues*. Cheltenham, UK: Edward Elgar.

——(2001). International cooperation for sale. *European Economic Review* 45, 1835–50.

Birnbaum, M. (2013). Europe consuming more coal. *The Washington Post*, February 7th, 2013.

Botteon, M. and C. Carraro (2001). Environmental coalitions with heterogenous countries: Burden sharing and carbon leakage. In A. Ulph (ed.), *Environmental Policy, International Agreements, and International Trade*. New York: Oxford University Press, 38–65.

Carraro, C., J. Eyckmans and M. Finus (2006). Optimal transfers and participation decisions in international environmental agreements. *Review of International Organizations* 1(4), 379–96.

d'Aspremont, C., A. Jacquemin, J.J. Gabszewicz and J. Weymark (1983). On the stability of collusive price leadership. *Canadian Journal of Economics* 16, 17–25.

Ellerman, A.D. and A. Decaux (1998). Analysis of post-Kyoto CO_2 Emissions trading using marginal abatement cost curves. MIT Joint Program on the Science and Policy of Global Climate Change Report #40.

Ellerman, A.D., H. Jacoby and A. Decaux (1998). The effect on developing countries of the Kyoto Protocol and CO_2 emissions trading. MIT Joint Program on the Science and Policy of Global Climate Change Report #41.

Finus, M. (2003). Stability and design of international environmental agreements: The case of transboundary pollution. In H. Folmer and T. Tietenberg (eds.), *International Yearbook of Environmental and Resource Economics, 2003/4*. Cheltenham, UK: Edward Elgar, pp. 82–158.

Fuentes-Albero, C. and S. Rubio (2010). Can international cooperation be bought? *European Journal of Operational Research* 202, 255–64.

Kolstad, C. (2007). Systematic uncertainty in self-enforcing international environmental agreements. *Journal of Environmental Economics and Management* 53, 68–79.

Kuik, O. and R. Gerlagh (2003). Trade liberalization and carbon leakage. *The Energy Journal* 24(3), 97–120.

McGinty, M. (2007). International environmental agreements among asymmetric nations. *Oxford Economic Papers* 59(1), 45–62.

——(2010). International environmental agreements as evolutionary games. *Environmental and Resource Economics* 45, 251–69.

——(2011). A risk-dominant allocation: maximizing coalition stability. *Journal of Public Economic Theory* 13(2), 311–25.

Pavlova, Y. and A. de Zeeuw (2013). Asymmetries in international environmental agreements. *Environment and Development Economics* 18(1), 51–68.

Revkin, A. (2013). Tough truths from China on CO_2 and climate. *New York Times*, February 26th, 2013.

Weikard, H-P. (2009). Cartel stability under an optimal sharing rule. *The Manchester School* 77 (5): 575–93.

10 Managing dangerous anthropogenic interference

Decision rules for climate governance

Richard B. Howarth and Michael D. Gerst

Introduction

In 1988 the World Conference on the Changing Atmosphere endorsed a strategy of phased reductions in greenhouse gas emissions aimed at protecting future generations from harm and abating the uncertain but potentially catastrophic impacts of climate change (World Meteorological Organization 1988). From that time to the present, the values and normative judgments behind this proposed strategy have been controversial amongst economists. Manne and Richels (1992: 1), for example, quipped that "the greenhouse debate is short on facts and long on rhetoric … [A]ctivists – many with backgrounds in the physical sciences – point to the potential for disastrous long-term trends in the global climate." In a related vein, Nordhaus (1992, 2008) has long argued that deep reductions in greenhouse gas emissions would reduce social welfare in a manner inconsistent with the achievement of economic efficiency. In this chapter, we shall explore the current status of this long-standing debate, drawing in part on our previous work (Howarth 1998; Gerst *et al.* 2013) in the context of the broader and evolving literature.

One starting point for discussion is Siegfried von Ciriacy-Wantrup's (1968) book *Resource Economics: conservation and policies*, which advanced the concept of "safe minimum standards" (SMSs) as an approach to managing long-run environmental risks. According to Ciriacy-Wantrup, the degradation of ecosystems is caused by activities such as the extractive use of natural resources and the discharge of effluents that, in many cases, generate relatively modest, short-run economic benefits. Such activities, however, can lead to the irreversible loss of ecosystem services that are potentially essential to the livelihoods and flourishing of future generations. Ciriacy-Wantrup understood that the futurity of environmental benefits pointed inexorably to issues of uncertainty, since the preferences of future generations, the path of technological change, and the response of ecosystems to anthropogenic disturbances are all fundamentally unknowable. Accordingly, he favored an approach that mandates the conservation of environmental resources unless the costs are judged to be intolerable.

Bishop (1978) elaborated Ciriacy-Wantrup's approach and provided it with a game-theoretic foundation built around the use of the maximin criterion as a basis for rational decision-making under uncertainty. Analogous frameworks are provided by Perrings' (1991) work on the "precautionary principle" and the concept of "strong sustainability" in ecological economics (Neumayer 2003). Advocates of strong sustainability, for example, emphasize the moral premise that the natural environment should be viewed as a public trust resource that belongs jointly to the members of present and future generations (Brown 1998). In this perspective, actions that impose uncompensated, catastrophic risks on future generations are morally proscribed. Under such circumstances, the right to utilize environmental resources then carries with it the duty to conserve the resource base for the benefit of posterity (Howarth 2007).

The SMS/strong sustainability framework can be criticized for its perceived imprecision. There is typically no clear point at which the costs of environmental conservation become "intolerable." Instead, decision makers must reach informed, ultimately subjective judgments that are translated into concrete policies and institutions. An example is provided by the 1992 United Nations Framework Convention on Climate Change (UNFCCC), which calls for the stabilization of greenhouse gas concentrations to "prevent dangerous anthropogenic interference" with Earth's climate that would threaten agriculture, biodiversity, and the achievement of sustainable development.

As operationalized under the Copenhagen Accord of 2009, this language is now interpreted as requiring that the long-run increase in mean global temperature caused by human action should be limited to no more than 2°C relative to the pre-industrial norm. This interpretation, of course, remains controversial. Authors such as Hansen *et al.* (2008) argue that a more stringent temperature limit would be required to avert major risks related to sea-level rise and the large-scale loss of unique species and ecosystems. This argument is sobering, in that many analysts now believe that achieving the 2°C target will become unfeasible in the absence of urgent actions to reduce greenhouse gas emissions. Such actions seem unlikely given the current state of play in international climate negotiations, where there is little agreement concerning emissions reductions targets or the distribution of burdens between parties.

Critiques have also been advanced by economists such as Nordhaus (1992, 2008), whose Dynamic Integrated Climate-Economy (DICE) model is a touchstone of the literature on climate change policy. Nordhaus reasons that environmental resources are a form of capital that should be managed based on the same criteria as those applied in private-sector investment decisions. In particular, he argues that observed economic data implies that investors discount future net monetary benefits at a rate of about 6 percent per year, so that an action that would provide one dollar of benefits one century from today has a present value of just three-tenths of one cent. Because the benefits of climate change policies accrue over century-scale time horizons, this framework suggests that, in the short run, only modest steps are warranted to reduce greenhouse gas emissions. Given the optimal policies described in the 2007 version of DICE (Nordhaus 2008), mean global temperature would rise to a level of 3.5°C above the pre-industrial norm by the late twenty-second century.

Nordhaus' critique poses a deep challenge to advocates of the 2°C temperature limit and, more deeply, use of the "dangerous anthropogenic interference" concept as a foundation for climate governance. One response is provided by authors such as Stern (2007), who argues that the emphasis that Nordhaus places on market-based discounting techniques is inappropriate because it attaches too little weight to the well-being of future generations. In contrast, the Stern Review is based on the assumptions that: (a) individual utility is logarithmic with respect to consumption; and (b) approximately equal weight should be attached to present and future utility. Given realistic assumptions about the baseline rate of economic growth, this approach suggests that aggressive climate policies are socially optimal.

A second response is to note that the Nordhaus model abstracts away from uncertainty in calculating the optimal trajectory of greenhouse gas emissions. To be sure, Nordhaus (2008) conducts a sensitivity analysis on the parameters of DICE as a way to explore the implications of uncertainty. Ackerman *et al.* (2010) extend this analysis to allow for the risk of catastrophic impacts, showing that aggressive climate policies can be derived in this model, given parameter values that, although statistically unlikely, are well within the range of plausibility. Large welfare losses occur in scenarios where emissions abatement rates are low and climate damages are high. Intuitively, this suggests that precautionary policies might provide significant insurance benefits.

In a similar vein, Woodward and Bishop (1997) explore how the intuition associated with Safe Minimum Standards can be applied in a revised version of DICE. Their analysis proceeds from the assumption that there is a low but statistically unknowable probability that climate change has the potential to impose catastrophic costs, understood as a loss of 25 percent of economic output given a 3°C temperature rise (see Nordhaus 1994). Based on an axiomatic framework developed by Arrow and Hurwicz (1972), Woodward and Bishop then show that policy makers would opt for stringent reductions in greenhouse gas emissions if they applied the well-known maximin criterion as a basis for decision making under uncertainty. This work is updated and extended in a recent paper by Iverson and Perrings (2012), who consider a version of the "minimax regret" criterion in which decision makers balance the risks of both catastrophic impacts and excessive abatement costs.

Although suggestive, none of these studies closes the loop in reconciling the DICE model with the aggressive emissions reductions needed to achieve the 2°C warming limit set forth by the Copenhagen Accord. On the one hand, Stern's approach to discounting has been criticized as ad hoc because it appears to suggest that public policies should be based on rates of time preference that differ from those employed by private households in the context of the market (Anthoff *et al.* 2009). This seems inconsistent with the notions of consumer sovereignty and revealed preference in ways that demand deeper grounding and justification.

In parallel, Woodward and Bishop's very insightful analysis is best understood as a thought experiment that explores the implications of one particular decision criterion under a stylized set of assumptions. The contribution of Ackerman *et al.*

is to conduct sensitivity analysis in the context of a deterministic model, which stops short of characterizing how policy makers do or should make rational decisions under conditions of uncertainty.

In the present discussion, we approach this set of issues from two distinct angles. First, we appeal to a modeling framework developed by Howarth (1998) to reinterpret the debate between Nordhaus (2008) and Stern (2007) over the so-called "pure rate of time preference," which is supposed to represent the respective weight that decision makers attach to the welfare of present and future generations. Second, we describe some results from the *et al.* stochastic version of the DICE model of Gerst *et al.* (2013), focusing especially on the analysis of decision makers' attitude towards time and risk. While Nordhaus (2008) has advocated a "policy-ramp" approach that favors deferring major action on climate change into the future, here we reach a different conclusion – stringent policies are warranted if one attaches sufficient weight to either the rights and interests of future generations or the precautionary benefits of climate stabilization.

Nordhaus, Stern, and future generations

Our first task is to understand how issues of time preference and discounting are depicted in the debate between Nordhaus (2008) and Stern (2007). At an operational level, both analysts assume that decision makers aim to maximize the social welfare function:

$$W = \sum_{t=0}^{\infty} N_t c_t^{1-\eta} / (1-\eta) / (1+\rho)^t \tag{1}$$

through the choice of climate change policies. In this setup, N_t is the world's population at a sequence of dates $t = 0, 1, 2, \ldots$, while c_t is the level of per capita consumption. The models considered by Nordhaus and Stern involve detailed descriptions of the factors that drive economic growth, the generation of greenhouse gas emissions, the impacts of greenhouse gas concentrations on future climate conditions, and the costs and benefits of emissions abatement strategies. Climate damages are assumed to affect future welfare through their impacts on material consumption. The choice of optimal policies then reduces to the two parameters embodied in the social welfare function.

Nordhaus (2008) assigns a value of $\eta = 2$ to the first parameter, the elasticity of marginal utility. Conceptually, this represents the rate at which the incremental satisfaction derived from increased consumption diminishes as the overall level of prosperity rises. In parallel, he chooses a value of $\rho = 1.5$ percent per year for the second parameter, the pure rate of time preference. He justifies this calibration based on its empirical realism. In particular, this specification replicates both the 2.25 percent per year consumption growth rate and the 6 percent annual return on private capital investment that he infers from his analysis of real-world economic data.

Stern's (2007) analysis departs from that of Nordhaus in several key respects. First, Stern works with the PAGE model developed by Hope *et al.* (1993).

Although DICE and PAGE are broadly similar in their assumptions concerning the costs and benefits of climate change policies, they differ in terms of their modeling strategies and level of aggregation. Significantly, PAGE assumes that baseline economic output follows a fixed path over time, while DICE models economic growth as driven endogenously by capital investment. Most important, however, are Stern's assumptions concerning the two preference parameters that characterize the social welfare function. In particular, Stern assumes that $\eta = 1$ while $\rho = 0.1$ percent per year. Each of these assumptions serves to increase the weight attached to costs and benefits that accrue to members of future generations.

Stern's lower value for η entails that, in a growing economy, the marginal utility derived from monetary costs and benefits falls relatively slowly over time, thereby leading to a lower monetary discount rate (see Broome 2008). Values of η between 1 and 2 are commonly employed in the economics of climate change and are viewed by many analysts to be plausible and supported by the available empirical evidence. Earlier versions of DICE, for example, shared Stern's view that the elasticity of marginal utility should be assigned a value of $\eta = 1$ (Nordhaus 1992).

Stern's bolder move is to assume a near-zero rate of pure time preference. Unlike Nordhaus' emphasis on descriptive realism, Stern argues that equal weight should be attached to the well-being of present and future human beings, but that a low, positive value for ρ is warranted based on the probability that the human species will be rendered extinct by a planetary catastrophe such as a large asteroid collision.

In terms of policy, the difference between Nordhaus and Stern concerning the value of these two parameters is a matter of considerable consequence. In the base calibration of the 2007 version of the DICE model, the optimal path allows greenhouse gas emissions to increase from 8.4 to 11.7 billion tonnes of carbon per year over the course of the next century, leading to a long-run temperature increase of 3.5°C relative to the pre-industrial norm. Now suppose that we revise DICE to:

1 maintain the rates of capital investment and baseline economic growth that Nordhaus identifies as both empirically realistic and socially optimal; but
2 employ Stern's preference parameters to choose the optimal path for greenhouse gas emissions reductions.

Under these assumptions, the optimal level of emissions is limited to 5.4 billion tonnes per year in 2015 and 0.3 billion tonnes in the year 2115. A rapid phase-out of high-carbon energy technologies limits the long-run increase in mean global temperature to just 1.7°C.

This result leaves us with an important and well-known paradox. Although Nordhaus' approach is consistent with the observed rate of economic growth, it attaches relatively little weight to the welfare of future generations, which some analysts argue is morally objectionable. Stern addresses this problem by using a low rate of time preference. Yet Stern's approach seems inconsistent with people's observed decisions concerning the desired level of capital investment. In particular, lower discount rates are used in evaluating climate policies than private household decisions. The question thus arises as to how and in what sense this is morally justified.

Overlapping generations

One potential answer to this problem is provided by Howarth (1998). The starting point for this analysis is to restructure the optimal growth framework employed by both Nordhaus and Stern, according to which society's preferences are depicted through appeals to a hypothetical, infinitely-lived agent. Instead, Howarth works with an overlapping generations model of an intertemporal, competitive economy in which individuals make savings decisions to maximize the utility they derive over the course of their finite lifespans (see also Gerlagh and van der Zwaan 2001). More formally, suppose that a person born at date t enjoys the consumption levels c_{yt} in youth and c_{ot+1} in old age. Here, time periods are measured in spans of 35 years to reflect a generational temporal scale, starting in the early twenty-first century. As Howarth shows, the utility function:

(2) $$U_t = 1\text{n}(c_{yt}) + 1\text{n}(c_{ot}+1)/1+\delta)$$

replicates both the observed rate of economic growth and market returns to capital investment when the rate of individual time preference attains a value of $\delta = 0.5$ percent per year, or 19 percent per 35-year time interval.

Using this model, Howarth then considers the implications of two normative frameworks for the evaluation of climate change policies. First is the Kaldor-Hicks criterion, in which the marginal cost of controlling emissions is equated with the ensuing discounted marginal benefits, setting the discount rate equal to the market return on private-sector capital investments. The resulting outcome is Pareto efficient, involving an allocation where enhancing the welfare of one generation is impossible without rendering at least one other generation worse off. Not surprisingly, this criterion supports only modest reductions in greenhouse gas emissions given its reliance on a relatively high, market-based discount rate. This invites the criticism that this outcome is inconsistent with the achievement of intergenerational fairness, since it permits the imposition of large, uncompensated environmental costs on future generations so that the present generation can avoid relatively low climate mitigation costs.

Second, Howarth considers a governance framework in which: (a) decisions concerning personal consumption and investment are made by private individuals in the context of competitive markets; while (b) decisions concerning climate change policies are made according to the principles of classical utilitarianism, attaching equal weight to the welfare of each present and future person. In technical terms, this involves maximizing the social welfare function:

(3) $$SWF = \sum_{t=0}^{\infty} N_t U_t$$

in which N_t represents the number of persons born at date t.

Why might this decoupling between individual investment decisions and social decisions concerning climate policy be politically and morally appropriate? One answer is that, by its very nature, climate policy involves a conflict between two types of rights that pertain to the use of a common pool resource. On the one hand, members of the present generation have a prima facie right to use energy and conduct other activities that generate greenhouse gas emissions in their pursuit of the good life. On the other hand, future generations have a presumptive right to protection from uncompensated harms (Shue 1999). It is this conflict between perceived positive and negative rights that puts climate governance in the public rather than the private sphere, necessitating a form of collective action. Aiming to maximize the total utility derived from climate policies is then one way of arriving at solutions that all affected parties would have reason to accept as fair and legitimate. In effect, this approach resolves conflicts between rights by minimizing the utility losses (and hence human harms) caused by rights violations.

In contrast with the Kaldor-Hicks criterion, basing climate change policy on the principles of classical utilitarianism favors immediate and aggressive cuts in greenhouse gas emissions, replicating the early findings of Cline (1992) and anticipating the later work of Stern. A key point is that the choice between these competing decision frameworks boils down to a set of core value judgments that go beyond the choice of a particular parameter – the pure rate of time preference – in a model that does not distinguish between the concepts of individual time preference and those of moral judgments regarding the respective weight that should be attached to the interests of present and future persons. There are strong reasons to work with overlapping generations models of the type described in this section, which explicitly distinguish between these categorically distinct aspects of values and preferences.

Accounting for uncertainty

We have seen that the DICE model supports stabilizing mean global temperature at a level below the 2°C threshold established by the Copenhagen Accord when it is recalibrated to attach sufficient weight to the future well-being. This outcome, however, seems to conflict with a commitment to the achievement of economic efficiency as gauged by the Kaldor-Hicks criterion. Since different observers reach different judgments concerning the appropriate moral basis for climate policy, the appeal to classical utilitarianism is suggestive but not decisive.

The discussion above is also inconclusive in the following sense. The DICE model represents rational decision making based on the contrary-to-fact assumption that people have perfect foresight regarding future social, economic, and environmental conditions. In reality, however, concerns about uncertainty are at the very heart of debates over climate change. This is seen in the concept of "dangerous anthropogenic interference" and the literature on safe minimum standards, the precautionary principle, and strong sustainability as referenced in the introduction. The question is then how to extend DICE to value appropriately the risk abatement benefits generated by climate change policies.

Insights on this matter are provided by the provocative work of Weitzman (2009). As Weitzman notes, considerable uncertainty exists concerning both climate change damages and the measurement of climate sensitivity – i.e., the level of climate change that would arise given a doubling of greenhouse gas concentrations. Until the 2000s, it was common to believe that this parameter attained a value between 1.5ºC and 4.5ºC with a mean value of around 2.5ºC (Intergovernmental Panel on Climate Change 1996). Recent scientific research, however, suggests a less certain picture. Roe and Baker (2007), for example, conclude that climate sensitivity follows a fat-tailed statistical distribution, with a 20 percent probability of exceeding 5.0ºC. As Weitzman notes, this level of uncertainty implies a low but characterizable risk of catastrophic climate impacts. If decision makers were sufficiently risk averse, the value of climate stabilization might in principle be infinite, at least in the simplified thought experiment developed in Weitzman's paper.

Pinning down and analyzing this effect in DICE requires attention to several factors that are addressed in a recent paper by Gerst *et al.* (2013). First, the base version of DICE abstracts away from uncertainty, yet the value of precautionary actions to stabilize climate depends on the interplay between uncertainties related to baseline economic growth, the pace of technological change, greenhouse gas emissions abatement costs, and climate change damages. To address these factors, Gerst *et al.* develop a stochastic version of DICE that:

1 replaces the deterministic growth model at the core of DICE with a stochastic Lucas–Mehra–Prescott framework in which baseline economic growth follows a random-walk statistical process (see Lucas 1978; Mehra and Prescott 1985; Barro 2006);
2 employs Nordhaus' representation of the uncertainties in each parameter of DICE based on the assumption that these parameters are normally and independently distributed; and
3 employs Roe and Baker's (2007) statistical distribution for the model's climate sensitivity parameter.

This approach is in one sense conservative. Unlike Woodward and Bishop (1997) and Gerst *et al.* (2010), this analysis does not assume a high degree of uncertainty concerning the climate change damages that would arise given a specified increase in mean global temperature. This is a point where Nordhaus' work has been critiqued for potentially understating the true level of uncertainty (Ackerman *et al.* 2009).

A crucial part of the Gerst *et al.* (2013) model is its representation of decision makers' attitudes towards time and risk. Drawing on the work of Barro (2006) and Ding *et al.* (2012), the model depicts preferences using an expected social welfare function in which the expectations operator $E[\cdot]$ calculates the probability-weighted mean over uncertain future outcomes:

(4) $$W = E\left[\sum_{t=0}^{\infty} N_t (c_t^{1-\eta} - c_m^{1-\eta}) / (1-\eta) / (1+\rho)^t\right].$$

In this specification, the pure rate of time preference is assigned a value of ρ = 1.5 percent per year, matching the value employed in the base version of DICE. Gerst *et al.* depart from Nordhaus, however, in assuming that the elasticity of marginal utility attains a value of $\eta = 5.6$ as compared with Nordhaus' value of $\eta = 2$.

What is the justification for this change in parameters? In brief, the Gerst *et al.* calibration is based on the assumption that private households make sequentially rational investment decisions under conditions of uncertainty, facing a 1 percent per year rate of return on safe capital investment with an average return of 6 percent per year on risky investments. Ding *et al.* (2012) show that explaining the high returns people demand on risky capital assets implies that they have a high degree of risk aversion, noting that the parameter η represents the coefficient of relative risk aversion in the context of this model. Nordhaus' value of $\eta = 2$ implies that the rates of return on safe and risky investments should be almost identical. The value of $\eta = 5.6$, in contrast, is consistent with data on both investment returns and stochastic movements in the level of economic output. Thus Gerst *et al.* interpret this value as a measure of people's true risk aversion. (See Atkinson *et al.* [2009] for a similar estimate based on a large-sample preference elicitation survey.)

Finally, equation (4) includes the parameter c_m, which reflects the level of consumption that is minimally sufficient to meet people's subsistence needs. For the purposes of analysis, Gerst *et al.* assign a value of $228/person/year to this parameter based on the work of Ahmed *et al.* (2007). This specification implies that utility moves towards zero as consumption approaches the subsistence level, with the economy terminating in rare scenarios in which consumption dips below this minimum. This utility function is therefore bounded, contrasting with Weitzman's (2009) analysis, in which outcomes that are unboundedly negative have an important bearing on the results.

Gerst *et al.* describe and evaluate a family of policy scenarios that differ in terms of the timing and stringency of controls on greenhouse gas emissions. For the purposes of the present discussion, some key results are as follows.

- In the absence of emissions controls, the model predicts that mean global temperature will increase by a median value of 7.5°C through the year 2400, with a 5 percent probability of exceeding 16°C.
- This climatological risk translates into very substantial economic risks. In the absence of emissions controls, the model calculates a 3.3-in-1000 probability of an economic collapse, defined as a case in which catastrophic climate impacts drive per capita consumption all the way down to the subsistence level.
- The risk of an economic collapse can be reduced to less than 1-in-10,000,000 through the pursuit of sufficiently aggressive climate stabilization measures.
- Climate stabilization is a high-value social investment. Of the various scenarios considered by Gerst *et al.*, the highest level of social welfare occurs when the rate of controls on greenhouse gas emissions increases to 50 percent

in the year 2050 and 86 percent in the year 2100. In comparison with the no-controls baseline, this scenario generates net benefits equivalent to a permanent 377 percent increase in per capita consumption at every point in time and each uncertain state of nature.

- These control rates are roughly twice as large as those judged to be optimal in the base version of DICE. They are sufficient to limit the median increase in mean global temperature to 2.9°C, overshooting the 2°C target from the Copenhagen Accord.

Conclusions

In this chapter we have explored how alternative decision rules for climate governance are tied to the prevention of "dangerous anthropogenic interference" with Earth's climate as defined under the UN Framework Convention on Climate Change. On the one hand, limiting greenhouse gas emissions to the levels required to meet the 2°C temperature target appears to be socially optimal if sufficient weight is attached to the welfare of future generations. As Stern (2007) points out, the costs of achieving climate stabilization are relatively moderate, averaging about 1 percent of gross economic output over the next several decades. Integrated assessment models suggest that this initial investment is likely to generate much larger benefits accruing to posterity.

In addition, we believe that the Gerst *et al.* (2013) model sheds light on the long-standing debate over the role of safe minimum standards, the precautionary principle, and the concept of strong sustainability in environmental policy analysis. Gerst *et al.*'s finding that climate stabilization generates very substantial net benefits echoes Ciriacy-Wantrup's concern that the degradation of biophysical systems poses low-probability, catastrophic risks to members of future generations, and that these risks might trump the costs of conservation, given a prudent approach to risk management.

Given its limitations and simplifying assumptions, it would be a mistake to interpret the Gerst *et al.* analysis as providing sharp recommendations on the optimal timing and stringency of reductions of greenhouse gas emissions. Above, we have noted this study's conservatism concerning the uncertainties pertaining to the functional relationship between mean global temperature and the costs of climate change damage. In addition, it is worth noting that the model in question fails to provide separate parameters that pertain to decision makers' risk aversion, inequality aversion, and ambiguity aversion – i.e., their willingness to choose decision options that hold the potential for negative outcomes with poorly-defined probabilities (see Atkinson *et al.* 2009; Traeger 2009). That said, the analysis may contribute to the establishment of a methodological bridge between Nordhaus' (2008) DICE model and the value judgments and epistemic position implicit in the work of Ciriacy-Wantrup and his intellectual successors.

Acknowledgements

We thank Mark Borsuk, the editors, and an anonymous referee for helpful comments and suggestions.

References

Ackerman, F., S.J. DeCanio, R.B. Howarth and K. Sheeran (2009). Limitations of integrated assessment models of climate change. *Climatic Change* 95, 297–315.

Ackerman, F., E.A. Stanton and R. Bueno (2010). Fat tails, exponents, extreme uncertainty: simulating catastrophe in DICE. *Ecological Economics* 69, 1657–65.

Ahmed, A., R. Vargas-Hill, L. Smith, D. Wiesmann and T. Frankenberger (2007). *The World's Most Deprived: characteristics and causes of extreme poverty and hunger.* Washington: International Food Policy Research Institute.

Anthoff, D., R.S.J. Tol and G.W. Yohe (2009). Risk aversion, time preference, and the social cost of carbon. *Environmental Research Letters* 4, 024002.

Arrow, K.J. and L. Hurwicz (1972). An optimality criterion for decision-making under ignorance. In C.F. Carter and J.L. Ford (eds.), *Uncertainty and Expectations in Economics.* Oxford: Blackwell.

Atkinson, G., S. Dietz, J. Helgeson, C. Hepburn and H. Sælen (2009). Siblings, not triplets: social preferences for risk, inequality and time in discounting climate change. *Economics E-Journal* 3, 2009–26.

Barro, R.J. (2006). Rare disasters and asset markets in the twentieth century. *Quarterly Journal of Economics* 121, 823–66.

Bishop, R.C. (1978). Endangered species and uncertainty: the economics of a safe minimum standard. *American Journal of Agricultural Economics* 60, 10–18.

Broome, J. (2008). The ethics of climate change. *Scientific American* 298, 97–102.

Brown, P.G. (1998). Toward an economics of stewardship: the case of climate. *Ecological Economics* 26, 11–21.

Ciriacy-Wantrup, S.V. (1968). *Resource Economics: conservation and policies.* Berkeley: University of California Press.

Cline, W.R. (1992). *The Economics of Global Warming.* Washington: Institute for International Economics.

Ding, P., M.D. Gerst, A. Bernstein, R.B. Howarth and M.E. Borsuk (2012). Rare disasters and risk attitudes: international differences and implications for integrated assessment modeling. *Risk Analysis* 32, 1846–55.

Gerlagh, R. and B.C.C. van der Zwaan (2001). Overlapping generations versus infinitely-lived agent: the case of global warming. *Advances in the Economics of Environmental Resources* 3, 287–313.

Gerst, M.D., R.B. Howarth and M.E. Borsuk (2010). Accounting for the risk of extreme outcomes in an integrated assessment of climate change. *Energy Policy* 38, 4540–48.

——(2013). The interplay between risk attitudes and low probability, high cost outcomes in climate policy analysis. *Environmental Modelling and Software* 41, 176–84.

Hansen, J.E., M. Sato, P. Kharecha, D. Beerling, V. Masson-Delmotte, M. Pagani, M. Raymo, D.L. Royer and J.C. Zachos (2008). Target atmospheric CO_2: where should humanity aim? *Open Atmospheric Science Journal* 2, 217–31.

Hope, C., J. Anderson and P. Wenman (1993). Policy analysis of the greenhouse effect: an application of the PAGE Model. *Energy Policy* 21, 327–38.

Howarth, R.B. (1998). An overlapping generations model of climate-economy interactions. *Scandinavian Journal of Economics* 100, 575–91.

——(2007). Towards an operational sustainability criterion. *Ecological Economics* 63, 656–63.

Intergovernmental Panel on Climate Change (1996). *Climate Change 1995: the science of climate change.* New York: Cambridge University Press.

Iverson, T. and C. Perrings (2012). Precaution and proportionality in the management of global environmental change. *Global Environmental Change* 22, 161–77.

Lucas, R. (1978). Asset prices in an exchange economy. *Econometrica* 46, 1429–45.

Manne, A.S. and R.G. Richels (1992). *Buying Greenhouse Insurance: the economic costs of CO_2 emission limits.* Cambridge: MIT Press.

Mehra, R. and E. Prescott (1985). The equity premium: a puzzle. *Journal of Monetary Economics* 15, 145–61.

Neumayer, E. (2003). *Weak versus Strong Sustainability: exploring the limits of two opposing paradigms*, Cheltenham, UK: Edward Elgar.

Nordhaus, W.D. (1992). An optimal transition path for controlling greenhouse gases. *Science* 258, 1315–19.

——(1994). Expert opinion on climatic change. *American Scientist* 82, 45–52.

——(2008). *A Question of Balance: weighing the options on global warming policies.* New Haven: Yale University Press.

Perrings, C. (1991). Reserved rationality and the precautionary principle: technological change, time and uncertainty in environmental decision making. In R. Costanza (ed.), *Ecological Economics: The science and management of sustainability*, New York: Columbia University Press.

Roe, G.H. and M. Baker (2007). Why is climate sensitivity so unpredictable? *Science* 318, 629–32.

Shue, H. (1999). Bequeathing hazards. In M.H.I. Dore and T.D. Mount (eds.), *Global Environmental Economics: equity and the limits to markets.* Oxford: Blackwell.

Stern, N. (2007). *The Economics of Climate Change.* Cambridge, UK: Cambridge University Press.

Traeger, C.P. (2009). Recent developments in the intertemporal modeling of uncertainty. *Annual Review of Resource Economics* 1, 261–86.

Weitzman, M.L. (2009). On modeling and interpreting the economics of catastrophic climate change. *Review of Economics and Statistics* 91, 1–19.

Woodward, R.T. and R.C. Bishop (1997). How to decide when experts disagree: uncertainty-based choice rules in environmental policy. *Land Economics* 73, 492–507.

World Meteorological Organization (1988). The changing atmosphere: implications for global security. Conference statement, Toronto, Canada.

11 Exclusive approaches to climate governance

More effective than the UNFCCC?

Steinar Andresen

Introduction: purpose and scope

Given the lack of progress along the inclusive UNFCCC track, there have been increasing calls for more exclusive or regional approaches as supplements or alternatives to the global approach (Victor 2011). Theoretically, as well as based on lessons from other regimes like WTO and the climate regime, it may be argued that reaching agreement is easier within smaller groups of countries (Olson 1965; Greenspan Bell and Ziegler 2012). The point seems particularly relevant within this issue area, as a handful of key actors are responsible for well above 50 percent of the world's total greenhouse gas (GHG) emissions.

In this chapter I will not delve much into the arguments in favor of more exclusive approaches, as this has been done by several other analysts. Political and academic discussions have tended to concentrate on the merits and shortcomings of the various approaches. In general, European researchers (and policy makers) have favored the UN approach, whereas US academics (and policy makers) have favored the "regime-complex" (or the "club") approach (Biermann *et al.* 2010; Keohane and Victor 2011). It is now time to move on and focus on the effectiveness of the more exclusive approaches: "The real test is whether these other forums are achieving their core purpose of reducing GHG emissions" (Greenspan Bell and Ziegler 2012: 38).

On this background, this chapter analyzes the achievements of some of the major existing exclusive approaches, empirically testing whether they can be said to be more effective. This is a necessary first step for judging whether these approaches are superior to the UN track.

After presenting the analytical perspective on measuring and explaining the effectiveness of international environmental institutions, I examine what has been achieved along the UNFCCC track over time, as a baseline for comparison. In the main section of the chapter, I present and discuss some of the major exclusive approaches to emerge thus far. First, an account will be given of probably the substantially most important supplement or alternative to the UNFCCC, the Asia Pacific Partnership on Clean Development and Climate Change (APP) and its subsequent (partial) transition into a broader framework. Second, I present an account of the Major Economies Forum on Energy and Climate (MEF), the G20 and a few other relevant forums. In conclusion I briefly discuss the merits and shortcomings of the various approaches, and offer some thoughts on future developments.

The effectiveness of environmental institutions

The effectiveness of international environmental institutions has been extensively studied over the past two decades (see Miles *et al.* 2002; Andresen 2013). Most emphasis has been on problem-solving effectiveness, with less attention paid to issues like equity and fairness (see Chapter 2 of this book). In this present chapter, the main focus is on problem-solving effectiveness but the question of legitimacy will also be touched upon. A consensus has emerged that effectiveness, the dependent variable, can be *measured* in terms of output, outcome, and impact (Underdal 2002). Output is essentially potential effectiveness, as it concerns the rules and regulations emanating from the regime, for example the Kyoto Protocol. We would normally expect stricter rules (in terms of ambition, legal commitments and compliance) to lead to higher effectiveness, but there is no guarantee that this will happen. Therefore we also need the outcome indicator, which focuses on actual achievements made. Here it is necessary to establish a causal link between the institution in question and behavior by key target groups. For example, the dramatic fall of emissions in today's "economies in transition" has not been caused by the climate regime but by economic recession following in the wake of the demise of the Soviet Union. The impact indicator is the most important one as it seeks to establish the extent to which the problem has been solved by the regime in question. For example, has the climate regime been able to stabilize GHG concentrations in the atmosphere at a level that would prevent dangerous anthropogenic interactions with the climate system? Unfortunately, due to the influence of a host of intervening variables, this approach is fraught with methodological challenges and is therefore less used. Perhaps, in seeking to measure the achievements of "softer institutions" like the global conferences on sustainability, we should also introduce less rigorous or less demanding criteria, like learning and agenda-setting (Andresen 2007). Such criteria may be more applicable to the more exclusive approaches, as these have existed for a relatively short time and are characterized by a soft-law approach.

As to *explaining* effectiveness, various approaches have been suggested (Mitchell 2010). However, there is agreement that both the institution as such, as well as non-regime attributes like the characteristics of the issue-area, can make a difference to what is achieved. Moreover, analysts agree that the fact that climate change is a very "malign" issue is a main factor in explaining why progress has been so slow. More interesting from an analytical as well as a policy-making perspective is the extent to which the problem-solving ability of the institution makes a difference (Underdal 2002). This begs the question of the significance of institutional design. Can the way an institution is designed be of significance for its effectiveness? Some analysts claim that this is the case and argue, for example, that more can be achieved by establishing a World Environmental Organization

(WEO), compared to the existing rather unruly system (Biermann and Bauer 2005). Others argue that when vested interests are strong, key actors tend to refrain from action due to the high costs incurred, and that these perceptions cannot be removed by clever institutional design (Najam 2003). How does this play out regarding climate change architecture? Is it possible to conclude, from empirical evidence, whether the exclusive approach has been superior to the inclusive approach, or at least that the former has a higher potential for problem-solving in the long run? Or is it the malignancy of this problem that is decisive for the lack of overall progress, irrespective of approach?

The UNFCCC regime: more effort but lower effectiveness

The year 2012 should have been a year of "climate celebration" as it was the twentieth anniversary of the United Nations Framework Convention on Climate Change (UNFCCC). But there was not much celebration at the 18th Conference of the Parties (COP) in Doha during the fall of 2012. Stalemate and gridlock have characterized the UN-based process over the last decade. Although efforts by climate diplomats have intensified significantly in terms of frequency of meeting, progress has been extremely limited. In fact it may even be argued that the process has moved backwards, as there are fewer countries with specific legally binding international obligations today than in 2005 when the Kyoto Protocol came into force (Andresen *et al.* 2014). Until a new regime is supposed to be in place by 2020, there are no international commitments whatsoever regarding some 85 percent of the emissions from the large majority of countries. The main reason for the stalemate lies in the deep-seated conflicts between the North and the South. As a rather rare example of equity concern, the developing countries were exempted from any "hard" commitments in the UNFCCC and the Kyoto Protocol. However, the world is now a very different place than it was 20 years ago. Due to strong economic growth, the emerging economies of the South are now among the major emitters: China is by far the biggest emitter, accounting for more than a quarter of the world's greenhouse gas emissions. The developed countries have pressed these emerging economies to take on commitments, but generally to little avail. However, at the 2011 COP it was argued that "the firewall" between the North and the South in terms of commitments had been broken down (Bodansky and Rajamani 2012) – but in all likelihood, this does not mean that the emerging economies will take on internationally legally binding commitments. Rather, the post-2020 UN regime will probably be a bottom-up pledge and review regime, representing a farewell to the top-down, legally binding Kyoto approach (Bodansky and Rajamani 2012). Judging from the (bottom-up) pledges so far, there is no chance of staying within the 2°C goal agreed in 2009.

Counter to the goal of the UN climate regime, emissions have climbed steadily ever since it was established. In a problem-solving perspective, the effectiveness of the climate regime is bound to be low, whatever indicator is used. The flexible mechanisms of the Kyoto Protocol (output) were quite innovative, but they have not delivered in terms of emission reductions. Joint implementation has been little used; and indeed, it has been argued that the CDM has brought an increase in emissions (Lund 2013). The quota trading system of the Protocol has never been applied, although it may have spurred similar schemes in the EU and elsewhere. For various reasons, however, these measures have not led to significant reductions in emissions. Actually, we do not know how much of a difference the climate regime has made compared to a business-as-usual scenario. Probably emissions would have been somewhat higher in its absence – but so far, the ups and downs of the world economy as well as switches in energy base unrelated to climate policies, have been far more important for climate gas trajectories (Andresen *et al.* 2014). To put it bluntly, the continuous and sharply rising GHG emissions show the failure of 20 years of UN multilateral diplomacy.

More inclusive approaches: have they delivered?

For a long time the UNFCCC was the only game in town, but this is no longer the case. Considering the failure of the UN approach, it is not surprising that alternatives or supplements have been suggested and also initiated. One line of argument has been that it is much easier to reach agreement in smaller groups of major economies (and emitters): a "club approach" or a mini-regime. Bottom-up conditional pledges by key emitters are said to be both more realistic as well as more effective compared to the random top-down approach of the UN regime (Victor 2011). The APP has distinct similarities with this approach. Another club approach is the Major Economies Forum on Energy and Climate (MEF), which encompasses 17 of the largest emitters. As these states account for more than 80 percent of global emissions, it may be argued that this is more instrumental than the unruly 190+ UN approach. Although the G8 and subsequently the G20 were originally set up for other purposes, they have also recently become involved in climate issues.

The emergence of an increasing number of institutions focusing on climate change has given rise to the study of climate change through the prism of "regime complexes." According to Keohane and Victor (2011), international institutions can be placed on a continuum between comprehensive regulation and fragmented arrangements. Along this continuum, these authors position regime complexes as a more fragmented set of institutions than traditional regimes. A regime complex is "an array of partially overlapping and non-hierarchical institutions governing a particular issue-area" (Raustiala and Victor 2004: 279), marked by the absence of an overall architecture or hierarchy (Keohane and Victor 2011). The formation of such a regime complex can be explained in part by the failure of the comprehensive UN climate regime. This development has also been driven by the fact that key actors like the USA, and the emerging economies, as well as large sectors of the business community, also want to work through institutions other than the UN.

Still, it is difficult to see why the massive diplomatic efforts and the comprehensive UN approach are regarded as similar to many of the "mini-lateral" efforts with only few parties. Thus, I will argue that there is more of a hierarchy present than argued by Keohane and Victor. The regime complex approach is fruitful from an analytical perspective, as it encompasses today's complex realities. It is more difficult to follow the argument that the regime complex per se should be more effective than the UN regime: "policy-makers keen to make international regulations more effective, a strategy focused on managing a regime complex may allow for more effective management of climate change than large political and diplomatic investments in efforts to craft a comprehensive regime" (Keohane and Victor 2011: 7). This complex already exists, but thus far little in the way of emissions reductions has resulted from it; how and by whom it should be managed to become more effective will need to be specified in order to substantiate such an assumption.

Be that as it may, a large number of competing or supplementary institutions to the UN regime have been in operation ever since the 2002 Johannesburg Summit, mostly in the form of partnerships of various kinds, often with some kind of public–private participation. Abbott (2012) notes that 69 such initiatives have been launched. Apart from those already mentioned, there are the Carbon Sequestration Leadership Forum, the International Partnership for a Hydrogen Economy, the Renewable Energy and Energy Efficiency Partnerships, and the Climate and Clean Air Coalition. Some researchers have tended to regret this fragmentation, on the grounds that it may undermine the UNFCCC process, lead to forum-shopping, and contribute to unclear rules and regulations. Others have praised this development, not least because non-state actors often play a key role in many of these partnerships. This has been seen as a way of bypassing the state centric UNFCCC process, thereby contributing to a more effective approach. Unfortunately, most analyses of these partnerships have applied a "process perspective," giving an account of these vibrant and often creative activities without paying attention to effectiveness: whether the partnerships actually lead to reduced emissions (Abbot 2013; Greene 2013).

The Asia Pacific Partnership (2006–2011): More learning than emission cuts?

On July 28, 2005, China, India, South Korea, Japan, Australia, and the USA agreed to set up a regional institution, the Asia Pacific Partnership on Clean Development and Climate Change (APP). The partnership was launched on January 12, 2006, and Canada joined in 2007 as the seventh member. In other words, some of the main emitters joined forces, Kyoto partners as well as non-Kyoto ones. Equally important, this was a blend of developed as well as key developing countries: potentially a highly potent and effective club. The founders were the USA and Australia. Both governments had rejected the Kyoto Protocol, and the two heads of state shared a conservative ideology and sought supplements or alternatives to the UN approach. The initiative came as a surprise to those not involved in the process, and the timing appeared well chosen – a few months after the Kyoto Protocol came into force (Karlsson-Vinkhuyzen and van Asselt 2009). At the time of its creation, the APP countries collectively represented close to 50 percent of the world's population, global GDP, energy consumption, as well as

GHG emissions (Peay 2007). Since then, their share of GHG emissions has risen significantly due to the sharp increases on the part of China and India.

Officially the APP was set up as a supplement to the UN regime, but it has been documented that its founders saw it more as a potential alternative (McGee 2010). Its architecture was almost diametrically opposed to the UN regime. In addition to its exclusive nature, there were no legally binding commitments or targets; a bottom–up, technology-oriented approach was adopted; and it was based on the philosophy that environmental quality improves with economic growth. Civil society organizations were not invited to the APP meetings. In contrast to the economy-wide approach of the Kyoto Protocol, the APP represented a sectoral approach. It was driven by public–private partnerships in eight task forces. During its existence, it launched three partnerships in energy-supply sectors (cleaner fossil energy, power generation and transmission, renewable energy and distributed generation) and five partnerships in energy-intensive sectors (aluminum, buildings and appliances, cement, coal mining, and steel). Each of the task forces developed an action plan setting out the main objectives and identifying activities and projects to be implemented. Some 175 projects were launched by the APP. Each task force was led by a chair and co-chair selected from among the partner countries. The main decision making body was the Policy and Implementation Committee (PIC), which oversaw the Partnership as a whole, guided the task forces, and periodically reviewed their work. The PIC was composed of senior government representatives. The USA had the dominant position in the institutional set-up, chairing the PIC from its inception. In contrast, India and China were the only partners not to hold the chair of a task force (Karlsson-Vinkhuyzen and van Asselt 2009). Three ministerial high-level meetings were arranged.

During the two or three first years of its existence the APP received considerable attention. Not surprisingly, the ENGO community as well as all other "progressive" climate actors were deeply opposed. The most common criticism was that it was an attempt by the Bush administration to weaken the UN process. Some also argued that it was doomed to fail, as it had no binding emissions reductions and no targets. In the academic discourse, most attention focused on explaining the establishment of the APP, the interplay between it and the UNFCCC, and the significance of the APP as a soft-law instrument (Karlsson-Vinkhuyzen and van Asselt 2009; McGee 2010). No attention was paid to the extent to which the approach introduced by the APP could in fact lead to reductions in emissions. As it became clear that the APP posed no real threat to the UNFCCC, policy makers and scholars turned their focus elsewhere. And then, in April 2011, the APP formally terminated its work, although several individual projects continued in other forums.

No comprehensive study on the problem-solving effectiveness of the APP has been conducted. It may also be argued that evaluating the Partnership in this perspective would be unreasonable, as it closed down so soon. However, some research has been conducted that may shed light on its performance. An evaluation based on interviews with APP stakeholders has been prepared for the Centre for

European Policy Studies (CEPS) (Fujiwara 2012). Here I will not go into the methodology used in the survey; suffice it to say that respondents from the two major developing countries, China and India, represented a very small share of total respondents; moreover, no response rates are given. We therefore do not know how representative the survey is. Still, a look at some of the main findings can provide a crude indicator of perceptions of the APP's achievements and challenges. In the survey, 55 percent of the respondents say they were satisfied with the results of their participation. A large majority mentioned major benefits from information sharing (90 percent) and networking (83 percent). The score was much lower on more specific measures, like access to existing technology (31 percent), market access (21 percent), access to new technologies (17 percent), and access to financing (14 percent). Other analysts of the APP have also underlined the lack of financing as a major performance bottleneck (McGee 2010).

The CEPS survey indicates that performance is considered good regarding the less demanding (and less important) indicators, whereas scores are much lower on indicators important for reducing emissions. This finding is supported by other studies as well. According to an evaluation by Kanie *et al.* (2013) of the APP and other partnerships focusing on technology, the added value of APP is questionable. There was no mechanism to evaluate GHG emissions reduction, and many of the activities were based on business-as-usual cooperation within the sectors. Few activities focused on technology development and transfer, although that was supposed to be the primary function of the APP. Other analysts have noted the positive aspect of bringing industry from many nations together to talk about best practices for energy efficiency, but point out the lack of funding for technical assistance: the APP was the "right idea (bringing industry to the table) but an essential ingredient, climate ambition, was lacking" (Ken Purvis, email to the author, April 19, 2013).

Although problem-solving effectiveness was low, the APP may still have had some value in indicating new ways for dealing with climate change. The fact that such diverse and major emitters decided to work pragmatically together is itself no small achievement, and stands in stark contrast to the sterile ideological "blame-game" taking place under the UN umbrella. Moreover, the APP was a major public–private partnership, implementing specific projects in a range of sectors. As such it was a prime example of the sectoral approach promoted by some actors in the UN negotiations, and it may gain increasing importance in the post-2020 UN regime. The same may well be the case for both the "soft" nature of APP cooperation as well as for the application of the bottom-up approach. Thus, the APP may prove quite important regarding learning and diffusion of new ideas and approaches.

Why was the APP terminated? It may have lost traction due to the shift in government in the USA and in Australia to a more pro-UN policy, and President Obama has been more inclined to focus on the broader MEF. Another viewpoint: "It was closed down, mainly, because of the view that it wasn't really getting things done. I suspect too, that the view that it was an excuse not to engage with Kyoto didn't help APP fate. The APP always struck me as a good idea in principle but badly implemented in practice" (David Victor, email to the author, April 19, 2013).

What about the broader institutional framework to which parts of the APP were transferred? Three of the eight task forces continued under a new partnership, the Global Superior Energy Performance Partnership (GSEP). This partnership has a stronger focus on environmental performance, and the aim is to expand participation to a global scale (Fujiwara 2012). The objective of the GSEP is to reduce energy use by encouraging industrial facilities and commercial buildings to pursue continuous improvements in energy efficiency and by promoting public–private partnerships for cooperation on specific technologies. The GSEP membership is significantly broader than the APP as it includes such key actors as the EU, Russia, South Africa and Mexico as well as some smaller developed countries. This indicates that the political controversy characterizing the APP has been transcended, which is promising from the perspective of joint problem-solving. The GSEP was a result of an initiative of the Clean Energy Ministerial Meeting in 2010, and it has been accepted as a task group under the International Partnership for Energy Efficiency Cooperation (IPEEC). This partnership accounts for 75 percent of global GDP. However, the initiative has not been particularly visible; according to one source that has followed the process closely, so far little has come out of this process (Norichika Kanie, email to author, April 12, 2013).

Other exclusive initiatives: mostly discussion clubs?

Another US initiative under the George W. Bush administration was the establishment of the Major Economies Meeting in 2007, representing an informal discussion forum for the world's 17 largest economies, responsible for some 80 percent of total GHG emissions. This initiative was *de facto* endorsed by the Obama administration through the establishment of the Major Economies Forum on Energy and Climate (MEF) in 2009. Obama has been pushing this initiative as a supplement to the UN process. Meetings have been rather frequent, with the 15th and latest as of this writing held in April 2013. Like the IPEEC, the MEF has focused on the development of clean technologies. However, the initiative has received very limited attention, probably because nothing of substance has come out of this process. Judging from the short summaries from the meetings, it seems fair to see it as a discussion club where broad principles negotiated within the UNFCCC are often taken up. The first more practical task of the MEF was launched at the April 2013 meeting, when participants agreed to focus on improving energy performance in buildings. A working group will be established to work out the details (Chair's Summary, April 2013). It has been argued that "Given the stalemate in the UN climate negotiations, the best arena to strike a workable deal is among the members of the Major Economies Forum" (Roberts and Grosso 2013: 1). In principle this sounds like a good idea, but as yet there have been no indications that this forum will actually play such a role.

Other more exclusive but broader forums like the G8 and the G20, as well as the World Economic Forum, have also focused on climate change recently. The

G8 has regularly issued statements on the matter since 2005 and was the first forum to suggest the 2°C goal, but more recently the broader and more representative G20 has become more important in this regard. The G20, created as the Group of Twenty Finance Ministers and Central Bank Governors in 1999, was intended as a forum for cooperation on financial matters. Its membership represents 90 percent of global GDP, 80 percent of global trade and two-thirds of the world's population. The group gained increased significance in the wake of the financial crisis in 2008; since then, heads of state have participated and it has become the world's major economic forum. After the G20 turned its attention to energy and climate change, it might be a more appropriate forum than the UNFCCC, as deals can more easily be brokered in such more confined forums. Detailed analysis carried out on the attention paid by the G20 to climate change, however, indicates otherwise (Bruynicks and Happaerts 2013). Not surprisingly, frequent references are made to financial institutions, and other analysts have claimed that the G20 has been important in securing financing at COP15 (Kim and Chung 2012). Climate change is usually referred to in terms of energy policies, with emphasis on fuel subsidies and clean energy technology; and the G20 has engaged closely with stakeholders, particularly business and industry. There was a rise in the number of climate-related statements in the run-up and aftermath to COP15 in 2009. However, the increase in the G20 discourse on climate change has not translated into concrete commitments. Only one specific enforceable commitment has been identified: the phasing-out of fuel subsidies (Bruynicks and Happaerts 2013: 8). Controversy exists on the extent to which this one commitment has been effectively implemented, although some achievements have been made. Although the potential of the G20 should not be underestimated, so far the group has been better at setting ambitious goals than implementing them (Barbier 2010).

The G20, like the MEF, emphasizes its support for the UNFCCC process, indicating that its members do not see this forum as an alternative to the UN process. That supports my initial assumption that there exists more of a hierarchy in the climate regime complex than suggested by Keohane and Victor (2011). Moreover, the most important questions are never addressed in these G20 statements, due to the strong differences of interests among members (Bruynicks and Happaerts 2013). In other words, differences of interest do not disappear even if there are fewer members and less bureaucracy. The G20 would appear to have had limited influence on the UNFCCC as well as on the issue of climate change more generally.

A final potentially important exclusive forum is the Climate and Clean Air Coalition (CCAC), launched in February 2012 by UNEP and six countries, among them the USA and Canada. The goal is to reduce the emission of black carbon, or soot. Recent research has shown that this is a much bigger contributor to climate change than previously thought, and is now regarded as the second most important source of global warming. Of particular note is the fact that it is very short-lived, so that reductions have an immediate effect. Emissions following from black carbon represent a problem primarily in developing countries, due not least to inefficient means of cooking; serious health problems are also a result. The CCAC initiative may represent a win–win approach; its membership has expanded rapidly, to nearly thirty countries from both North and South, but no BASIC countries have yet joined. Financing has been modest, and the initiative has been met with skepticism from the "green" community, not least since it was spearheaded by the USA and Canada, neither of them known for proactive climate policies. The CCAC has a potential to reduce black-carbon emissions, but it is too early to say whether it will make a real difference.

Conclusions: political conflicts cannot be removed by new design

The UN climate regime has proven ineffective in a problem-solving perspective, so it is therefore understandable that more complex or more exclusive approaches have been suggested. However, several such institutions and forums have existed for quite some time already. This brief empirical examination of the achievements made by some of the most important ones indicates that they do not offer a panacea for dealing more effectively with climate change. The malignancy of the issue is so severe it cannot be removed by clever institutional design alone. Still, in a learning perspective, these institutions and forums may have had some influence, by showing that a more pragmatically-oriented bottom-up sector approach may be more workable than the traditional legally-binding top-down approach. Whether this will lead to more effective agreements, however, remains a still-unanswered empirical question.

Although the UNFCCC has not been effective, it scores high on legitimacy, thanks to its inclusive nature. This study has also indicated that the other forums acknowledge the UNFCCC as the main and legitimate body for dealing with issues of climate change. In turn, that indicates that the climate-change regime complex is actually more hierarchical than some analysts have held, and that the UN will stand as the main forum for the foreseeable future. However, much of the initial skepticism to other forums has now lessened. This is fortunate, because these supplements are also sorely needed if progress is ultimately to be achieved.

References

Abbott, K.W. (2012). The Transnational regime complex for climate change. *Environment and Planning C: Government and Policy* 30(4), 571–90.

——(2013). Constructing a transitional climate change regime: bypassing and managing states. *Social Science Research Network (SSRN)*, February 9, 2013.

Andresen, S. (2007). The effectiveness of UN environmental institutions. *International Environmental Agreements: Law, Politics and Economics* 7(4), 317–36.

——(2013). International regime effectiveness. In R. Falkner (ed.), *The Handbook of Global Climate and Environment Policy*. London: Wiley, pp. 304–19.

Andresen, S., N. Kanie and P.M. Haas (2014). Actor configurations in the climate regime: the states call the shots. In N. Kanie, S. Andresen and P.M. Haas (eds.), *Improving Global Environmental Governance Best Practices for Architecture and Agency.* London/New York: Routledge, pp. 175–96.

Barbier, E.B. (2010). A global green recovery, the G20 and international STO cooperation in clean energy. *STI Policy Review* 1(3), 1–15.
Biermann, F. and S. Bauer (eds.) (2005). *A World Environmental Organization: solution or threat for effective international environmental governance.* Aldershot: Ashgate.
Biermann, F., P. Pattberg and F. Zelli (eds.) (2010). *Global Climate Governance Beyond 2012: architecture, agency and adaptation.* Cambridge, UK: Cambridge University Press.
Bodansky, D. and L. Rajamani (2012). The evolution and governance architecture of the climate change regime. *Social Science Research Network (SSRN),* October 28, 2012.
Bruynicks, H. and S. Happaerts (2013). Negotiating in the new world order: emerging powers and the global climate change regime. Paper, ISA Annual Convention, San Francisco, April 3–6, 2013.
Chair's Summary (2013). Meeting of the leader's representatives of the major economies forum on energy and climate, April 12, 2013.
Fujiwara, N. (2012). *Sector-specific activities as the driving force towards a low-carbon economy from the asia-pacific partnership to a global partnership.* CEPS (Center for European Policy) Policy Brief No. 262, January 2012.
Greene, J. (2013). Order out of chaos: public and private rules for managing carbon. *Global Environmental Politics* 13(2), 1–25.
Greenspan Bell, R. and M. Ziegler (eds.) (2012). *Building International Climate Cooperation: lessons from the weapons and trade regimes for achieving international climate goals.* Washington, DC: World Resources Institute.
Kanie, N., M. Suzuki and M. Iguchi (2013). Fragmentation of international low-carbon technology governance: an assessment in terms of barriers to technology development. *Global Environmental Research* 17(1), 61–70.
Karlsson-Vinkhuyzen, S. and H. van Asselt (2009). Introduction: exploring and explaining the Asia Pacific Partnership on clean development and climate. *International Environmental Agreements: Politics, Law and Economics* 9(3), 195–211.
Keohane, R. and D. Victor (2011). The regime complex for climate change. *Perspectives on Politics* 9(7), 7–23.
Kim, J.A. and Suh-Yong Chung (2012). The role of the G20 in governing the climate change regime. *International Environmental Agreements: Politics, Law and Economics* 12(4), 361–74.
Lund, E. (2013). *Hybrid Governance in Practice: public and private actors in the Kyoto Protocol's Clean Development Mechanism.* Ph.D. thesis, Lund University, Sweden.
McGee, J. (2010). *The Asia Pacific Partnership and Contestation for the Future of the International Climate Regime.* Submitted for Doctor of Philosophy Graduate School of the Environment, Macquarie University, Sydney, Australia.
Miles, E., A. Underdal, S. Andresen, J. Wettestad, J.B. Skjærseth and E. Carline (2002). *Environmental Regime Effectiveness: confronting theory with evidence.* Cambridge, MA: MIT Press.
Mitchell, R. (2010). *International Politics and the Environment.* London: Sage Publications.
Najam, A. (2003). The case against a new international environmental organization. *Global Governance* 9(3), 367–84.
Olson, M. (1965). *The Logic of Collective Action: public goods and the theory of groups.* Cambridge, MA: Harvard University Press.
Peay, S.A. (2007). Joining the Asia-Pacific Partnership: the environmentally sound decision? *Colorado Journal of Environmental Law and Policy* 18 (Spring), 477.

Raustiala, K. and D. Victor (2004). The regime complex for plant genetic resources. *International Organization* 58(2), 277–309.

Roberts, T. and M. Grosso (2013). *A Fair Compromise to Break the Climate Impasse.* A paper: Global Views 40, Washington, DC: The Brookings Institution.

Underdal, A. (2002). One question, two answers. In E. Miles, A. Underdal, S. Andresen, J. Wettestad, J.B. Skjærseth and E. Carline, *International Environmental Regime Effectiveness: confronting theory with evidence,* Cambridge, MA: MIT Press, pp. 3–45.

Victor, D. (2011). *Global Warming Gridlock: creating more effective strategies for protecting the planet.* Cambridge, UK: Cambridge University Press.

Young, O. (2013). Does fairness matter in international environmental governance? Creating an effective and equitable climate regime. In T. Cherry, J. Hovi and D. McEvoy (eds.) (2014), *Toward a New Climate Agreement: conflict, resolution and governance.* London: Routledge.

12 Bottom-up or top-down?

Jon Hovi, Detlef F. Sprinz and Arild Underdal

Introduction

Politicians, bureaucrats, scholars, and environmentalists alike have expressed their disappointment that 20 years of UN negotiations have failed to produce an effective climate agreement. This disappointment has led to suggestions that the idea of a "top-down" treaty design, aiming for a science-based "optimal" solution to the climate change problem, be abandoned in favor of a "bottom-up" design, whereby countries pledge targets for emissions reductions or limitations according to what they see as economically, politically, and administratively feasible at home. This chapter probes the pros and cons of these suggestions in the context of international climate mitigation agreements.

To do so, we must consider a wide spectrum of group sizes, ranging from "one" to "universal." At one end, we find countries embarking upon an ambitious "*Energiewende*" (energy transition) more or less regardless of what others do. Germany at least shows economic and political feasibility at the stage of conceptualizing this policy, although it is premature to assess its ability to implement practical measures required in the coming decades. At the other end, we find the often hoped for universal-participation approach to combating global environmental problems in the spirit of a "global problem → global solution" argument.

Purely unilateral action without coordination will be unlikely to achieve ambitious global climate mitigation goals. Seemingly, strong domestic demand must exist in major countries in order to undergo national energy transformations regardless of what other countries do. Such unilateral and unconditional transformations may spread by horizontal diffusion, much as beliefs or innovations spread. In most countries, however, domestic demand for ambitious unilateral mitigation measures tends to weaken in times of economic stress and to become increasingly contingent on reciprocity as mitigation costs increase.

On the other hand, universal participation will be unlikely to materialize (e.g., the Holy See did not ratify the UNFCCC, despite the normative value that ratification could have had for over a billion followers). Moreover, some countries have smaller mitigation weights than others. Liechtenstein, San Marino, and Tuvalu may be signatories to the Kyoto Protocol, and Liechtenstein even has a

GHG freeze obligation, yet these countries are largely inconsequential for overall global GHG emissions. Thus it seems appropriate to focus on the relatively few countries (e.g., the G20) accounting for 65–80 percent of emissions. Their conduct will largely determine the future atmospheric concentration of GHGs.

Between the two ends of the spectrum, we find the EU, a conglomerate of countries lacking clear federal or unitary-actor character but having institutional capacity far greater than the typical IGOs. The EU position in international climate negotiations is subject to the internal coordination challenge; however, once a joint decision has been taken and codified, the EU can enforce it vis-à-vis the member states.

In this chapter, we limit the scope of analysis to international negotiations and agreements ($N \geq 2$). In the third section we consider the pros and cons of top-down and bottom-up processes of regime establishment and management, and of top-down and bottom-up architectures in the fourth section. We then consider the prospects for combining bottom-up and top-down elements in the fifth section. To prepare for this analysis, however, we first must clarify in the following section what we mean by "top-down" and "bottom-up" in this context.

Definitions

In a fundamental sense, *all* international agreements are bottom-up (Barrett 2012: 30); international agreements require the consent of each participating state. According to this view, top-down architectures presuppose a centralized institution able to authoritatively distribute and enforce rights and obligations among lower-level units. The international state system lacks such a centralized institution; hence it is anarchic. Therefore the role of international agreements is not only to distribute rights and obligations among states, but also to provide incentives to entice states to become signatories and comply with their commitments.

This conceptualization will not be pursued further here. Instead, we focus on two other ways of using the distinction between top-down and bottom-up. The first concerns different processes of regime creation. A top-down process is centralized and thus resembles the UNFCCC negotiations, whereby all UN member countries negotiate with a view to agreeing on a global climate treaty that will allocate rights and obligations to all (UNFCCC) countries. In contrast, a bottom-up process is decentralized (Lutsey and Sperling 2008); for example, it could start with uncoordinated regional initiatives that eventually become connected and perhaps integrated into a global regime.

A second way of using the distinction concerns different types of treaty architectures.[1] A (completely) top-down architecture may be defined as an agreement that: (1) fully coordinates and integrates policies (e.g., targets and timetables are determined jointly by mutually consenting parties); and (2) has all the world's countries as participants. In contrast, a (pure) bottom-up architecture is one where: (1) each country's target and timetable are intially decided unilaterally; and (2) only a small subset of the world's countries participates in an international agreement.

In a top-down architecture, the focus will typically be on "what needs to be done." For example, targets and timetables may be science based, insofar as they reflect scientific advice about what the targets should be and how quickly they should be reached. Moreover, each participating country's commitment is made conditional on other participating countries also making commitments.

In contrast, a bottom-up architecture focuses on "whatever makes sense and is feasible" in each participating country. The set of sensible and feasible measures may or may not depend on other countries' commitments. Targets and measures are determined by each participating country, depending on what it considers politically, economically, and administratively feasible at home. These individually determined targets are then reported to an international registry. We consider bottom-up architectures where at least two countries coordinate their mitigation policies.

Processes of regime establishment and management

Earth system science sees climate change as a systemic (global) challenge requiring an equally comprehensive response. Most policy makers and stakeholders seem to have adopted this perspective. If we combine a systemic understanding of the problem with a notion of good governance as involving broad stakeholder participation in democratic processes, a simple and clear procedural recipe emerges. Any major effort affecting people worldwide should be developed and implemented by the global community. With minor modifications, the UNFCCC negotiations pursue this "global problem → global solution" approach.

By standards of appropriateness (legitimacy), the logic of this argument seems quite compelling. Assessed in terms of problem-solving effectiveness, however, the "global problem → global solution" approach (and its main procedural tool, global conference diplomacy) is not the obvious choice.

Let us nevertheless begin with its bright side. First, global conference diplomacy often serves as an effective tool for raising awareness, setting political agendas, and simultaneously focusing the attention of governments and stakeholders worldwide on the same problem. Many governments, IGOs, and NGOs invest considerable time and energy preparing for high-visibility conferences such as the UN Conference on Environment and Development in Rio in 1992. These preparations involve efforts to improve general understanding of the problem and of alternative policy options. At this early stage, environmental ministries, agencies, and NGOs will often be able to influence the initial description and diagnosis of the problem and to shape political agendas (see e.g., Seyfang and Jordan 2002). Sometimes these efforts lead to unilateral upgrading of a country's policies and/or its institutional capacity.

Second, committees and expert groups provide important arenas for learning. In these settings, learning tends to be an asymmetrical affair whereby laggards learn from prestigious frontrunners. Indeed, some achievements that decision makers and observers attribute to international agreements may be as well or better understood as instances of policy diffusion, facilitated and supported by international institutions.

Third, even when they fail to reach agreement on new substantive measures, global environmental conferences often succeed in establishing institutional arrangements that can enhance the capacity for future problem solving. Although easily dismissed as a poor substitute for "the real thing," some investments in institution building seem to pay off handsomely later. In particular, institutions established to build a more solid platform of science-based knowledge for future negotiations can provide important services. IPCC is the obvious example in the climate change domain, but similar bodies help move cooperation forward in many other areas as well (see e.g., Breitmeier *et al.* 2011).

Finally, global conference diplomacy usually generates, for many of the people involved, positive stakes in its own success. As pointed out by Richardson (1992: 175), "The attention . . . focused on the government's response would generate pressure to raise its level." Moreover, working together for years may cause interpersonal networks to develop and could foster a sense of common purpose and a common will to "succeed" (or, at least, to recover sunk costs).

These four advantages are particularly important in the early stages. As negotiations progress towards considering substantive policy programs and specific measures, at least five limitations of the "global problem → global solution" approach tend to become increasingly evident.

First, the default decision rule of consensus leaves the burden of proof with any party working for policies or practices of a more ambitious nature, and empowers any opposed, significant party to act as veto player. In the activity systems to be regulated, however, some parties are clearly more important than others. An effective climate change mitigation program can, in principle, be established and implemented by a fairly small coalition of major actors (Victor 2011). The "global problem → global solution" approach is not well attuned to this world of realpolitik.

Second, global conferences – in particular, plenary sessions with political leaders in the spotlight – provide fertile ground for ideological posturing. Global climate change is characterized by stark asymmetries between countries' "guilt" in causing the problem and their capacity to alleviate it on the one hand, and socio-ecological vulnerability to climate change on the other. In such a domain, the risk of deadlock over basic principles and beliefs becomes imminent.

Third, more intense competition in international trade, finance, and politics strengthens countries' concerns about relative losses and gains. Over the past decade, economic growth rates have been significantly higher in major emerging economies such as China and India than in North America and Western Europe. Suffering from the consequences of financial turmoil, high rates of unemployment, and other economic woes, many Western countries have become less willing to undertake costly mitigation measures, at least as long as their increasingly successful competitors do not reciprocate. The static dichotomy between Annex I countries and the rest of the world is insensitive to these "tectonic" shifts in the global economy.

Fourth, the risk of bargaining deadlock is further aggravated by the UN system of caucusing and its internal coalition dynamics that tend to foster polarization rather than accommodation (see e.g., Sunstein 2002). States and delegations that stand out as the most articulate and committed advocates for the core interests and values of their group will often speak for that group.

Finally, global mega-conferences are plagued by overwhelming complexity and sprawling agendas. The number of delegates increased fourteen-fold from COP1 in 1995 to COP15 in 2009 (Schroeder *et al.* 2012: 835). As UN conference diplomacy allows in a wide range of stakeholder organizations and groups, focused negotiations tend to drown in a cacophony of voices and side events. Consequently, work to invent integrative solutions and design effective and politically feasible collective action programs will increasingly have to be undertaken elsewhere.

The understanding of climate change as a systemic challenge highlights its global public goods (or bads) nature. This is clearly one essential feature of the challenge, but equally important for understanding climate change politics is the fact that human influence on the climate system can be traced to a very wide range of activities, including all those that generate GHG emissions or involve significant land cover change. Some of these activities are local, regional, or national in scope, and each generates its own particular configuration of stakes for involved parties. An effective mitigation policy must somehow penetrate these activity systems and bring about transformative change. To succeed, it must also be politically feasible.

This conception of the climate change challenge points towards a different procedural approach. As Victor (2011: 6) states, "It starts with what nations are willing and able to implement." Moreover, it recognizes power configurations as a critical determinant of outcomes and therefore "starts with the interests of the most powerful countries" (Victor 2011: 265).

A fairly wide range of bottom-up procedures can meet Victor's two requirements. Interpreted as a general prescription, a bottom-up procedure seems to build on at least five key assumptions. First, most nations are indeed willing and able to take *some* meaningful mitigation measures. Second, which specific measures they will be willing and able to take will vary significantly, but at least some countries will go for more ambitious measures if allowed to choose what makes sense given their particular circumstances. Third, most nations' willingness to contribute will somewhat depend on what important others do. Even in a bottom-up approach, international cooperation will therefore be important in facilitating and enforcing voluntary exchanges of conditional commitments. Fourth, particularly concerning brokering and managing collective action, more can be achieved by taking advantage of the capabilities and services of existing international institutions. A number of functional and regional organizations initially established for other purposes (such as the IMO and the EU) have over the years built fairly effective systems for managing activities that are important sources of emissions and thus also important as potential targets for mitigation policies. Club-like world forums, notably the G8 and G20, may also play important roles in bringing climate change concerns to bear on discussions about "more urgent" problems of international trade, finances, and security. Fifth, although such a differentiated bottom-up approach will most likely lead to a "clumsy" patchwork of different measures

(Verweij *et al.* 2006), it can – in the aggregate – be expected to achieve more than the "global problem → global solution" approach ever will.

Of these five assumptions, the last seems most vulnerable. A number of bilateral and mini-lateral processes could plausibly come up with specific measures that – considered individually – would be more effective than are their functional equivalents created through global conference diplomacy. Such achievements constitute a necessary but not a sufficient condition for concluding that a bottom-up approach is superior to its top-down counterpart. The critical question is whether the aggregate impact of multiple "small-N" processes adds up to a more significant overall achievement. To answer that question, we must examine the two types of institutional architectures as well.

Institutional architectures

At least four arguments seemingly favor a bottom-up architecture. First, something is arguably better than nothing. The UNFCCC negotiation process has thus far failed to produce a top-down architecture that comes anywhere near solving the climate change problem (e.g., in the sense of reaching the +2°C target). Its main accomplishment, the Kyoto Protocol (Kyoto 1), suffers from serious deficiencies. Although nearly all countries in the world ratified it (the United States is a major exception), the 36 countries with legally binding targets (after Canada's withdrawal in 2011) are currently responsible for less than 20 percent of global emissions and committed to reducing their emissions by only about 5 percent on average. Hence, while inspired by scientific advice from the IPCC, Kyoto's ambition level fell well short of what the IPCC considered (and considers) necessary to avoid "dangerous anthropogenic climate change."[2] If Kyoto's achievements have been moderate, the UNFCCC has not been any more successful in its ambitions to negotiate a more effective successor agreement. Understandably, therefore, many observers, disappointed with the lack of progress, see a bottom-up architecture as more promising.

Second, because a bottom-up architecture decentralizes responsibility for formulating targets, it permits each participating country to pledge whatever it is able and willing to deliver (constrained by what other participating countries find acceptable). As each country presumably has the best knowledge of its own political, economic, and administrative capacity for reducing or limiting emissions, a bottom-up architecture should have some advantages concerning formulating feasible targets.

Third, if it is true that a bottom-up architecture will result in feasible targets, it should also produce comparatively high compliance rates. Limited compliance may result from two main sources: lack of capacity to comply, and lack of will to comply (e.g., Hovi and Underdal, forthcoming). A main motive for choosing a bottom-up architecture is precisely to ensure that targets reflect the capacity and will in each country to reduce or limit emissions. Thus, other things being equal, we should reasonably expect nationally determined targets to entail higher compliance rates than internationally negotiated targets.

Finally, to the extent that a bottom-up architecture entails targets that match each country's capacity and will to reduce emissions, the need for compliance enforcement should be limited. International agreements must be enforced either by the parties themselves (in a decentralized fashion) or by some institution that the parties unanimously erect and empower; thus global negotiations on compliance enforcement systems are subject to the "law of the least ambitious programme" (Underdal 1980; see also Hovi and Sprinz 2006). Concerning compliance enforcement, the least enthusiastic parties will likely be countries expecting to find compliance difficult or costly. Such countries may well attempt to prevent the erection of a potent enforcement system, insist on a loophole, or (failing those) decline to participate in the agreement (Downs *et al.* 1996). Hence a potent compliance enforcement system will be unlikely to receive unanimous consent precisely in those contexts in which it is highly needed; indeed, the stronger the incentives for non-compliance, the less likely that a consensus on potent compliance enforcement will emerge (e.g., Hovi *et al.* 2009). Therefore, a bottom-up architecture is attractive in part because it should make potent compliance enforcement less vital.

Given these upsides of a bottom-up architecture, what are the downsides? First, is something necessarily better than nothing? Of course, if we must choose between an agreement that achieves at least some emissions reductions, and no agreement at all, something is presumably preferable to nothing. However, the alternative to a weak agreement now may not be no agreement ever; it could be a stronger agreement later on. If so, perhaps we should not prefer just something now.

Consider Kyoto. Despite President George W. Bush's repudiation in 2001, other countries moved on without the United States, hoping that countries that did not participate initially might join later. We now know that this hope came to nothing. Although Australia ratified in 2007, Canada withdrew in 2011, so the number of countries participating with a binding target was the same when Kyoto 1 expired in 2012 as when it entered into force in 2005. The United States not only stuck by its decision to stay out; "Kyoto" was eventually a reviled term in US political debate. Thus, if anything, Kyoto made the United States less inclined to participate in global climate cooperation. We shall never know what would have happened if Kyoto had been scrapped following President Bush's 2001 repudiation; however, it is difficult to completely dismiss the possibility that an alternative climate agreement, with the United States as a member, might then have emerged.

Second, while a top-down architecture provides member countries an incentive to adopt ambitious targets by making commitments conditional, a bottom-up architecture frees them to choose whether to make their commitment conditional. For example, the Copenhagen Accord includes both conditional and unconditional commitments. Moreover, while many environmental top-down architectures have very limited ambitions, members of a bottom-up agreement are free to choose their targets unilaterally. Hence some countries may be tempted to pledge little more than business as usual, and thus a bottom-up architecture may well result in very shallow targets indeed.

Third, because bottom-up architectures are typically shallow, they can largely do without compliance enforcement. But if they can do without compliance enforcement simply because countries in such architectures pledge only little or even no change in their behavior, a limited need for compliance enforcement is hardly a good sign (Downs *et al.* 1996).

Fourth, suppose member countries in a bottom-up architecture actually do pledge targets requiring significant behavioral change. Will they invariably comply with these targets? The history of climate change policies displays a long list of nationally determined targets for emissions reductions that never even came close to being fulfilled (Barrett 2008: 240–41). Some of these targets may well have been so-called stretch goals intended to "inspire people to reach for a difficult-to-obtain objective" (Lempert 2009: 64); however, in retrospect others seem more like empty promises. Compliance enforcement may thus be necessary even in (ambitious) bottom-up architectures.

Finally, if a bottom-up architecture does not require significant behavioral change, one might wonder whether and why an agreement is needed at all. Why not leave the pledges for emissions reductions or limitations fully to each country? Through what mechanisms (if any) can a bottom-up architecture induce member countries to adopt stricter targets than they would adopt unilaterally? One possibility is that by bundling together, like-minded countries might benefit because of similar ambitions, information exchange, diffusion of best practices, and joint research. Another possibility is that bottom-up agreements "can pilot ideas, encourage action and pull societies into a new space to which others can aspire" (King and Steiner 2011). However, it is less clear precisely how this pull is supposed to work, why we should expect it to be created or reinforced through cooperation, and why we should expect it to work better bottom-up than top-down.

We have seen that both bottom-up and top-down approaches have strengths as well as weaknesses. While the third section suggested that bottom-up *processes* for regime creation may allow for somewhat more enthusiasm than the current UNFCCC system does, the fourth section presented a slightly less optimistic view concerning bottom-up *architectures*. Does this leave us with options to pursue that fall into neither of the two previous categories? The next section considers this question, with particular attention to club goods and differentiation of governance functions.

Club goods and functional allocation

Club goods are defined as goods that: (1) display little rivalry in consumption, that is, rising resource use only marginally reduces individual benefits;[3] and (2) exclusion can be achieved at low cost (Ostrom *et al.* 1994: 7). Weischer *et al.* (2012: 187–88) argue that a (transformational) club must generate significant benefits that are exclusively for its members. Moreover, all members must benefit and the benefits must be generated so as to respect existing law. Clubs therefore include only a proper subset of the universe of potential members; for example, in the context of climate change, clubs will have (considerably) fewer members than the UNFCCC does. Examples abound both in the literature on clubs for low-carbon urban development (e.g., Dirix *et al.* 2013) and in the literature on regional versus global agreements (e.g., Asheim *et al.* 2006). Clubs offer one possible way to combine select elements of bottom-up and top-down perspectives to climate governance. They include aspects of bottom-up in that membership is limited. And they include aspects of top-down insofar as the club coordinates and integrates policies among its members. Both David Victor (2011) and Robert Nordhaus (at the IIASA 40th Anniversary Conference in 2012[4]) have suggested that club goods may be avenues for decentralized, yet coordinated, mitigation efforts. For Nordhaus the question is one of choosing an optimal tariff to approach the welfare level afforded by the cooperative solution; in contrast, Victor focuses on technology options for mitigation and for geoengineering.

Victor (2011) suggests a nonbinding "pledge and review" club that allows for contingent commitments. A contingent commitment can enhance other parties' incentives to contribute (more), but the extent to which it does so will depend on: (a) the "exchange rate" offered; and (b) the credibility of the contingency itself. While the motivation for international clubs such as GATT and the WTO is the potential gains from trade (Pareto-superior outcomes), Victor argues that both "general benefits" and "other benefits . . . tailored to each member" exist in the context of climate change mitigation (2011: 250). While the latter benefits seem difficult to describe with precision, general benefits take the form of "access to carbon markets in the enthusiastic nations for CDM-like offset trading as well as access to government-to-government funding for emission abatement projects, adaptation, and capacity building" (Victor 2011: 249–50). These benefits may provide incentives for participation in mitigation action. Two questions arise: (1) the degree to which a club good can remain a club good; and (2) whether the goals we may expect a club to pursue could not also be pursued by uncoordinated unilateralism.

Rivalry in consumption is limited in the examples suggested by Victor, yet drawbacks exist concerning the feasibility of exclusion at low costs. Clearly, groups of governments can exclude other governments from their emissions trading system by not allowing markets to be formally linked, but to some extent the world market for offsets from the Clean Development Mechanism and Joint Implementation works as a slow proxy for linking emissions trading markets. Thus exclusion fosters indirect solutions. Government funding for abatement activities in partner countries is often bilateral and sometimes multilateral. Yet we should keep in mind that such investments are already happening anyway – regardless of whether there is a club much smaller in size than the UNFCCC's membership. Moreover, if governments club together to develop in common highly efficient technologies that are inexpensive to use (Hovi *et al.* 2009: 23), it may be difficult to practice exclusion later on. If such technologies were extremely inexpensive to use, then hampering diffusion might not only be politically impossible, it might also lead to mandatory licensing at low costs in developing non-club member countries (as happened regarding HIV and other medicines in some developing countries in order to bolster the market for locally produced, generic medicine). If potential club members anticipate such mandatory licensing or removal of the exclusive access to the technology, they may abstain from joining the fund at the outset.

Furthermore, it is not obvious why countries should form technology clubs. Financing a scientifically challenging project such as the International Thermonuclear Experimental Reactor is currently scheduled to cost around €15 billion,[5] essentially a cost item that the EU or a major country (say the United States, Russia, China, India, or Brazil) could finance from a single year's budget. The scope of the challenge must be reasonably daunting to justify the creation of a club. The search for a renewable electricity solution that solves intermittency and achieves electricity production costs at half the present rate of hard coal per kWh may be such a challenge; yet a national advanced market commitment (i.e., a guaranteed minimum purchase once the specifications have been honored on commercial scale applications) or an X-Prize could be viable options in wealthy countries (Sprinz 2009). Perhaps the best scope for technology clubs would be basic development of advanced technologies, where deep uncertainty exists concerning which technology will ultimately prevail (Barrett 2009: 73–74). Yet investment in basic research implies uncertain returns on investment at the firm level. Where deep uncertainty exists concerning which technology will ultimately prevail, investing collectively in a very large portfolio of technologies might be better than having a single firm or country develop a technology by itself (Barrett 2009: 73–74). In this case, investing in a share of a very large portfolio of technologies might make particular sense for countries. The same caveat about ex-post exclusion applies also in this case: a very successful component of the portfolio might be up for mandatory licensing, thereby undermining the incentives to join the club – except for charitable reasons. Given the traditional argument for club goods mentioned in the beginning of this section, Victor's argument appears theoretically appealing, yet may encounter a range of practical limitations.[6] The same applies to proposals for transformational 2ºC clubs by Weischer *et al.* (2012). They outline a range of theoretically convincing reasons why potential benefits can be generated by clubs. Yet the ability to exclude is solved by way of definition, and optimism regarding creating such clubs rests on analogies to the GATT/WTO and the European Union – both of which have been in deep trouble for several years.

Olson (1971) distinguishes between exclusive and inclusive collective goods. Jointness of supply is present for inclusive goods and may increase their provision with rising membership (Olson 1971: 38–43), but is not present for exclusive goods (e.g., reducing production to demand higher prices in a cartel). The concept of inclusive goods can be related to the example of the "30 percent club," which supported the development of the Helsinki Sulphur Protocol as part of the CLRTAP regime in UNECE member countries in the mid-1980s. This club announced a minimum policy ambition to which prospective club members had to adhere (Sprinz 1992). Perhaps this notion of minimum agreed effort, or similarity in policy effort by members, is implicitly hoped for by Victor (2011). In this context, Weischer *et al.* review a fair number of dialogue fora and implementation groups in the field of climate policy. They conclude, "[t]here are no clubs for which the level of ambition is the membership criterion, in the senses that a country would have to demonstrate a certain track record [on climate change, authors] or agree to specific commitments regarding emissions reductions as a condition for being admitted to the club" (Weischer *et al.* 2012: 183).

Overall, club goods in the context of international negotiations are about domestic policy coordination around shared common goals and minimum agreed efforts. Many arguments favoring club goods do not practically apply (at least not easily) to climate change mitigation. The irony may be that very costly undertakings (such as the German *Energiewende*) lend themselves to exclusion, yet find few wholehearted followers in the short run, while clubs finding low-cost solutions may suffer from the anticipation that non-members cannot be excluded from the fruits of club activity.

An alternative way to consider options that qualify neither as top-down nor as bottom-up perspectives is to allocate functions according to the comparative advantages of these perspectives.[7] First, climate change mitigation efforts could benefit from having an ultimate goal that allows a long-term focus. The Copenhagen and Cancun meetings achieved this by agreeing on the goal to limit global mean temperature change to +2°C over pre-industrial levels. This also serves to operationalize the UNFCCC's ultimate goal (United Nations 1992, Art. 2). Second, tasks that need international standardization (such as monitoring, reporting, and verifying emissions, impacts, and relevant research surrounding climate change) are best undertaken with a bird's eye perspective. Third, coordinating national mitigation efforts may be best left to smaller groups of countries. Even in top-down architectures for burden sharing, countries can easily eschew taking on obligations (as most countries have done under the Kyoto Protocol), use exit clauses (as Canada did in 2011), or go for unannounced, but intentional noncompliance to extract additional resources from other countries in the future.

Clubs may serve to enhance global cooperation if they include stronger institutions than those available at the global level. For example, the EU not only facilitated the Kyoto Protocol's entry into force (by inducing Russia's ratification); it probably also enhanced the EU member countries' Kyoto compliance (through developing an internal policy scheme involving the EU Emissions Trading Scheme and other measures).

The EU likely appears more attractive to many non-members than most other clubs. Indeed, several countries would no doubt be willing to adopt stricter climate policies if it would help them gain EU membership. Obviously, however, the EU would be unlikely to grant membership to very many of these countries, regardless of their climate policies.

The EU seems, however, to be an exceptional case. Other past and present clubs, such as the Asia-Pacific Partnership, the Climate and Clean Air Coalition, the G20, and the Major Economies Meeting, have records on climate change mitigation that so far have been no more impressive than those of the UNFCCC (see Andresen, Chapter 11 in this volume).

Conclusions

We conclude that neither the bottom-up nor the top-down strategies stands out as a dominant solution for developing climate change mitigation policies. Rather, our analysis indicates that each strategy has some distinct advantages over the other. The critical challenge for future research – and for policy makers and diplomats – is to find ways of combining top-down and bottom-up approaches according to their respective comparative advantages. This chapter offers some suggestions, but it also points to important feasibility constraints for such combinatorial approaches.

Acknowledgements

We would like to thank Geir Ulfstein for helpful comments and Frank Azevedo for excellent editorial assistance.

Notes

1 Because defining bottom-up and top-down architectures is difficult without specifying the broader category of agreements that one has in mind, we focus on agreements in the targets-and-timetables category.
2 In addition, Kyoto's enforcement system lacks potency (Barrett 2008).
3 This holds up to a point, beyond which congestion sets in and increasing resource use further reduces individual benefits significantly.
4 See http://conference2012.iiasa.ac.at/person.html?code=nordhaus&scode=s1s4 (accessed 30 May 2013) and IIASA/OeAW Public Lecture Series, Lecture 1, "Maastricht and Kyoto: A Tale of Two Treaties," www.iiasa.ac.at/web/home/about/events/upcoming events/iiasa-oeaw-25-october-lecture.en.html (accessed 30 May 2013).
5 See http://en.wikipedia.org/wiki/ITER (accessed 30 May 2013).
6 Given that climate change is a global force mechanism challenge at the stage of creating climate change, nobody can be excluded from the positive and negative global effects of mitigation. This is strikingly different from adaptation – which is essentially a private good in practice (and acknowledged as such by Victor [2011]).
7 See Sprinz (2010) for an early version of these ideas.

References

Asheim, G.B., C.B. Froyn, J. Hovi and F.C. Menz (2006). Regional versus global cooperation for climate control. *Journal of Environmental Economics and Management* 51(1), 93–109.

Barrett, S. (2008). Climate treaties and the imperative of enforcement. *Oxford Review of Economic Policy* 24(2), 239–58.

——(2009). The coming global climate-technology revolution. *Journal of Economic Perspectives* 23(2), 53–75.

——(2012). Credible commitments, focal points, and tipping. In R. Hahn and A. Ulph (eds.), *Climate Change and Common Sense. Essays in Honour of Tom Schelling*. Oxford: Oxford University Press.

Breitmeier, H., A. Underdal and O.R. Young (2011). The effectiveness of international environmental regimes: Comparing and contrasting findings from quantitative research. *International Studies Review* 13(4), 579–605.

Dirix, J., W. Peeters, J. Eyckmans, P.T. Jones and S. Sterckx (2013). Strengthening bottom-up and top-down climate governance. *Climate Policy* 13(3), 363–83.

Downs, G.W., D.M. Rocke and P.N. Barsoom (1996). Is the good news about compliance good news about cooperation? *International Organization* 50(3), 379–406.

Hovi, J., and D.F. Sprinz (2006). The limits of the law of the least ambitious program. *Global Environmental Politics* 6(3), 28–42.

Hovi, J., D.F. Sprinz and A. Underdal (2009). Implementing long-term climate policy: time inconsistency, domestic politics, international anarchy. *Global Environmental Politics* 9(3), 20–39.

Hovi, J. and A. Underdal (2014). Implementation, compliance, and effectiveness of policies and institutions. Forthcoming in D.F. Sprinz and U. Luterbacher (eds.), *International Relations and Global Climate Change*. Cambridge, MA: MIT Press.

King, D. and A. Steiner (2011). Is a global agreement the only way to tackle climate change? *The Guardian* 27 November 2011. Available from: www.theguardian.com/commentisfree/2011/nov/27/durban-climate-change-delivery (accessed 20 November 2013).

Lempert, R.J. (2009). Setting appropriate goals: a priority, long-term climate-policy decision. In R.J. Lempert, S.W. Popper, E.Y. Min, J.A. Dewar (eds.), *Shaping tomorrow today: near-term steps towards long-term goals*. Santa Monica, CA: Rand Corporation.

Lutsey, N. and D. Sperling (2008). America's bottom-up climate change mitigation policy, *Energy Policy* 36(2), 673–85.

Olson, M. (1971). *The Logic of Collective Action. Public Goods and the Theory of Groups.* Cambridge, MA: Harvard University Press.

Ostrom, E., R. Gardner and J. Walker (1994). *Rules, Games, and Common-pool Resources.* Ann Arbor: The University of Michigan Press.

Richardson, E.L. (1992). Climate change: problems of law-making. In A. Hurrell and B. Kingsbury (eds.), *The International Politics of the Environment*. Oxford: Oxford University Press.

Schroeder, H., M.T. Boykoff and L. Spiers (2012). Equity and state representations in climate negotiations. *Nature Climate Change* 2 (December), 834–36.

Seyfang, G. and A. Jordan (2002). The Johannesburg summit and sustainable development: how effective are mega-conferences? In O.S. Stokke and Ø.B. Thommesen (eds.), *Yearbook of International Co-operation on Environment and Development 2002–2003*. London: Earthscan.

Sprinz, D.F. (1992). *Why Countries Support International Environmental Agreements: the regulation of acid rain in Europe*. Ph.D. diss, Ann Arbor: Department of Political Science, The University of Michigan.

——(2009). Long-term environmental policy: definition, knowledge, future research. *Global Environmental Politics* 9(3), 1–8.

——(2010). The "sandwich solution" to global climate policy. In Deutsche Post, Delivering Tomorrow – Towards Sustainable Logistics. Bonn: Deutsche Post AG: 71–74.

Sunstein, C.R. (2002). The law of group polarization. *The Journal of Political Philosophy* 10(2), 175–95.

Underdal, A. (1980). *The Politics of International Fisheries Management: the case of the North-East Atlantic*. Oslo: Universitetsforlaget.

United Nations (1992). United Nations Framework Convention on Climate Change [Online]. New York, NY. Available from http://unfccc.int/resource/docs/convkp/conveng.pdf (accessed August 2013).

Verweij, M., M. Douglas, R. Ellis, C. Engel, F. Hendriks, S. Lohmann, S. Ney, S. Rayner and M. Thompson (2006). Clumsy solutions for a complex world: the case of climate change. *Public Administration* 84(4), 817–43.

Victor, D.G. (2011). *Global Warming Gridlock. Creating More Effective Strategies for Protecting the Planet.* New York: Cambridge University Press.

Weischer, L., J. Morgan and M. Patel (2012). Climate clubs: can small groups of countries make a big difference in addressing climate change? *Review of European Community & International Environmental Law* 21, 177–92.

Part III

Governance

Structures for a new agreement

13 Rethinking the legal form and principles of a new climate agreement

Geir Ulfstein and Christina Voigt

Introduction

In 2011 it was agreed to launch a new ad hoc Working Group on the Durban Platform for Enhanced Action (ADP) with a mandate "to develop a protocol, another legal instrument or an agreed outcome with legal force under the Convention applicable to all Parties" was launched[1]. The new negotiating process, which began in May 2012, is scheduled to end in 2015. The outcome shall come into effect and be implemented from 2020 onwards. However, while the door is pushed open for the formal discussion on questions of legal form, the choice of form is far from settled. The formulation in the ADP decision from Durban is marked by constructive ambiguity and requires further interpretation.

The negotiations of the new climate agreement are also marked by uncertainty about its substantive content, not least about if and how the principle of "common, but differentiated responsibilities and respective capabilities" should be incorporated in the future agreement. The UNFCCC and the Kyoto Protocol are based on a clear distinction between developed and developing states (Annex I and non-Annex states). But the basis for this distinction is becoming increasingly blurred by changing economic realities and trends towards steeper greenhouse gas emissions in advanced economies. Accordingly, there is a need to rethink how differentiation in responsibilities should be reflected in a new climate agreement.

Furthermore, the connection between the legal form and the substantive content should be examined. The Durban Platform commits states to negotiate an agreement with a legal status. But this does not necessarily mean that all elements of such an agreement need to be legally binding. It is therefore of interest to discuss which of the elements should be of a legally binding character, including whether differentiations between states may include the binding nature of their commitments.

This chapter begins with a discussion of the interpretation of the terminology in the ADP decision on the legal character of a climate agreement. In the next section we deal with the benefits of a legally binding agreement, with a particular focus on the binding character of the different elements of the agreement. Then we examine different approaches to differentiation between states in the agreement. Finally, we draw some conclusions about the substantive content and legal form of a future climate agreement.

The Durban Platform of Enhanced Action (ADP)[2]

Discussions on the legal form of a new agreement have weighed on the UNFCCC negotiation process for several years. The 2007 Bali Action Plan[3] set up the Ad-Hoc Working Group on Long-term Cooperative Action (AWG-LCA) with the purpose of reaching an "agreed outcome" on long-term cooperative action on climate change in 2009. The term "agreed outcome," however, did not provide clarity on the legal form or character of the outcome this process should produce, nor did the Bali decision contain a clear mandate to negotiate the legal character of such outcome. Since the Bali Action Plan, many Parties have repeatedly expressed their view that such outcome needs to be of legally binding character. Several Parties submitted proposals for various legally binding instruments under UNFCCC Article 17. These included different types of protocols suggested by Australia,[4] Japan,[5] Grenada,[6] Tuvalu,[7] and Costa Rica[8] and an Implementing Agreement from the United States.[9]

Decision 1/CP.17 limits the choice of legal form of a new climate agreement to three options: a protocol, another legal instrument, or an agreed outcome with legal force under the convention. Except for setting up these alternatives, the decision is silent on the legal character of that agreement, notably in the absence of the term "legally-binding."

A preliminary question is whether the qualifier "under the Convention" shall apply to all three options.[10] If answered in the affirmative, the qualifier excludes tracks separate from the convention, such as purely domestic or regional arrangements and laws.

Option 1 then reads to be "a protocol … under the Convention" and pertains to a protocol according to Article 17 UNFCCC. Option 2 is less straightforward. It would read as "another legal instrument … under the Convention." Legal instruments under the convention could refer to any of the legally binding instruments which the Conference of the Parties (COP) has the competence to adopt. These are – in addition to protocols – amendments (Article 15), annexes (Article 16) and amendments to annexes (Article 16). However, the COP may also adopt decisions. The majority view is that COP decisions are generally political in character and not legally binding. Legal bindingness may, however, be achieved in the exceptional circumstance that an explicit mandate by the treaty that establishes the COP (the Convention) empowers the COP to "regulate" or "legislate" through decisions. The Convention does not contain an explicit "regulatory mandate" for the COP.[11] This does not necessarily mean that COP decisions are devoid of legal effects. The decisions may have significance for the interpretation of substantive obligations of the states parties. They may also be binding at the internal level, for example decisions about the establishment of subsidiary bodies or the adoption of rules of procedure.[12] In the absence of an amendment to the Convention or a new protocol authorizing the COP to adopt binding decisions it is, however, difficult to see COP decisions as a relevant "legal instrument(s)."[13]

The third option is the least clear. The formulation "an agreed outcome with legal force under the convention" uses language which does not appear in the

Convention. It has been suggested that this option "seems to be designed to allow room for the negotiations to end with an outcome that doesn't take the form of the legal instruments expressively contemplated in the Convention, and yet is still 'under the Convention.'"[14]

The term "agreed outcome with legal force" is considered to be a more ambiguous option than the first two. This option conceivably gives more room for maneuver in negotiating the new climate agreement as it does not limit the result to one instrument. Rather, there could be a package of instruments including perhaps several COP decisions as well as a protocol or protocols.

The ambiguity in the formulation "agreed outcome with legal force" raises the same question as pertains to "legal instrument," namely whether a non-binding COP decision could qualify as such "outcome." An "agreed outcome with legal force" may well include COP decisions, but not as the *sole* result. Rather, one (or more) protocol(s) would need to be part of that package. The "agreed outcome" could, for example, consist of: (1) a protocol that in turn empowers the COP or its governing body to "regulate through decisions," similar to Article 17 of the Kyoto Protocol which instructs the COP to "define the relevant principles, modalities, *rules* and guidelines, in particular for verification, reporting and accountability for emissions trading"; (2) the decisions taken under that mandate; and (3) of aspirational decisions with a lesser degree of stringency and specificity.

Meanwhile, the possibility has been contemplated that "an agreed outcome with legal force" could also be interpreted as legal instruments embodied in domestic, rather than international, law.[15] Such "bottom-up" legal construction is reminiscent of an "implementing agreement" that allows for "legally binding approaches" based on Parties' municipal law.[16] This view, however, is difficult to maintain. The qualifying term "under the Convention" limits the range of options to those entailed in that international legal instrument, i.e., protocols, amendments and annexes. The difference to options 1 and 2, which envisage single instruments, lies in the fact that option 3 allows for the combination of a number of different instruments in a "package solution" as explained above and thereby creates greater flexibility.[17]

Finally, Decision 1/CP.17 instructs the ADP to complete its work no later than 2015 in order to adopt a protocol, another legal instrument or an agreed outcome with legal force at COP 21 and for it to come into effect and be implemented from 2020. COP decisions do not need to "come into effect." On the other hand, legally binding instruments such as a protocol would require certain domestic legislative processes to take place before they can enter into force and come into effect. The time gap of 5 years indicates that there is a difference between "adoption," "coming into effect" and "implementation" of the new climate agreement. In other words, the time frame indicates the necessity of a lengthy process for the "evolution" from adoption to implementation, reminiscent of a ratification process of an international agreement. Such a procedural time window would neither be necessary for purely domestic solutions, nor for decisions of the COP. Rather, the generous time frame provides yet another interpretative argument for the legally binding possibilities under the UNFCCC.

Which elements should be legally binding?

Introduction

There is no necessary connection between a legally binding instrument and the binding status of each of the elements of such an instrument. While the instrument can be of a legally binding nature, its provisions need not be binding.[18]

One example is the emissions reduction commitment in Article 4(2) of the UNFCCC. This provision is part of a binding instrument, but is arguably not a legally binding obligation.[19] Being part of a legally binding instrument means that this provision is also subject to the accepted principles of treaty interpretation, as reflected in Articles 31–33 of the Vienna Convention on the Law of Treaties (1969). Furthermore, violation of a legally binding instrument may entail stronger political condemnation. i.e., a heightened "shaming factor." Elements of a legally binding instrument may also more likely be implemented in national legislation.[20] But its breach of a non-binding provision will have no legal consequences at the international level.

In the following a functional approach will be applied. It will be asked which elements of a legally binding instrument should, in light of the international legal consequences, have a legally binding character. But it will also be examined whether such elements should be contained in the legal instrument itself, or be delegated to subsequent decisions by the COP.

The legal character of different elements

The legal consequences of including elements of a binding character may be different, depending on the character of the rule. For example, binding procedural rules may delegate powers to the COP to adopt legally binding – as opposed to political ("soft law") – decisions. Meanwhile, legally binding substantive obligations, such as emissions reduction commitments, may, in cases of violation, have legal consequences as defined in the relevant instrument, possibly including sanctions, they may also entitle other states parties to take legal action, such as countermeasures, under general international law. The different legal consequences will be discussed in relation to the pertinent elements of the legal instrument.

Objectives and principles

The objectives and principles for cooperation may be contained either in the preamble or in the articles – the operative provisions – of a legal instrument. While the preambular provisions are not binding as such, the operative provisions are binding at the outset.

However, the objectives and principles are by nature not immediately operational – they are meant to serve as a background for the operative provisions. They may be used in the interpretation of the operative provisions and serve as guidance in establishing the powers of the COP and other institutional organs. Decisions taken by the COP and other organs may even be considered invalid cooperation (*ultra vires*) if they contradict the objectives and principles for the cooperation. Whether such objectives and principles are contained in the preamble or in the operative provisions will not necessarily make much legal difference. But their inclusion in the operative provisions may give them a stronger political status.

Emissions reductions and limitations

Emissions reductions and limitations of greenhouse gases are at the core of a new climate agreement. The need for binding legal obligations depends on their strictness. But if the 2°C (or even 1.5°C) goal shall be reached, strict obligations would be required. In such case, the burdens on states would be significant and the temptation to circumvent – acting as a "free-rider" – would increase correspondingly. Both the need for national implementation and the "shaming factor" in case of violation would speak in favor of internationally binding emissions reductions and limitations. But are legally binding obligations desirable due to the legal consequences of violations?

In principle, it would not be necessary to have legally binding emissions reductions and limitations in order to impose binding consequences, such as sanctions, under a climate agreement (see below). Such legally binding consequences can be established on any factual or legal basis as defined by the state parties in the agreement. But states would hardly impose legally binding consequences for violations of political (soft law) commitments. Furthermore, the legal status may be of importance for allocation of benefits under the climate agreement, whether it pertains to rights to trade in quotas (see below), receipt of economic or technological incentives (see below) or representation by the relevant state party on elected subsidiary organs.

The use of legally binding emissions reductions and limitations would also allow states to apply general international law in the form of countermeasures and state responsibility against states that do not respect their obligations. Moreover, they may take them to available international courts outside the climate regime, such as the International Court of Justice. While other states will usually be reluctant to take such actions, they should not be ruled out. The whaling case brought by Australia and New Zealand against Japan is an example of states taking court action to enforce a multilateral regime for living resources.[21]

The use of legally binding commitments may also be of significance with respect to other international obligations. The international legal system is increasingly fragmented in terms of diverse regulations in different legal regimes, such as trade, human rights and the environment. It may be of importance that the commitments under the climate regime are legally binding in order to prevail in relation to other treaties, such as the international trade regime.[22]

Trade in emission quotas and other flexibility measures

A new climate agreement may contain different kinds of "flexibility mechanisms," from trading in emissions quotas to, say, different forms of credits for investment in environment-friendly technologies in other states, reforestation or reduced emissions from the forest or other sectors. Such measures may need a legally binding form in order to ensure effective transactions between the involved states. But a legally binding form is also essential for the connection with substantive obligations concerning emissions reductions and limitations. It is difficult to imagine that states should be credited with such transactions in relation to legally binding emissions reductions and limitations unless the transactions also have a binding legal basis.

Financial transfer and transfer of technology

Massive economic and technological support is necessary if the 2°C goal (or the 1.5°C goal) shall be reached. Financial commitments and transfer of technology by developed states would be part of a package deal.

Such arrangements may take the form of conditions for implementation of commitments by developing states, as is arguably the case of Article 4.7 UNFCCC.[23] This would mean that non-fulfilment of the commitments would have legal consequences in the form of suspension of the substantive obligations by developing states. But, first, there may be a need to strengthen the obligations of the developed states through establishing them as unqualified legal obligations, i.e., not only as having the consequence of suspending the obligations of developing states. Furthermore, a legally binding status may be desirable in order to apply legal consequences in cases of violations (see next subsection).

Non-compliance procedures, including incentives and sanctions

The legal status of the enforcement measures adopted by the Compliance Committee of the Kyoto Protocol has been disputed. Article 18 of the Protocol establishes that "any procedures and mechanisms under this Article entailing binding consequences shall be adopted by means of amendment to this Protocol." It may thus be argued that none of the consequences adopted by the Committee are binding in the absence of such amendment.[24]

Incentives should obviously be part of a compliance regime. But it would generally seem that stricter commitments would require use of binding legal sanctions as a means to prevent "free-riders."[25]

The need for legally binding consequences should, however, be assessed in relation to the character of the substantive commitments and the relevant consequences. For example, it may be asked whether legally binding sanctions should be applied in relation both to violations of substantive obligations and to commitments to provide financial and technological resources. Furthermore, the consequences may be diverse, spanning everything from a duty to provide a plan on how to bring about compliance to the imposition of financial penalties. The decisive criterion for establishing legally binding consequences should be the need for establishing the sanctions as a separate legal obligation on the relevant states – in addition to the obligations taken on as part of the climate agreement as such.

It should also be mentioned that competence to adopt legally binding consequences, including sanctions, may have repercussions on the institutional features and procedures of the organs dealing with such issues. First, there needs to be assurance that such organs are credible in terms of assessing both relevant facts and legal commitments. Moreover, the imposition of legally binding consequences should only occur if relevant due process guarantees are respected. This may speak in favor of also establishing the organs and their procedures in a legally binding form.

Regulations in the agreement vs. delegation to the COP

The establishment of an institutional framework in the form of a COP and possible other organs is key to ensuring dynamic cooperation and developing the substantive commitments over time – possibly all the way to 2050. The Marrakesh Accords is a good example of wide-reaching decision making by the COP serving as the meeting of the Parties (CMP).

It is necessary to balance which substantive obligations should be contained in the agreement itself and what could be delegated to decisions by the COP. But to the extent that legally binding regulations through COP decisions are needed, the delegation of powers to the COP must be contained in legal provisions in the agreement.

The procedures for decision making could be left to subsequent adoption by the COP as Rules of Procedure. But such delegation to the COP may, in the absence of consensus, prevent their adoption – as we have seen under the UNFCCC.[26] It might also be that a more sophisticated system for decision making should be adopted as part of a climate agreement. Different procedures could be developed for different kinds of decisions. For example, new substantive commitments could require consensus but not formal amendment and a ratification procedure. Other decisions, e.g., on trading and other flexibility mechanisms, could require forms of qualified majority. This could be combined with systems of weighted voting. The Enforcement Branch of the Kyoto Protocol's Compliance Committee provides an example by its membership from Annex I and non-Annex I countries and regional representation, as well as Small Island Developing States (SIDS), and its requirement of a three-quarters majority, including a majority among both Annex I and non-Annex I countries.

In short, a future climate agreement raises not only the question about its legal form, but also the legally binding character of its elements. Furthermore, the consequences of violations of such binding elements must be determined. Finally, a choice must be made between which elements should be included in the agreement itself and what can be delegated to decision making by the COP.

Equity and differentiation

The success of the negotiations under the ADP will, among other things, depend on a common understanding of equitable sharing of efforts. While equity is a principle of the Convention, its meaning and scope remain contentious. In general terms, equity refers to the quality of being impartial, fair and just. In the international climate discourse, equity and fairness are used interchangeably.[27] The broad understanding is that the new agreement must reflect states' different "situations", whether they are the stage of development, economic means, risk (exposure and vulnerability) of climate impacts, contributions to increasing GHG concentrations in the atmosphere (historical, current and future trends), technological capacities, etc. States differ in many regards and these differences must be duly reflected in the architecture of the new agreement.

At the same time, the design of an equitable regime poses a paramount challenge to the traditional structure of public international law, which is based on the sovereign equality of states. States are supposed to be treated equally, where as a starting point, the same rights and duties apply to all. There is, however, growing understanding that in order to treat states equally, their differences and national circumstances must be taken into account. Differentiation and positive discrimination (e.g., through affirmative action) is necessary in order to treat different states on an equal basis.

Since the 1972 Stockholm Conference, the making of international environmental treaties has changed from providing identical treatment to all contracting states to providing for differential treatment of developing countries, based on concepts of cooperation and solidarity. The idea is to bring about effective – rather than formal – equality among de facto unequal states and to ensure the participation of all countries in international environmental agreements.

With the 1992 Rio Declaration, a specific form of differential treatment has found its way into international environmental law-making. The Declaration's Principle 7 reads:

> States shall cooperate in a spirit of global partnership to conserve, protect and restore the health and integrity of the Earth's ecosystem. In view of different contributions to global environmental degradation, States have common but differentiated responsibilities. The developed countries acknowledge the responsibility they bear in the international pursuit of sustainable development in view of the pressures their societies place on the global environmental and of the technologies and financial resources they command.

An equitable approach to the climate challenge is often understood as an approach based on the principle of common but differentiated responsibilities and respective capabilities (CBDRRC) as mentioned in Article 3 UNFCCC, which – unlike the Rio Declaration – says that "parties should protect the climate system … on the basis of equity and in accordance with their common but differentiated responsibilities and respective capabilities."

This principle has so far mainly been interpreted by major developing countries in the context of the Rio Declaration and used as an argument for demanding that developed countries should acknowledge their historic contributions to the increased concentration of greenhouse gases in the atmosphere and accordingly "take the lead" in climate mitigation and finance.[28] The other side of this argument is that developing countries, having historically contributed marginally to the current concentrations, should not be required to take on mandatory commitments to reduce their GHG emissions. The CBDRRC principle is reflected in the "firewall" between developed (Annex-I) and developing (non-Annex) countries and is the main reason for the current difficulties in finding a truly equitable solution.

Differentiation has so far been along the dividing line of "developed" and "developing" states and according to the historical contributions of developed countries to environmental degradation, as well as the capability of developed countries to engage in cost-intensive environmental mitigation action. These factors (criteria) have led to substantively stronger obligations of developed countries, with developing countries having no, or milder, obligations as well as entitlements to both financial transfers and transfer of technology and know-how from developed countries. Based on these criteria, "positive discrimination" in favor of developing countries has led to highly asymmetric environmental obligations coupled with mechanisms for capacity building, and transfer of financial resources and technology as well as compliance assistance.

The immense current global challenges commonly faced by all states can only be tackled by taking cooperative and large-scale remedial action. However, the factual preconditions under which states are acting still differ considerably. Yet the "landscape of similarities and differences" has changed since its perception more than 40 years ago. The world is characterized now by different disparities in resources and capabilities. The antagonistic dividing line between developed and developing countries is becoming increasingly blurred. Any of the two groups, if they can be identified at all,[29] is no longer homogeneous but marked by stark internal differences. In addition, the international landscape undergoes changes and fluctuations. Any attempt at categorization might be insufficient to capture such dynamism. For that reason there needs to be more differentiation[30] and differentiation should be flexible and dynamic and only granted on a temporary basis.

At the time of the conception of the Convention, this "grouping" of states might have been a suitable reflection of the state of the world. But 20 years on, it isn't. Not only have a number of developing countries become economies with strong growth, they have also increased their emissions and partly overtaken the highest emitting developed countries – and such trends are continuing.[31] A differentiation of responsibility in addressing the climate challenge that rests entirely on historic contributions is not only not fair – it is not effective. In short, it is dangerous because it will miss out those emissions that cause global temperatures to tilt by 4°C or more. Moreover, the traditional binary differentiation does not address the heterogeneity that is found within the group of developing nations as well as developed countries. The question is thus how to reflect in the new agreement the different "situations" of states and their dynamic development. The change in the world between 1992 and 2013 tells us that such differentiation cannot be static; it needs to be based on dynamic and flexible parameters, which allow the structure of the agreement to evolve as the world evolves.

When the ADP was adopted in Durban, the decision was free from references to developed/developing or Annex I/non-Annex I countries. Such wording was celebrated as breaking down the traditional firewall between the two major groups. But already the decision from Doha, in December 2012, had Parties acknowledging that the ADP "shall be guided by the principles of the Convention."[32] While explicit attempts to include the principle of common but differentiated responsibilities and respective capabilities, or to refer to the climate paragraphs of the outcome document of Rio+20, were not successful, the principles of the Convention are also guiding principles for the ADP.[33]

The argument put forward in this chapter is that the principles of the Convention must apply, but they are not static. Their principled character is a means to adjust and adapt the Convention to changing political circumstances. Principles of law provide a necessary means by which law can develop in a dynamic fashion that is responsive to today's problems. Principles are "an authoritative recognition of a dynamic element on international law."[34] Law is a continuing process and principles provide for a "welcome possibility for growth,"[35] in which capacity they also contribute to the development of international law. Principles are never "finished products." From their identification to the final determination of the principles' content in a particular context is a "continuing process."[36]

While only one particular interpretation of the CBDRRC principle (i.e., differentiation based on historic contributions) has so far dominated the climate negotiations, the time has come for more countries to put forward their (evolved) understanding of that principle. In Doha, a number of countries came forward with their understanding that the CBDRRC principle is reflective of a certain dynamism in the climate regime because it is not set in stone but evolves over time. In the ongoing ADP negotiations, many Parties therefore observed "that the principles [of the Convention] are not rigid, and should be applied in a dynamic and evolving manner taking into account national circumstances, changing economic realities and levels of development."[37] "Principles … need to be forward-looking and take into account what the world might look like in 2020."[38] Some Parties also stated that the application of the Convention

> should be adapted in order to improve its vitality and relevance in the modern world and in order to enable it to become a modern instrument to address climate change. It was pointed out that the Convention has evolved, and will continue to evolve over time, and thus the manner in which the principles apply also needs to evolve.[39]

Yet finding reliable, but flexible and dynamic, criteria or means to define various (groups of) "equals" and allocating rights and responsibilities accordingly, may be an impossible task in multilateral environmental treaties. Making instead, alternatives to criteria-based differentiation should be explored. As such an alternative, we suggest self-differentiation within a spectrum of options/contributions.

As a starting point, the question of how to treat differentially is not the same as the question of how to "group" parties. Traditionally, differentiation has been made along the fault line of "developed" and "developing" countries. This antagonistic dividing line is particularly difficult to maintain in a climate context for some of the fastest growing developing countries. Differentiation should thus be flexible and dynamic and only be granted on a temporary basis.

In terms of giving flexibility to differentiations, the 1991 Volatile Organic Compounds (VOC) Protocol to the Convention on Long-Range Transboundary Air Pollution (LRTAP VOC Protocol) stands out as an interesting approach. It establishes specific targets and timetables that commit Parties to control and reduce their emissions of VOCs. In order, however, to reflect the need for differentiation based on a Party's emissions and particular geographic and demographic circumstances, the VOC Protocol offers Parties three ways in which to meet the emission reduction requirement. Upon signature or ratification, a Party must choose one of these options. While the first option is open to all Parties, the availability of the other two options depends on particular criteria and circumstances.

The first option to achieve emission reductions is for (any) Party to

> take effective measures to reduce its national annual emissions of VOCs by at least thirty per cent by the year 1999, using 1988 levels as a basis or any other annual level during the period 1984 to 1990, which it may specify upon signature of or accession to the present Protocol.[40]

The second option is only available to a Party whose annual emissions contribute to tropospheric ozone concentrations in areas under the jurisdiction of one or more other Parties, and where such emissions originate only from an area under its jurisdiction that is specified as a tropospheric ozone management area (TOMA) under Annex I to the Protocol.[41] A Party that chooses this way shall

> as soon as possible and as a first step, take effective measures to: (i) Reduce its annual emissions of VOCs from the areas so specified by at least 30 per cent by the year 1999, using 1988 levels as a basis or any other annual level during the period 1984–1990, which it may specify upon signature of or accession to the present Protocol; and (ii) ensure that its total national annual emissions of VOCs by the year 1999 do not exceed the 1988 levels.[42]

The third way is only available to Parties whose national annual emissions of VOCs in 1988 were lower than 500,000 tonnes and 20 kg/inhabitant and 5 tonnes/ km^2. Such a Party "shall, as soon as possible and as a first step, take effective measures to ensure at least that at the latest by the year 1999 its national annual emissions of VOCs do not exceed the 1988 levels."[43]

Furthermore, no later than 2 years after the Protocol entered into force, each Party was required to apply "appropriate" national or international emission standards to new stationary and new mobile sources based on "best available technologies which are economically feasible" (BATEF). No later than 5 years after the entry into force, in those areas in which national or international tropospheric ozone standards are exceeded or where transboundary fluxes originate or are expected to originate, each Party must apply BATEF to existing stationary sources in major categories.[44]

In carrying out their obligations, Parties are invited to give highest priority to reduction and control of emissions of the substances with the greatest photochemical ozone creation potential. A further innovative requirement is that states must ensure that they do not substitute toxic and carcinogenic VOCs, and those that harm the stratospheric ozone layer, for other VOCs.[45]

The success of the LRTAP is partly due to its restricted regional scope and its comparatively small and homogeneous group of Parties. It has nevertheless served as a model for the UNFCCC and the Vienna Convention on the Protection of the Ozone Layer.

One lesson from the VOC Protocol for the climate negotiations is that by giving Parties the (free) choice of option, based on certain criteria – or no criteria – for the availability of that particular option, the Parties themselves can find their "group" in some form of "self-differentiation." For example, as option 3 of the VOC Protocol illustrates, certain emission intensity criteria (e.g., t/capita or t/km^2), overall emission amounts, or other economic, demographic and/or geographic criteria can be included in the design of options.

In addition, the options could differ either in substance (higher baselines or different base years or reference levels, higher emissions caps or less stringent targets) or in form (flexibility for implementation, supplementarity, time frame for implementation etc.) or in both. For example, some options might be equal in substance but differ in the time frame for implementation periods (so-called grace periods as in the Montreal Protocol[46]). Moreover, some options could be linked to either providing or receiving financial, technological and scientific support and assistance. This could be fine-tuned with the particular design of the various options. One example could be the choice between two particular options, where one includes stronger emission reduction targets but weaker obligations to provide financial support, whereas the other option contains the opposite: more moderate emission reduction targets coupled with significant financial transfer obligations. Such an "optioning" approach, coupled with criteria for the availability of particular options by otherwise free choice, could open for much needed flexibility in the design of new international environmental agreements, particular climate change.

While the optioning approach could be a feasible way forward, it is important to keep the architecture dynamic. Dynamic elements should thus be integrated into the options or complement them (see VOC Protocol, Art. 3). Dynamic elements in terms of technological responses are, for example, the requirement of using "best available technologies" or "best practices." Dynamism in terms of substance can also be maintained by: (1) adoption of a "critical loads" approach that allows for upward adjustments; (2) regular review of the appropriateness of the targets in the light of latest scientific findings; (3) automatic strengthening of commitments at given intervals, or adjustments on the basis of available scientific, environmental, technical and economic information (see Montreal Protocol, Art. 6); or (4) review of the level of ambition by an assessment expert panel.

The point is that there will not be one type of differentiation that "fits all" and covers all the very different circumstances and situations of Parties. Rather, it will be the right combination or "mix" of substantive commitments, incentive structures, entitlements, procedural requirements, etc., that will be crucial for the success of a new agreement. A well-designed and fine-tuned "catalogue" of options (with differing commitments, contributions, and/or entitlements), which Parties can choose from upon signature or ratification, might be a feasible way forward, reflecting the diversities of a globalized and interconnected world in the sophisticated design of a comprehensive agreement.

Conclusions

The negotiations of the new climate agreement face challenges both regarding its legal form and its principles. The Durban Platform establishes a framework by the requirement that the agreement shall be "a protocol, another legal instrument or an agreed outcome with legal force under the Convention applicable to all parties." While the formulation is marked by "constructive ambiguity," it provides guidance and allows states to choose between the distinctive options of adopting a new protocol, amendments to the UNFCCC (or annexes or amendments to annexes), or a package including delegation of powers to the COP to adopt legally binding decisions.

But the future climate agreement also raises the question of the legally binding character of each of its elements. This should be decided on the basis of the international legal consequences of choosing a legally binding form of the relevant element, i.e., a functional approach. It would generally seem that the stringency and complexity required of a new agreement would militate in favor of legally binding elements. However, the need for a dynamic regime suggests that extensive decision-making power should be delegated to the COP.

It is also necessary to rethink the relationship between the sovereign equality of states and the meaning of equity in light of the principle of common but differentiated responsibilities and respective capabilities (CBDRRC). The UNFCCC and the Kyoto Protocol make a sharp distinction between Annex I and non-Annex states. But the economy of different states, and their emissions, have become so heterogeneous that it is no longer desirable to uphold this binary division. There is a need for more differentiation, flexibility and dynamism. This could include the possibility that states choose between different options of substantive commitments ("self-differentiation"), incentive structures, etc., combined with dynamic criteria such as using "best available technology", or a "critical loads" approach that allows for upward adjustment of commitments.

Such an approach would be less prescriptive than an (unrealistic) top-down structure but more ambitious than mere bottom-up approaches, while leaving necessary freedom and flexibility to the Parties.

Acknowledgement

This chapter was supported by the Research Council of Norway through its Centres of Excellence Funding scheme, project number 223274.

Notes

1 Decision 1/CP.17, paras 2 and 4.
2 The following paragraphs draw upon: C.Voigt (2012), The legal form of the Durban platform agreement: seven reasons for a protocol; *Ethics, Policy & Environment* 15(3), 276–82.
3 Decision 2/CP.13.
4 Draft protocol to the Convention prepared by the Government of Australia for adoption at the fifteenth session of the Conference of the Parties, FCCC/CP/2009/5 (6 June 2009).
5 Draft protocol to the Convention prepared by the Government of Japan for adoption at the fifteenth session of the Conference of the Parties, FCCC/CP/2009/3 (13 May 2009).
6 Proposed protocol to the Convention submitted by Grenada for adoption at the sixteenth session of the Conference of the Parties, FCCC/CP/2010/3 (2 June 2010).
7 Draft protocol to the Convention presented by the Government of Tuvalu under Article 17 of the Convention, FCCC/CP/2009/4 (5 June 2009).
8 Draft protocol to the Convention prepared by the Government of Costa Rica for adoption at the fifteenth session of the Conference of the Parties, FCCC/CP/2009/6 (8 June 2009).
9 Draft implementing agreement under the Convention prepared by the Government of the United States of America for adoption at the fifteenth session of the Conference of the Parties, Note by the secretariat, FCCC/CP/2009/7 (6 June 2009).
10 The issue of the legal form of a new agreement "under the convention" is separate from the question of whether all principles and provisions of the convention should apply equally to that agreement. The latter question is not discussed here.
11 The UNFCCC only empowers the COP to "make, within its mandate, the decisions necessary to promote the effective implementation of the Convention" and to "exercise such other functions as are required for the achievement of the objective of the Convention." Article 7.2 UNFCCC.
12 R.Churchill and G. Ulfstein (2000), Autonomous institutional arrangements in multilateral environmental agreements: a little noticed phenomenon in international law; *American Journal of International Law* 94(4), 623–60.
13 As mentioned above, many states made clear the need for a legally binding agreement. As one climate expert noted: "Given the gathering momentum towards a legally binding instrument in the lead-up to the Durban Climate Conference, it would be safe to assume that the majority of countries that negotiated the Durban Platform, however, did not intend this [to consider COP decisions "legal instruments"] to be the case." See L. Rajamani (2012), The Durban Platform for enhanced action and the future of the climate regime; *International Comparative Law Quarterly* 61(2), 506.
14 J.Werksman (2011), Q&A: The legal aspects of the Durban Platform text; *WRI Insights,* 14 December 2011.
15 Rajamani (2012), 507. D. Bodansky (2012), The Durban Platform negotiations: goal and options; *Harvard Project on Climate Agreements*, Viewpoint, June 2012, 3; and D. Bodansky (2012), *The Durban Platform: issues and options for a 2015 agreement*, Center for Climate and Energy Solutions, December 2012.
16 Rajamani (2012), 507. See also India's submission to the ADP, where it is noted that "an agreed outcome with legal force" need not have the legal form of a protocol or a legal instrument; it could be an outcome that derives legal force from municipal or

international law." Indian Submission on ADP Work Plan, FCCC/ADP/2012/Misc.3, 30 April 2012, 33.

17 Moreover, a solution based on domestic climate laws may not satisfy the requirements of universal applicability required by Decision 1/CP.17. In order to be "applicable to all Parties," the outcome must adhere to a certain degree of consistency and coherence in form and character, even if not in substance. This element pays respect to the insistence of the US and others on legal symmetry. Universality of application thus prevents a solution based on (many different) domestic instruments. This argument is further supported by the need to strengthen "the multilateral, rules-based regime under the Convention" as expressed in the third paragraph of the Preamble to Decision 1/CP.17.

18 D. Bodansky (2012), *The Durban Platform: issues and options for a 2015 agreement.* Center for Climate and Energy Solutions; available at SSRN: http://ssrn.com/abstract=2270336, 3.

19 D. Bodansky (1993), The United Nations Framework Convention on Climate Change: A commentary; *Yale Journal of International Law* 18, 451–558.

20 D. Bodansky (2010), *The Art and Craft of International Environmental Law.* Cambridge, MA: Harvard University Press, p. 179.

21 ICJ (2010), *Whaling in the Antarctic (Australia v. Japan) Application.* Application instituting proceedings 31 May 2010.

22 J. Werksman (2012), Compliance and the use of trade measures. In J. Brunnée, M. Doelle, and L. Rajamani (eds.), *Promoting Compliance in an Evolving Climate Regime.* Cambridge, UK: Cambridge University Press, pp. 262–86.

23 L. Rajamani, Developing countries and compliance in the climate regime. Ibid., p. 383.

24 G. Ulfstein and J. Werksman (2005), The Kyoto compliance system: towards hard enforcement. In O.S. Stokke, J. Hovi, and G. Ulfstein (eds.), *Implementing the Climate Regime: International Compliance.* London: Earthscan, pp. 57–8.

25 D. Bodansky (2011), *The Art and Craft of International Environmental Law*, pp. 236–37.

26 In the absence of formally adopted rules of procedure, the Climate Change Convention applies draft rules contained in UN Doc. FCCC/CP/1996/2. Draft Rule 42 on majority voting, however, is not applied. See Report of the Conference of the Parties on its Fifth Session, UN Doc. FCCC/CP/1999/6, paragraph 14.

27 J. Ashton and X. Wang (2003), Equity and climate: in principle and practice. In J.E. Aldy (ed.), *Beyond Kyoto: advancing the international effort against climate change.* Pew Center on Global Climate Change, 62.

28 Arts. 3, 4.2 and 4.7 UNFCCC.

29 See, for example, the classification by the International Monetary Fund (IMF) in "advanced economies" and "emerging market and developing economies." International Monetary Fund, *World Economic Outlook*, October 2012, 179. The IMF notes: "This classification is not based on strict criteria, economic or otherwise, and it has evolved over time" (p. 177). According to the United Nations Statistics Division, "there is no established convention for the designation of 'developed' and 'developing' countries or areas in the United Nations system." See UNSD, *Composition of macro geographical (continental) regions, geographical sub-regions, and selected economic and other groupings (fn. C)*, revised 11 October 2012, available at: http://unstats.un.org/unsd/methods/m49/m49regin.htm#ftnc (accessed: August 2013).

30 See also J. Pauwelyn (2013), The end of differential treatment for developing countries? Lessons from the trade and climate regimes, *RECIEL* 22(1), 29–41. Pauwelyn argues for more differentiation and further subdivisions.

31 See the overview of annual global carbon budgets and trends updated by the Global Carbon Budget, available at: www.globalcarbonproject.org/carbonbudget/index.htm (accessed: August 2013).

32 Decision x/CP.18.

33 Summary of the Roundtable Under Workstream 1, ADP1, part 2, Doha, Qatar, November-December 2012, Note by the Co-Chairs, 7 February 2013 (ADP. 2012.6. Informal Summary), 2.

34 J. L. Brierly (1963), *The Law of Nations: an introduction to the international law on peace*. Oxford: Clarendon Press, p. 63.

35 See M. Bos (1977), The recognized manifestations of international law, twentieth *German Yearbook of International Law*, 42.

36 Ibid.

37 Summary of the Roundtable under Workstream 1, ADP1, part 2, Doha, Qatar, November-December 2012, Note by the Co-Chairs, 7 February 2013 (ADP. 2012.6. Informal Summary) at p. 3.

38 Ibid.

39 Ibid.

40 Art. 2(2)(a) VOC Protocol. This option has been chosen by Austria, Belgium, Estonia, Finland, France, Germany, Netherlands, Portugal, Spain, Sweden and the United Kingdom, with 1988 as base year; by Denmark with 1985; by Liechtenstein, Switzerland and the United States with 1984; and by Czech Republic, Italy, Luxembourg, Monaco and Slovakia with 1990 as base year.

41 Art. 2(2)(b) VOC Protocol. Annex I specifies TOMAs in Norway (base year 1989) and Canada (base year 1988). The total Norwegian mainland as well as the exclusive economic zone south of 62°N latitude in the region of the Economic Commission for Europe (ECE), covering an area of 466,000 km^2, is a TOMA.

42 Art. 2(2)(b) VOC Protocol. This has been chosen by Bulgaria, Greece, and Hungary.

43 Art. 2(2)(c) VOC Protocol.

44 Art. 3 VOC Protocol.

45 Art. 5 VOC Protocol.

46 Montreal Protocol on Ozone Depleting Substances, Article 5, which allows certain developing countries to postpone for ten years compliance with their obligations under the protocol.

14 Technology agreements with heterogeneous countries

Michael Hoel and Aart de Zeeuw

Introduction

There is a large literature showing that international environmental agreements focusing only on reducing emissions, such as the Kyoto Protocol on climate change, cannot be expected to achieve much (for example, Barrett 1994; Finus 2003). One alternative to such a comprehensive international environmental agreement is to focus instead on technological improvements in order to reduce abatement costs. A sufficiently large reduction in abatement costs might induce countries to undertake significant emission reductions. Even without an explicit general agreement on emission reductions, some agreement leading to lower abatement costs as a consequence of the R&D agreed upon might result in a broad reduction of emissions. This is the background for proposals of a climate agreement on technology development (for example, Barrett 2006; Hoel and de Zeeuw 2010). The present chapter discusses this issue in more detail, emphasizing the fact that countries differ with respect to their valuation (or willingness to pay) for reducing greenhouse gas emissions.

The previous literature on the relationship between technological development and international environmental agreements considers different aspects. De Coninck *et al.* (2008) argue that agreements should focus on technology because technology is essential for handling the problems, technology is already part of environmental policies anyway, and some important countries only want to discuss this type of agreement. Moreover, hold-up problems may arise if technology choice precedes agreements on emission reductions. Buchholz and Konrad (1995) put out a warning that countries have incentives to choose and commit themselves to bad technologies before they enter the negotiations for an agreement, because they may then be able to shift the burden of emission reductions onto other countries that have lower costs. However, Battaglini and Harstad (2012) show that in a dynamic context where both the size and length of the agreement are endogenous, this hold-up problem may actually be beneficial. The idea is that the hold-up problem generated by a short-term agreement is a credible threat off the equilibrium path and reduces the incentives to free-ride. Another hold-up problem arises in Goeschl and Perino (2012), who connect a regime of international property rights to an international environmental agreement. They show a hold-up effect from the anticipation of rent extraction by the innovator that induces a reduction in abatement commitments in an agreement.

The main reason that international environmental agreements are not expected to achieve much is that large agreements with large possible gains of cooperation are not stable in the sense that free-rider incentives dominate the incentives to cooperate. Benchekroun and Ray Chauduri (2012) show that eco-innovations can reduce the stability, using a farsighted stability concept. Buchner and Carraro (2005) use the FEEM RICE (Regional Integrated model of Climate and the Economy) model to assess whether technology agreements perform better than agreements on emission reductions. They show that technology agreements are usually more stable but not necessarily more environmentally effective. Nagashima and Dellink (2008) use the STACO (Stability of Coalitions) model to show the effects of spillovers of existing technology on international environmental agreements: global emission reductions increase, of course, but the stability of the agreement hardly changes.

To focus on the technology aspect, we assume that there is no cooperation on emission reductions. However, countries may in various ways cooperate on the development of new, climate friendly technology that reduces the costs of abatement. This means that the agreement is on R&D expenditures, for example in the form of joint ventures, and not on emission reductions. We model this very crudely, by assuming that abatement costs are a decreasing function of the total amount of R&D expenditures by a group of cooperating countries. Formally, we consider a three-stage game. In the first stage, each country decides whether or not it wants to belong to a coalition of countries that is undertaking R&D aiming to reduce abatement costs. In the second stage, the coalition decides on its amount of R&D (and how to share this cost among its members). Finally, in stage three all countries (coalition members and outsiders) decide on how much to abate. The decisions at this final stage are made non-cooperatively but the decisions are of course influenced by the previous decision of the coalition. Note that it is possible that a group of countries decides in stage two to lower the costs of abatement so much that all countries decide to abate in the final stage. Moreover, in the model we assume marginal costs over the relevant range of abatement to be constant. Hence at this stage each country either chooses zero abatement or some fixed amount of abatement. This decision may differ across countries, since they are assumed to have different valuations of emission reductions: Each country abates if, and only if, the cost of abatement does not exceed the country's valuation of emission reductions. The basic question is how far the coalition wants to go in its investments in R&D. The higher the investments, the lower the costs of abatement and the higher the number of countries that switch to the climate-friendly technology.

The model

We consider a world consisting of N countries each having the same potential amount of abatement (in absolute, e.g., physical terms) using the new technology. This identical amount of abatement is normalized to 1. The abatement potential could, for instance, be all the emissions within a specific sector in each country, e.g., the production of electricity.

Abatement decisions are made non-cooperatively, with each country choosing to abate if, and only if, the cost of doing so does not exceed the country's valuation of the corresponding emission reduction. Countries are assumed to be heterogeneous with respect to these valuations, denoted by v_i for country *i*. Countries are indexed so that $v_1 \geq v_2 \ldots \geq v_N \geq 0$.

The cost of abating (at the amount 1) in each country is given by *c*(*M*), where *M* is the amount of total R&D expenditures by all countries. Knowledge created by R&D is hence considered to be a perfect public good. We make the following assumptions on *c*(*M*):

$$c(0) > v_1, 0 > c'(M) > -1.$$

The inequality $c(0) > v_1$ means that without any R&D, no country will abate. Abatement costs are assumed to be declining in total R&D *M*, but *c*(*M*)+(*M*) is increasing in *M*. This last condition implies that no country will undertake R&D unilaterally in order to reduce its abatement costs.

Consider a coalition of *k* countries investing *M* in the development of new technology. Define *m*(*M*) as the number of countries satisfying $v_i \geq c(M)$. Clearly, *m*(*M*) is (non-strictly) increasing in *M*.

If all coalition members abate once the technology is developed, we have $m(M) \geq k$. However, as we will see in the next section, there may be equilibria where $m(M) < k$, i.e., only some of the coalition members abate, although they all participate in the financing of the new technology. The reason they participate in the coalition is that they obtain benefits from other countries' abating due to the developed technology.

We assume that a coalition of *k* countries consists of the countries that benefit most from the coalition. These are the countries with the highest valuations of abatement, i.e. countries 1, 2, …, *k*. The benefit to the coalition of *k* countries of one unit of abatement is hence

$$W(k) = \sum_{i \leq k} v_i$$

which increases as *k* increases (and is only defined for integer values of *k*).

Using the definitions above, it is clear that the benefit to the coalition members of *m*(*M*) countries abating is *m*(*M*)*W*(*k*). The investment cost of the coalition is *M*. The abatement cost of the coalition is *kc*(*M*) if all members abate, and *m*(*M*)*C*(*M*) otherwise. The payoff to a coalition of *k* countries that optimizes its amount of R&D is hence

$$(1) \qquad V(k) = \max_M \{m(M)W(k) - (\min[k, m(M)])c(M) - M\}$$

Notice that $V(k) \geq 0$ for all *k*, since the coalition always has the option of setting *M*=*0* and obtaining *m*(*0*)=*0* (from our assumptions about the valuations v_i and the cost function *c*(*M*)) and hence *V*(*k*)=0.

To have a non-trivial equilibrium we assume that $V(1)=0$ and $V(k)\geq 0$ for sufficiently high values of $k\leq N$. For these values of k, $V(k)$ is strictly increasing in k. This is easiest to see by treating k as a continuous variable instead of an integer. Applying the envelope theorem to (1) then gives us

(2) $V'(k) = m(M)W'(k)$ for $m(M) < k$

(3) $V'(k) = m(M)W'(k) - c(M)$ for $m(M) > k$

Since $W'(k) > 0$ it is immediately clear that $V'(k) > 0$ for $1 < m(M) < k$. For the case $m(M) > k$ we must have $v_k \geq c(M)$, which together with $W'(k) = v_k$ and $m(M) > 1$ implies $V'(k) > 0$.

The optimization problem defined by (1) gives M as a function of k. From our assumptions and the discussion above, it follows that $M(k)=0$ for sufficiently low values of k but $M(k)>0$ for $k\geq k^*$, where k^* is some threshold not exceeding N. The coalition size k^* is a coalition size satisfying the conditions for internal stability (see, for example, d'Aspremont *et al.* 1983; Barrett 1994). No country will want to leave a coalition of size k^*, since all countries will receive a payoff of zero when the size of the coalition becomes k^*-1 (due to $M(k^*-1)=0$).

For k^* to be the largest possible stable coalition, it must be true that for any coalition larger than k^* at least one country will benefit from leaving the coalition. Consider a coalition k larger than k^*. The total payoff to the coalition may be written as

(4) $V(k) = \{m(M(k-1))W(k) - (\min[k, m(M(k-1))])c(M(k-1)) - M(k-1)\} + \varepsilon(k)$

where $\varepsilon(k)$ is the loss in payoff from choosing $M(k-1)$ instead of the optimal value $M(k)$. This will typically be a small number, as the payoff function is flat at the top.

The abatement decision of each country is independent of whether or not the country is a member of the coalition. The gain from leaving the coalition for country i is therefore simply its saved investment costs. However, by leaving the coalition it also obtains a loss in the form of its share of $\varepsilon(k)$. Formally, country i is hence better off in the coalition than outside if

(5) $\alpha_i M(k-1) < \varepsilon_i(k)$

where α_i is country i's share of the investment costs (with $\Sigma_i \alpha_i=1$) and $\varepsilon_i(k)$ are some numbers satisfying $\Sigma_i \varepsilon_i(k)=\varepsilon(k)$. To have a stable coalition no country must be able to gain from leaving. Hence, inequalities of type (5) must hold for all members. Summing up these inequalities we obtain the following condition for coalition stability:

(6) $M(k-1) < \varepsilon(k)$

If this inequality holds, it is possible to find α_i's satisfying $\Sigma_i \alpha_i = 1$ that make (6) hold for all i, hence making the coalition stable.

Clearly, the inequality (6) holds for k^* defined above, since $M(k^* - 1) = 0$ by the definition of k^* and $\varepsilon(k^*) = V(k^*) > 0$. Can we have a stable coalition for values of k above k^*? We cannot rule out this possibility. However, typically $\varepsilon(k)$ will be "small," implying that this will only occur if $M(k-1)$ is sufficiently small. In the rest of this chapter we assume that the valuations v_i and the cost function $c(M)$ have properties implying that the only stable equilibrium is k^* as defined above (i.e., the lowest integer giving $M(k) > 0$).

The equilibrium $\{k^*, M(k^*), m(M(k^*))\}$ will of course depend on all valuations v_i and on the cost function $c(M)$. We start by considering how the equilibrium coalition size depends on the cost function.

Coalition size and the cost function

Let the cost function be given by $c(M) + \alpha g(M)$ where $g(M) \geq 0$ with a strict inequality for some M, and the parameter α is equal to 0 initially. An increase in α is thus equivalent to some positive shift in the cost function.

Inserting $c(M) + \alpha g(M)$ into (1) and differentiating with respect to α gives (using the envelope theorem)

$$(7) \qquad \frac{dV(k)}{d\alpha} = -\left(\min\left[k, m(M)\right]\right) g(M) \leq 0$$

In other words, any positive shift in the cost function will either leave the function $V(k)$ unchanged (if $g(M) = 0$) or it will decline (if $g(M) > 0$). If $V(k)$ is unchanged there will be no change in the coalition size. If, however, it is reduced, the maximal value of k giving $V(k) = 0$ will increase. If this change is sufficiently large, the equilibrium coalition size will increase. We can thus conclude that to the extent that the equilibrium coalition size is affected by the cost function $c(M)$, it is larger the higher the position of the cost function.

The result above is quite intuitive. A higher investment cost required to achieve some level of abatement (i.e., some countries abating) reduces the benefits of a coalition trying to achieve this level of abatement, since there will be more investment costs to cover. Hence the coalition loses from such a cost increase. The coalition can of course adjust its ambition with respect to abatement, but this will only reduce the loss, not eliminate it. The loss to the coalition means that more countries are needed in the coalition in order for the members to have a positive net benefit from being coalition members.

Coalition size and valuations of abatement

An increase in some or all of the valuations v_i can be represented by a positive shift in $W(k)$. This shift is introduced by replacing $W(k)$ by $W(k) + \beta f(k)$, where $f(k) \geq 0$ with a strict inequality for some k, and the parameter β is equal to 0 initially. A shift in the valuations will generally also affect the number of countries who want to abate at any given cost $c(M)$. Hence the function $m(M)$ also gets a positive shift to $m(M)+\beta r(M)$, where $r(M)\geq 0$. Inserting $W(k)+\beta f(k)$ and $m(M)+\beta r(M)$ into (1) and differentiating with respect to β gives (using the envelope theorem)

$$(8)\quad \frac{dV(k)}{d\beta}=m(M)f(k)+W(k)r(M) \text{ for } k<m(M)$$
$$\frac{dV(k)}{d\beta}=m(M)f(k)+[W(k)r(M)-c(M)r(M)]r(M) \text{ for } k>m(M)$$

which is non-negative for all k, since $W(k)>c(M)=v_{m(M)}$ for $k>m(M)$.

In other words, any positive shift in some of the valuation parameters will either leave the function $V(k)$ unchanged or it will increase. If $V(k)$ is unchanged there will be no change in the coalition size. If, however, it is increased, the maximal value of k giving $V(k)=0$ will decline. If this change is sufficiently large, the equilibrium coalition size will decline. We can thus conclude that to the extent that the equilibrium coalition size is affected by a valuation parameter v_i, it is smaller the larger this valuation parameter.

The result above is quite intuitive. Higher valuations of abatement among coalition members increase the benefits of the coalition for any given amount of abatement. Moreover, higher valuations of abatement may induce more countries to abate for any given abatement cost; this will also be beneficial to the coalition. The increased benefits to the coalition countries mean that fewer countries will be needed in the coalition in order for the members to have a positive net benefit from being coalition members.

Investment and abatement

The number of countries abating will be lower the higher is the cost $c(M)$, and for any given abatement cost $c(M)$ the number of countries abating will be higher the higher the valuations of the countries. However, changes in either the cost function $c(M)$ or the countries' valuations of abatement will generally change the equilibrium value of M. As we shall see in the next section, it is not obvious in which direction M moves, and it is therefore not possible to say how the equilibrium abatement depends on the valuations and the cost function for the general case. To be able to shed some light on this issue we therefore proceed by considering a special case of the general model used so far.

Model with two types

In the rest of this chapter we consider the special case of only two types of country: one with "high" valuation h of abatement and one with "low" valuation $l(<h)$ of abatement. There are $n\in[0,N]$ of the h-types. Compared with the notation above, we thus have $v_1=v_2=\ldots=v_n=h$ and $v_{n+1}=v_{n+2}=\ldots=v_N=l$.

From the cost function $c(M)$ we can find two endogenous investment levels M_1 and M_2 defined by $c(M_1)=h$ and $c(M_2)=l$, as illustrated in Figure 14.1. The investment level M_1 is just sufficient to make all h-types abate, while $M_2(> M_1)$ is just sufficient to make all countries abate.

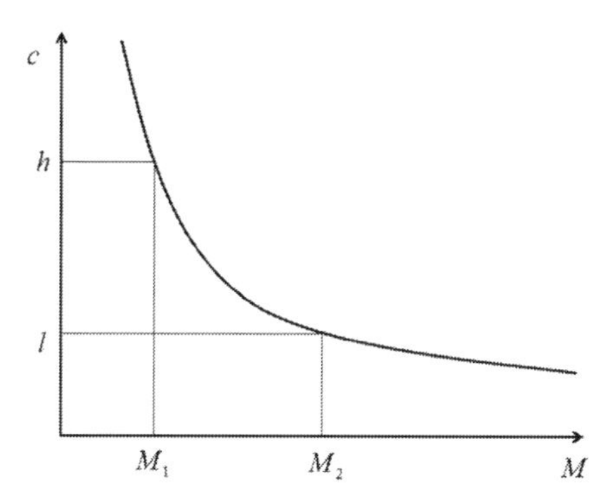

Figure 14.1 The cost function

A coalition of k countries has three relevant options. The first option is trivial; it is characterized by zero investment and hence no abatement. The two non-trivial options are to invest M_1 and achieve abatement by the n h-countries (henceforth called partial abatement), or to invest M_2 and achieve abatement by all countries (henceforth called full abatement).

Consider first the case of investing M_2, and hence achieving full abatement. The payoff to the coalition depends on whether k is smaller or larger than n, and is given by

(9) $V^F(n,k)=k[Nh-l]-M_2$ for $k\leq n$

(10) $V^F(n,k)=n[Nh-l]+(k-n)[Nl-l]-M_2$ for $k>n$

The curve for $V^F(n,k)$ is increasing in k in the (k,V) space, with a kink at $k=n$. At $k=n$ the slope of the piecewise linear curve drops from $Nh-l$ to $Nl-l$.[1]

If the coalition instead invests only M_1 it only achieves partial abatement. As in the case above, the payoff depends on whether k is smaller or larger than n, and is given by

(11) $V^P(n,k)=k[nh-h]-M_1$ for $k\leq n$

(12) $V^P(n,k)=n[nh-h]+(k-n)nl-M_1$ for $k>n$

The curve for $V^P(n,k)$ is increasing in k in the (k,V) space, with a kink at $k=n$. At $k=n$ the slope of the piecewise linear curve changes from $nh-h$ to nl.[2]

Given that a coalition maximizes its payoff, we get (ignoring the possibility of achieving 0 by not investing)[3]

(13) $\tilde{V}(n,k)=\max[V^F(n,k),V^P(n,k)]$

This payoff is piecewise linear and increasing in k, and typically has two kinks; one at $k=n$ and one at $V^F=V^P$. The stable coalition k^* is the smallest integer satisfying $\tilde{V}(n,k)\geq 0$.

The possible equilibria are illustrated in Figures 14.2–14.5. In all figures the V^F curve starts at $-M_2$ and increases with k, with a kink at $k=n$. The V^P curve

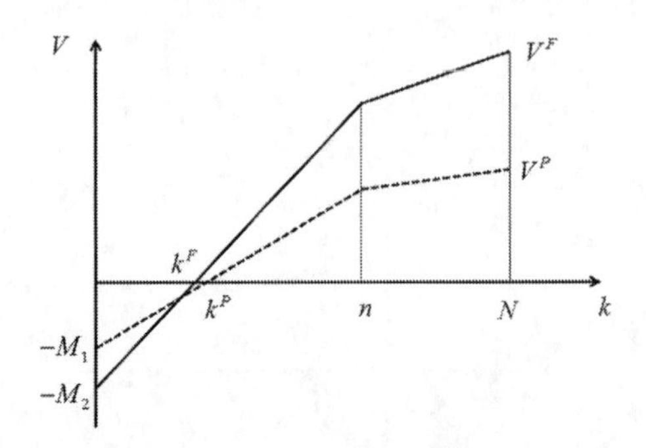

Figure 14.2 Coalition of only h-countries: full abatement

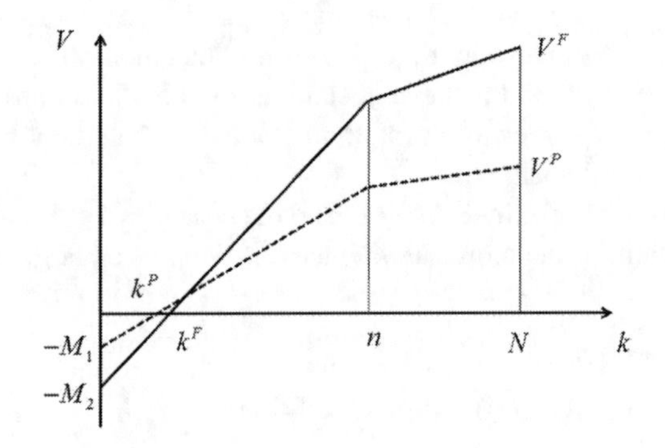

Figure 14.3 Coalition of only h-countries: partial abatement

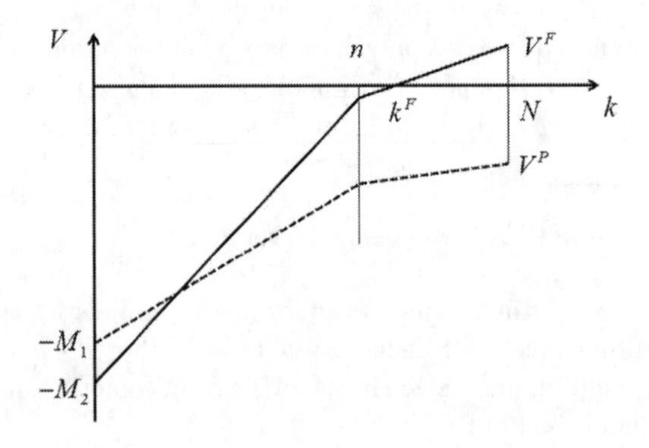

Figure 14.4 Coalition of all h-countries and some l-countries: full abatement

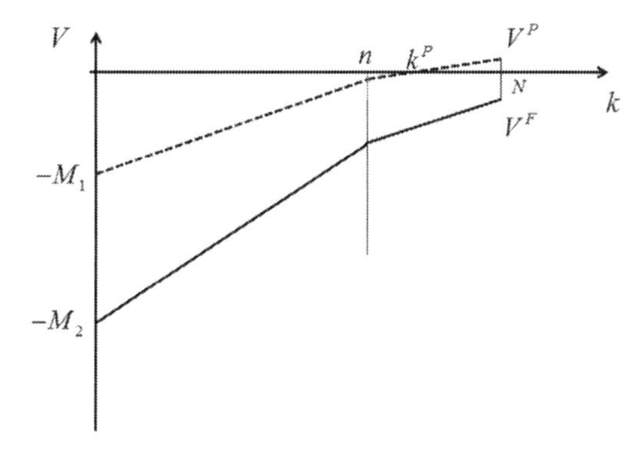

Figure 14.5 Coalition of all *h*-countries and some *l*-countries: partial abatement

starts at $-M_1$ and increases with k, with a kink at $k=n$. For all values of k the V^F curve is steeper than the V^P curve. The value function $\tilde{V}(n,k)$ is the piecewise linear curve equal to the maximum of V^F and V^P. Our assumption $V(N)>0$ implies that at least one of the two curves must intersect the horizontal axis at some k. We define the values k^F and k^P by $V^F(n,k^F)=0$ and $V^P(n,k^P)=0$, respectively. From (9) to (12) we hence have

$$(14) \quad k^F = \frac{M_2}{Nh-l} \text{ for } k \le n$$

$$(15) \quad k^F = n + \frac{M_2 - n(Nh-l)}{(N-1)l} \text{ for } k > n$$

$$(16) \quad k^P = \frac{M_1}{(n-1)h} \text{ for } k \le n$$

$$(17) \quad k^P = n + \frac{M_1 - n(n-1)h}{nl} \text{ for } k > n$$

From the previous section we know that an internally stable coalition is given by the smallest integer k^* satisfying $k^* \ge \min[k^F, k^P]$. In Appendix A we show that this is the only possible stable equilibrium under reasonable conditions.

Consider first Figure 14.2. In this figure we have $k^F < k^P$ and $k^F < n$. Hence the stable coalition k^* in this case consists only of *h*-countries, and they invest so much that full abatement is achieved.

In Figure 14.3 we have $k^P < k^F$ and $k^P < n$. Also in this case, therefore, the stable coalition k^* consists only of *h*-countries. However, in this case the coalition invests only M_1, so that only *h*-countries abate in equilibrium.

In Figure 14.4 $V^P(n,k) < 0$ for all k, and $k^F > n$. (The properties of the equilibrium would be the same if we had assumed instead $V^P(n,k^P)=0$ for some $k^P \in (k^F, N)$.) Hence the stable coalition k^* in this case consists of all *h*-countries and some *l*-countries, and they invest so much that full abatement is achieved.

In Figure 14.5 $V^F(n,k) < 0$ for all k, and $k^F > n$. (The properties of the equilibrium would be the same if we had instead assumed $V^F(n,k^F)=0$ for some $k^F \in (k^P, N)$.) Hence the stable coalition k^* in this case consists of all *h*-countries and some

l-countries, and the coalition invests only M_1, so that only h-countries abate in equilibrium.

The size of the coalition and, more importantly, the equilibrium amount of abatement depend on both the properties of the cost function $c(M)$ and the preference parameters (h, l, n). The next section discusses how properties of $c(M)$ and the preference parameters affect the equilibrium coalition size, while the determinants of the amount of abatement are discussed in the section after that.

Determinants of the coalition size

As explained in the previous section, the size of the coalition is determined by the intersection point between $\tilde{V}(n,k)$ and the horizontal axis, i.e., by the lowest of the values k^F and k^P. We start by considering how k^F and k^P are affected by a shift in the cost function.

From the analysis above we know that this will increase the equilibrium coalition size. This can also be seen directly from Figures 14.2–14.5: from the definitions of M_1 and M_2 it is clear that a positive shift in the cost function $c(M)$ will generally increase both M_1 and M_2. This affects the starting points of the curves for V^F and V^P, but not their slopes. It therefore immediately follows from Figures 14.2–14.5 that the equilibrium size k^* of the coalition must increase. Such a cost increase may therefore also move us from an equilibrium where only h-countries cooperate to an equilibrium where all h-countries and some l-countries cooperate.

The result above is quite intuitive. A higher investment cost required to achieve either partial or full abatement reduces the benefits of a coalition of any given size, since there will be more investment costs to cover. Hence, more countries are needed in the coalition in order for the members to have a positive net benefit from being coalition members.

Consider next a change in the preference parameters. Increasing n, h or l is equivalent to an increase in some v_is in the general case. It therefore follows from the analysis above that the equilibrium coalition size either remains unchanged or declines. To see whether the coalition size is independent of or increasing in n, h or l, we can use equation (8). We only study small changes that do not induce a switch from partial to full abatement or vice versa. The next section discusses switches between abatement regimes in more detail.

Consider first an increase in n. For a small coalition ($k<n$) this will increase the number of abating countries, hence $r(k)$ in (8) is positive. This means that $dV(k)/dn>0$, so that the equilibrium coalition size goes down. For a large coalition all countries abate, so $r(k)=0$. However, in this case the change in the valuation from l to h for one or more countries increases $W(k)$, i.e., $f(k)>0$. Therefore $dV(k)/dn>0$ also in this case, so that the equilibrium coalition size goes down.

Consider next an increase in h. Whatever the size of the coalition, this increases $W(k)$, since there are always some h-countries in the coalition. This means that $f(k)>0$, implying $dV(k)/dn>0$, so that the equilibrium coalition size goes down.

Finally, consider an increase in l. For a large coalition ($k>n$) this leads to a higher value of $W(k)$, i.e., $f(k)>0$. This means that $dV(k)/dl>0$, so that the equilibrium coalition size goes down. If the coalition is small ($k<n$), $W(k)$ is unaffected by an increase in l, so $f(k)=0$. Since both partial and full abatement are independent of the values of l under consideration, the number of countries abating is independent of l, hence $r(k)=0$. According to (8) this leaves $V(k)$ unchanged. However, the analysis leading to (8) only gives first-order effects. If the small coalition is investing M_2 in order to induce full abatement, its value function $V(k)$ will be unaffected by l if the coalition leaves its investment unchanged and higher than necessary. However, the increase in l implies that the investment needed for full abatement goes down. This gives the coalition a benefit, so that $V(k)$ in fact increases in this case.

To conclude, with one exception the optimal coalition size declines as a response to an increase in n, h or l. The exception is that an increase in l has no effect on a small coalition that is only investing enough to achieve partial abatement.

Determinants of abatement

This section discusses how properties of $c(M)$ and the preference parameters affect abatement. We start by considering the cost function.

The properties of the cost function $c(M)$ will obviously generally affect whether we get an equilibrium with full or only partial abatement. From Figures 14.2–14.5 or equations (14–17) we immediately see the following:

- If the cost function $c(M)$ changes so that M_1 increases while M_2 remains unchanged, k^P will increase while k^F will remain unchanged. Hence such a shift in the cost function may move us from an equilibrium with partial abatement to an equilibrium with full abatement.
- If the cost function $c(M)$ changes so that M_2 increases while M_1 remains unchanged, k^F will increase while k^P will remain unchanged. Hence such a shift in the cost function may move us from an equilibrium with full abatement to an equilibrium with partial abatement.
- If the cost function $c(M)$ changes so that M_1 and M_2 increase by the same amount, k^F and k^P will both increase, but k^P will increase most since the V^P curve is flatter than the V^F curve. Hence such a shift in the cost function may move us from an equilibrium with partial abatement to an equilibrium with full abatement.

We turn next to the preference parameters and start by considering n (the number of h-countries). First consider the situation in which the stable coalition consists only of h-countries. Partial abatement can only be an option for a coalition of h-countries if the number n of h-countries is sufficiently large so that the total benefits (net of abatement costs) $n(n-1)h$ are higher than the investment costs M_1. This implies that in this situation only values of n that are larger than n^* have to be considered, where $n^*>0$ satisfies $n^*(n^*-1)h=M_1$ or

$$n^* = \frac{1+\sqrt{1+4M_1/h}}{2}$$

The size of the stable coalition k^P is the number of h-countries that yields coalitional net benefits just covering the investment costs, so that $k^P = M_1/(n-1)h$ as given by (16). As shown above, a larger number n of h-countries reduces the size of the stable coalition. The reason is that the net benefits of partial abatement per country $(n-1)h$ increase so that fewer h-countries are needed in the stable coalition to just cover the investment costs M_1.

Full abatement provides net benefits $(Nh-l)$ to each of the h-countries. This implies that the size of the stable coalition k^F is the number of h-countries that yields coalitional net benefits just covering the investment costs M_2, so that $k^F = M_2/(Nh-l)$ as given by (14). Note that k^F is independent of the total number n of h-countries. The switch from partial abatement to full abatement occurs at the value of n where $k^F = k^P$. It is clear that this happens when the investment costs M_2 are sufficiently small, or when the valuation l is sufficiently large. It is interesting, however, that it will be harder to achieve this switch by lowering M_2 when the number n of h-countries gets larger. The reason is that the coalition has a lower incentive to induce full abatement because it receives more net benefits from partial abatement. The situation is depicted in Figure 14.6, where k^P and k^F are drawn as functions of n, and where n^* denotes the minimal number of h-countries needed to have partial abatement as an option for a coalition consisting of only h-countries:

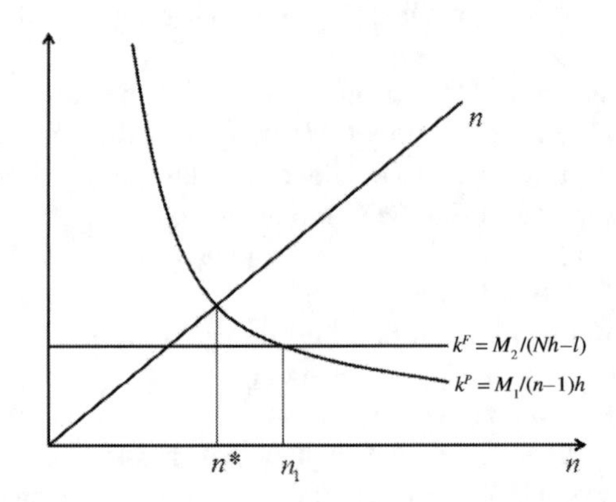

Figure 14.6 Switch from full abatement to partial abatement: only h-countries

The curves k^P and k^F only intersect for $n > n^*$ when

$$\frac{M_2}{Nh-l} < n^*$$

In fact we are employing the condition $k^P = k^F < n$. The point n_1 in Figure 14.6 is defined by the intersection of the curves k^P and k^F. From (14), (16) and $k^P = k^F$, we get

$$n_1 = 1 + \frac{(Nh-l)M_1}{hM_2} \Leftrightarrow M_2 = \frac{(Nh-l)M_1}{h(n_1-1)} \tag{18}$$

so that an inverse relationship between M_2 and n results. The lower M_2, the further to the right lies the switch point n_1. For $n > n_1$ we have that $k^P < k^F$, so that only partial abatement occurs. For $n^* < n < n_1$ we have that $k^F < k^P$, so that full abatement occurs.

Note that full abatement with only h-countries is also achieved for values of n below n^* as long as $M_2 < Nh - l)n$. Otherwise, some l-countries are needed in the stable coalition to cover the investment costs M_2.

Consider now the situation that the stable coalition consists of all h-countries and some l-countries. This situation is more complicated than the previous one. The switch points are determined by:

$$k^P = k^F \geq n, k^P = n + \frac{M_1 - n(n-1)h}{nl}, k^F = n + \frac{M_2 - n(Nh-l)}{(N-1)l}$$

Again, lowering the investment costs M_2 will move k^F below k^P and induce a shift from partial abatement to full abatement. However, the form of the relationship between M_2 and n that determines the switch points is not immediately clear, because both k^F and k^P decrease when n increases, as was seen in the previous section. It is shown in Appendix B that in this situation a decreasing relationship between n_1 and M_2 also holds and is given by

$$(19) \qquad n_1 = \frac{M_2 - (N-1)h - \sqrt{(M_2 - (N-1)h)^2 - 4(h-l)(N-1)M_1}}{2(h-l)}$$

Summarizing, we can draw a graph in the (n, M_2)-plane as illustrated in Figure 14.7.

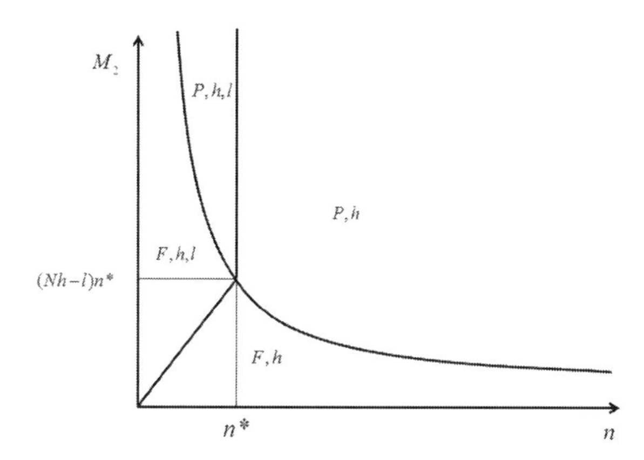

Figure 14.7 Switch from full abatement to partial abatement

When $n > n^*$ we get partial agreement with only h-countries in the stable coalition unless M_2 becomes sufficiently small to induce a shift to full abatement, as we have seen in the first part of the analysis above. Moreover, we have seen in that part of the analysis that fixing M_2 and decreasing n below n^* requires at some value of n to add l-countries to the stable coalition that achieves full abatement. This value is determined by $M_2=(Nh-l)n$. The upper-left part of the figure shows the switch points from partial abatement to full abatement in the case where the stable coalition consists of all h-countries and some l-countries. This curve was derived in Appendix B.

Note that Figure 14.7 does not contain the upper limits of n and M_2. From (11) it follows that in order to be able to achieve full abatement with two types of countries, M_2 needs to be smaller than $(Nh-l)n+(N-1)l(N-n)$. This line is upward sloping, starts at $N(N-1)l$ and meets the continuation of the line $(Nh-l)n$ at $n=N$. This observation implies that the area with full abatement in Figure 14.7 above this line does not occur, but the switches between the different areas remain intact.

Figure 14.7 shows that we need a sufficiently low investment level to make it worthwhile to invest to achieve full abatement, given the number of h-countries, which is to be expected. More interestingly, however, it also shows that given the investment level M_2, we need a sufficiently low number n of h-countries to achieve full abatement. Otherwise the stable coalition will prefer partial abatement.

Finally we need to say what happens to Figure 14.7 when the parameters M_1, h and l change. It is easy to see that an increase in M_1 only (meaning that the cost function only shifts out around h) moves the whole figure out: n^* increases and both curves, given by (19) and (20), shift out. This implies that the area where full abatement occurs becomes larger, which is to be expected since the investment costs of partial abatement are larger.

The effect of an increase in h, and therefore a decrease in M_1, is more complicated. The direct effect of an increase in h is that the lower part of the curve in Figure 14.7 shifts out, because n^* decreases, the slope of the line $(Nh-l)n$ increases and from (19):

$$\frac{\partial M_2}{\partial h}=\frac{\partial}{\partial h}\frac{(Nh-l)M_1}{h(n-1)}=\frac{lM_1}{h^2(n-1)}>0.$$

However, combined with the indirect effect of the decrease in M_1, the total effect of an increase in h is not clear.

The effect of an increase in l, and therefore a decrease in M_2, is also not clear. The direct effect of an increase in l is that the lower part of the curve in Figure 14.7 shifts in, n^* does not change and the slope of the line $(Nh-l)n$ decreases. However, this has to be interpreted for a lower M_2 and therefore the total effect of an increase in l is not clear.

For a given value of n, we have two possible values of abatement: full ($=N$) or partial ($=n$). When n varies, there is a larger range of possible values of abatement. This is illustrated in Figure 14.8, based on a given set of values for (M_1, M_2, h, l). For values of n up to n_1 there is full abatement, i.e., abatement equal to N. As n passes n_1 abatement drops to n_1. As n increases further toward N, abatement also increases toward N.

Notice that Figure 14.8, like the rest of the analysis above, was based on the assumption that the value function for the coalition is positive if the coalition is sufficiently large, i.e., $V(N)>0$. This is an implicit assumption on the sizes of the

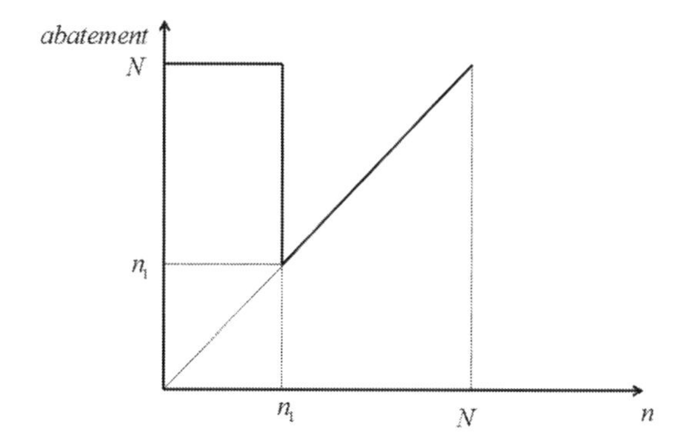

Figure 14.8 Full abatement if the number of *h*-countries is small

elements in the vector (M_1, M_2, *n*, *h*, *l*). Not all combinations of values will satisfy $V(N)>0$. For Figure 14.8 to be valid we must have $V(N)>0$ even if $n=0$. From (9)–(13) it follows that $V^P(0,N)<0$. To achieve $V(N)>0$ even if $n=0$ we must therefore have $V^F(0,N)>0$. From (10) we see that this holds if $M_2<N(N-1)l$. Since M_2 is higher the lower is *l*, this inequality is less likely to hold the lower is *l*. If the inequality does not hold, we will have zero abatement for *n* sufficiently low.

An equilibrium with positive abatement will occur when *n* is large enough to make either V^F $(n,N)>0$ or V^P $(n,N)>0$. The critical value of *n* for positive abatement, denoted n_0, is hence given by $n_0 = \min\left[n_0^F, n_0^P\right]$, where n_0^F and n_0^P are defined by $V^F(n_0^F, N)=0$ and $V^P(n_0^P, N)=0$. From (10) and (12) it follows that

$$n_0^F\left[Nh-l\right]+(N-n_0^F)\left[N-1\right]l-M_2=0$$
$$n_0^P\left[N_0^P h-h\right]+(N-n_0^P)n_0^P l-M_1=0$$

Solving, we obtain

$$n_0=\min\left[\frac{M_2-N(N-1)l}{N(h-1)}, \frac{-(Nl-h)+\sqrt{(Nl-h)^2+(h-l)M_1}}{2(h-l)}\right]$$

Figure 14.9 illustrates the case when the first of the two numbers in square brackets is the smaller of the two, while Figure 14.10 illustrates the opposite case.

A numerical illustration

To take a numerical example, suppose that $h=2$, $l=1$, $N=20$, and the cost function is $c(M)=\gamma/M$. With this cost function we have $M_1=\gamma/h$ and $M_2=\gamma/l$. From the previous section we know that Figure 14.8 is valid for $M_2<N(N-1)l$; inserting the numerical values for (*N*, *h*, *l*) from above gives $\gamma<380$. If, for example, $\gamma=250$, it follows from that n_1 in Figure 14.8 is equal to 10.8. In other words, abatement is equal to 20 for $n\in[0,10]$ and equal to *n* for $n\in[11,20]$. For higher values of γ we get either Figure 14.9 or Figure 14.10. If, for example, $\gamma=500$, we have Figure 14.9 with $n_0=6$ and $n_1=10.5$, and if $\gamma=1000$ we have Figure 14.10 with $n_0=15.1$.

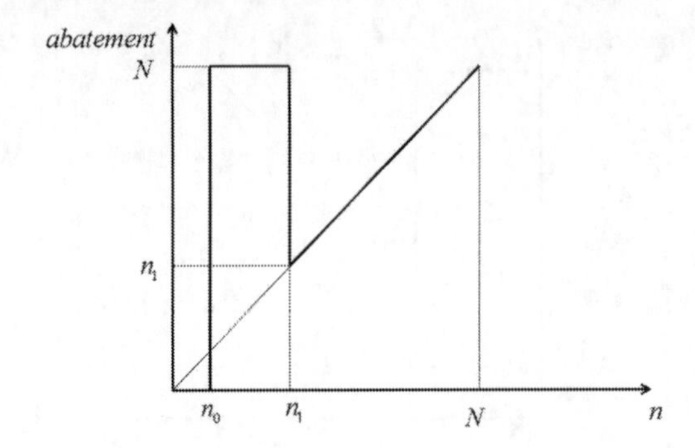

Figure 14.9 Full abatement if the number of *h*-countries is intermediate

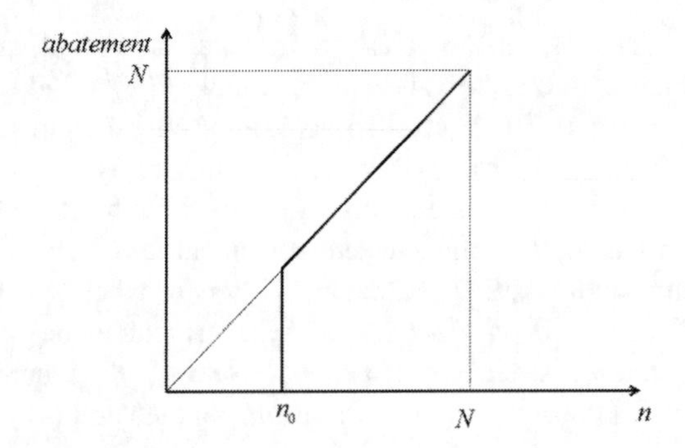

Figure 14.10 Never full abatement

Consider the case of $\gamma=500$ in more detail: For $n<6$ there is no investment in R&D and no abatement. For $n\in[6,10]$ there is a coalition of the *n* *h*-countries and some *l*-countries investing M_2, hence giving full abatement. For $n=11$ the coalition consists of the 11 *h*-countries and 3 *l*-countries, investing M_1 so that abatement is 11. For $n\in[12,20]$ the coalition will consist only of *h*-countries, investing M_1, giving abatement equal to *n*.

The effect on the critical values n_0 and n_1 of changes in the valuations *h* and *l* for the case of $\gamma=500$ is illustrated in Table 14.1. We immediately see that increasing the valuation for the *l*-countries increases the range of *n* giving full abatement, while increasing the valuation of the *h*-countries reduces this range.

Table 14.1 n_0/n_1 for different values of *h* and *l*

l ↓	*h* →	2	3
1		5/10.8	3/7.4
1.5		0/15.4	0/10.8

Concluding remarks

Even without any international agreement on emission reductions, significant emission reductions are possible if abatement costs are sufficiently low. In principle, future abatement "costs" could be negative, i.e., reducing emissions could give benefits even when the effect on the climate is ignored. This would be the case if a form of carbon-free energy with costs lower than the costs of fossil energy is discovered. A more likely scenario is that some future technology will give abatement costs that are positive, but sufficiently low that countries with a valuation of emission reductions exceeding some threshold will use this technology to reduce emissions. This is the situation we have analyzed in this chapter, with an emphasis on heterogeneity across countries with regard to their valuations of emission reductions.

If all countries have a sufficiently low valuation of emission reductions, there will be no emission reductions if such reductions have a positive cost. However, when some countries have a sufficiently high valuation of emission reductions, we have shown that there may be an equilibrium with a coalition of countries undertaking R&D in order to bring down abatement costs, and with some countries non-cooperatively adopting the new technology and hence reducing emissions. This implies that a focus on technology development in international environmental agreements may be successful in terms of emission reductions without the need for a broad participation in the agreement.

One of our results is that the equilibrium size of the coalition will be smaller (or unaffected) the higher is any country's valuation of emission reductions. However, the relationship between aggregate abatement and the countries' valuations of emission reductions is ambiguous. In the numerical illustration above, an increase in the number of high-valuation countries could either reduce or increase aggregate abatement. Increased valuation by the high-valuation countries could reduce abatement, while increased valuation by the low-valuation countries could increase abatement.

In our formal analysis we have ignored all types of uncertainty. In reality, the consequences of a given R&D expenditure for abatement costs, and hence total abatement, will be uncertain. However, introducing uncertainty will not change the analysis. In the general expression for a coalition's payoff, the terms for abatement costs and total abatement must simply be reinterpreted as expected values instead of being deterministic. The analysis for the general case will be unchanged by this reinterpretation. The details of the specific case analyzed above must be modified, but the main conclusions will remain valid.

Acknowledgements

We are grateful to FEEM, who provided us with excellent working conditions while we visited there in October 2012 to work on this chapter. While carrying out this research we have been associated with CREE – Oslo Center for Research on

Environmentally-friendly Energy. CREE is supported by the Research Council of Norway.

Appendix A: stable coalitions larger than k^*?

The condition for a coalition to be internally stable was given by equation (6). Consider first an integer K such that $V^F(n,k) > V^P(n,k)$ for both K and $K-1$. A coalition of size K cannot be internally stable, since the optimal investment (and amount of abatement) are identical for K and for $K-1$. Hence $\varepsilon(K)=0$, so that the condition for internal stability is violated.

Consider instead an integer K such that

$$V^F(n,K) \geq V^P(n,K)$$
$$V^F(n,K-1) < V^P(n,K-1)$$

In this case the value of $\varepsilon(K)$ is equal to the difference between the actual payoff $V^F(n,K)$ to the K countries in the coalition and what they would have gotten by investing M_1 instead of M_2. The latter is simply $V^P(n,K)$. Hence

$$\varepsilon(K) = V^F(n,K) - V^P(n,K)$$

Define k^{**} by $V^F(n,k^{**}) = V^P(n,k^{**})$, implying $K-1 < k^{**} \leq K$. It follows that

$$\varepsilon(K) = (K - k^{**}) \left[\frac{\partial V^F(n,k)}{\partial k} - \frac{\partial V^P(n,k)}{\partial k}\right]$$

From the definitions of $V^F(n,K)$ and $V^P(n,K)$ it follows that

$$\varepsilon(K) = (K - k^{**})[(N-n+1)h - 1] \; for \; k \leq n$$
$$\varepsilon(K) = (K - k^{**})[(N-n-1)l] \; for \; k > n$$

The condition for K to be stable is therefore (from [6])

$$M_1 < (K - k^{**})[(N-n+1)h - 1] \; for \; k \leq n$$
$$M_1 < (K - k^{**})[(N-n-1)l] \; for \; k > n$$

Since $K-k^{**}<1$, a necessary condition for the first inequality to hold is that the square bracket is larger than M_1. A necessary condition for this is in turn that $(N-n+1)h > M_1$, which means that a single h-country would be willing to pay M_1 in order to get $N-n+1$ countries to abate. A necessary condition for the second inequality to hold is that $(N-n-1)l > M_1$, which means that a single l-country would be willing to pay M_1 in order to get $N-n+1$ countries to abate.

We cannot rule out that a coalition of size $K > k^*$ is internally stable if N is sufficiently large. Although a collation of this size is stable by our formal definition, we believe that such coalition sizes are not very relevant from an economic point of view for the following reason: Assume a coalition of size $K > k^*$

is stable. Consider a very small change in either the cost function $c(M)$ or one of the parameters (h, l, n) such that k^{**} increases but remains below K. If such a change makes k^{**} sufficiently close to K, (6) will no longer hold, and K will no longer be a stable coalition. From an economic point of view, a coalition size that depends on integer properties in this manner does not seem to be of particular interest.

Appendix B: the switch points between regimes

The switch points in the second situation are determined by

$$k^P = k^F \geq n, k^P = n + \frac{M_1 - n(n-)h}{nl}, k^F = n + \frac{M_2 - n(Nh-l)}{(N-1)l}$$

This implies that the switch points are determined by the intersections n_l of the quadratic functions

$$f_1(n) = n(M_2 - (Nh-1)n), f_2(n) = (N-1)(M_1 - n(n-1)h).$$

Note that the function f_1 has roots in 0 and $M_2/(Nh-l)$ and the function f_2 has a root in n^* and is maximal in $n = ½$. The situation is depicted in Figure 14.11.

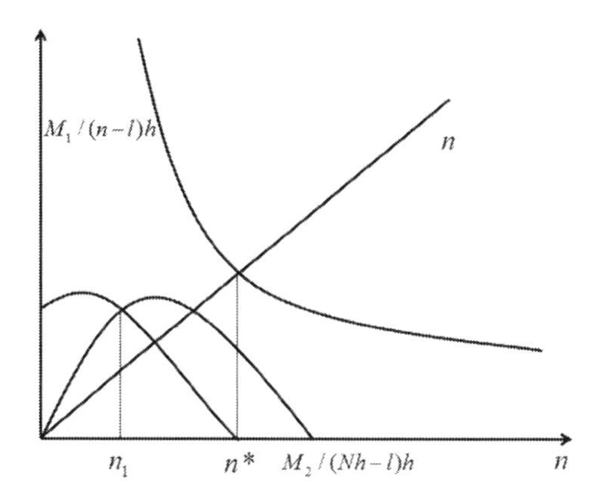

Figure 14.11 Switch from full abatement to partial abatement: both h- and l-countries

For $n_1 < n < n^*$ we have that $k^P < k^F$ so that only partial abatement occurs. For $n < n_1$ we have that $k^F < k^P$ so that full abatement occurs.

The difference of the quadratic functions is given by

$$f_2(n) - f_1(n) = (h-l)n^2 - (M_2 - (N-1)h)n + (N-1)M_1.$$

Since this difference is positive at $n=0$ and the slope is negative at $n=M_2/2(Nh-l)$, it can only have positive roots. The switch point n_l is the smallest root and it is given by

$$n_1 = \frac{M_2-(N-1)h-\sqrt{(M_2-(N-1)h)^2-4(h-l)(N-1)M_1}}{2(h-l)}$$

Note that for $M_2=(Nh-l)n^*$ it follows that $f_1(n^*)=f_2(n^*)=0$ so that $n_1=n^*$.

Since

$$\frac{\partial n_1}{\partial M_2} = \frac{\sqrt{(M_2-(N-1)h)^2-4(h-l)(N-1)M_1}-(M_2-(N-1)h)}{2(h-l)\sqrt{M_2-(N-1)h)^2-4(h-l)(N-1)M_1}}<0,$$

n_1 is decreasing in M_2. It approaches zero as M_2 goes to infinity.

Notes

1 Notice that if our previous assumption that $V(N)>0$ is to hold for all values of n we must have $V^F(0,N)>0$ i.e. $N(N-1)l>M_2$.

2 In the figures it is implicitly assumed that $nl<(n-1)h$.

3 Including the option of not investing gives us the value function defined above, i.e. $V(k)=\max[0,\tilde{V}(n,k)]$.

References

d'Aspremont, C., A. Jacquemin, J. Gabszewicz and J. Weymark (1983). On the stability of collusive price leadership. *Canadian Journal of Economics* 16(1), 17–25.

Barrett, S. (1994). Self-enforcing international environmental agreements. *Oxford Economic Papers* 46 (supplement 1), 878–94.

——(2006). Climate treaties and "breakthrough" technologies. *American Economic Review, Papers and Proceedings* 96(2), 22–25.

Battaglini, M. and B. Harstad (2012). Participation and duration of climate contracts. *Working Paper*, Princeton University.

Benchekroun, H. and A. Ray Chauduri (2012). Cleaner technologies and the stability of international environmental agreements. *Working Paper*, McGill University, Montreal.

Buchholz, W. and K.A. Konrad (1995). Global environmental problems and the strategic choice of technology. *Journal of Economics* 60(3), 299–321.

Buchner, B. and C. Carraro (2005). Economic and environmental effectiveness of a technology-based climate protocol. *Climate Policy* 4(3), 229–48.

de Coninck, H., C. Fischer, R.G. Newell and T. Ueno (2008). International technology-oriented agreements to address climate change. *Energy Policy* 36(1), 335–56.

Finus, M. (2003). Stability and design of international environmental agreements: the case of transboundary pollution. In H. Folmer and T. Tietenberg (eds.), *The International Yearbook of Environmental and Resource Economics 2003/2004: a survey of current issues.*, Cheltenham: Edward Elgar, pp. 82–158.

Goeschl, T. and G. Perino (2012). The climate policy hold-up: how intellectual property rights turn international environmental agreements into buyer cartels for abatement technologies. *Working Paper*, University of Heidelberg.

Hoel, M. and A. de Zeeuw (2010). Can a focus on breakthrough technologies improve the performance of international environmental agreements? *Environmental and Resource Economics* 47(3), 395–406.

Nagashima, M. and R. Dellink (2008). Technology spillovers and stability of international climate coalitions. *International Environmental Agreements* 8(4), 343–65.

15 International guidance for border carbon adjustments to address carbon leakage

Aaron Cosbey and Carolyn Fischer

Introduction

Economists have long argued that the most efficient way to reduce global carbon emissions would be to place a global price on carbon. However, in reality we see no global agreement in sight on either carbon targets or prices. Furthermore, the UN Framework Convention on Climate Change (UNFCCC) enshrines the principle of common but differentiated responsibility (CBDR), which explicitly recognizes that developing countries should not be expected to implement the same kinds of policies as developed countries. In an environment of decentralized climate policy making, for countries that do seek to implement market-based policies for reducing carbon emissions, a major concern will be the potential for international trade to undermine the effectiveness of carbon pricing, the competitiveness of domestic industries, and domestic political support for strong measures. As a result, these countries may seek complementary measures to cope with carbon leakage.

One such measure is border carbon adjustment (BCA): a measure applied to traded products that seeks to make their prices in destination markets reflect the costs they *would have* incurred had they been regulated under the destination market's greenhouse gas (GHG) emissions regime[1]. Applied to imports, the charge would reflect the GHG emissions associated with imported products and the price of emissions faced by comparable products in the destination market. Applied to exports, the adjustment would be a rebate of emissions charges levied in the country of origin. In a seamless system of globally applied BCA, this would be followed by border adjustment in the destination market, with the objective that all products in their destination markets should reflect domestic emissions prices.

While BCA sounds intuitively appealing, and seems to rest on simple principles, in fact the elaboration and implementation of such a regime is devilishly complicated. In no small part this is because the ideal scheme would be forced to serve multiple, often competing, objectives. What follows is a description of one permutation of such a regime, based on the results of a drafting exercise carried out by a small group dedicated to defining best practice in BCA (Cosbey *et al.* 2012).

BCA, of course, is not the only available policy tool to address the competitiveness and leakage issues that national climate policies can engender. The full list also includes special treatment of vulnerable sectors (e.g., free allocation of allowances or wholesale exclusion from the climate policy), international sectoral agreements (under which one or more countries agree to regulate sectors in a similar coordinated manner), and GHG intensity standards. While BCA would need to be weighed as an option against these alternatives, we do not conduct such a comparative assessment here. Our aim is narrower: to describe a BCA regime that can adequately meet four competing objectives:

- environmental effectiveness (reducing leakage);
- administrative feasibility, cost effectiveness;
- good governance; and
- respect for international legal obligations of the implementing state.

With these criteria in mind, we review key design principles for BCAs and suggest the more compatible options.

Motivation

The design of a BCA regime will vary depending on the motivation, and policy makers should be explicit about their goals. BCA can have at least three possible motivations:

1. **Reducing risks of leakage.** Leakage is an increase in GHG emissions in foreign jurisdictions resulting from climate policies taken in an implementing jurisdiction (Metz *et al.* 2007).
2. **Maintaining industry competitiveness.** This motivation concerns the loss of profits, market share, production, investment, and related jobs. Those losses could be due to industry relocating to jurisdictions with lower costs of compliance, industry losing market share to firms from such low-cost jurisdictions, or a diversion of new investment to those same jurisdictions.
3. **Leverage.** BCA, or the threat of BCA, might be used to encourage other countries to adopt policies to reduce GHG emissions.

Preventing leakage – an environmental motivation concerned with making domestic climate policies effective – is the only motivation we recommend as a basis for BCA. Any BCA regime will have to justify itself under the GATT's Article XX General Exceptions, since all BCA violates the GATT's Article III (national treatment) obligations not to discriminate between "like" domestic and foreign products. In the eyes of trade law, goods with very different carbon intensities of production are nonetheless considered "like" by most scholars. The grounds for an exception to the Article III obligations include legitimate environmental pursuits (e.g., conserving exhaustible natural resources; protecting human, animal or plant life or health).

Notably, these grounds do *not* include preserving competitiveness, which is a purely economic concern. Indeed, the WTO was created to counteract the beggar-thy-neighbor behavior that results from attempts to preserve domestic competitiveness through trade barriers. Although energy-intensive, trade-exposed (EITE) sectors may experience disproportionate effects from carbon pricing, it may be difficult to disentangle these effects from more general economic trends causing continued shifts of manufacturing activities from industrialized to emerging economies. Responding to competitiveness motivations through BCA would thwart legitimate economic drivers of comparative advantage and trade.

Leverage – the use of BCA to pressure other countries to take actions to reduce their emissions – is also an inappropriate motivation. Using BCA to encourage adoption of specific measures at the national level, as opposed to BCA aimed at the practices of producers, would undoubtedly conflict with CBDR. Regardless, the effectiveness of BCA for such purposes may be limited, relative to the controversy they would provoke. Export streams represent a small part of total country-level production within the relevant EITE sectors, and an even smaller share of the broader economy. Yet BCA is sufficiently divisive as to risk impairing efforts to achieve multilateral climate agreement. The intense controversy surrounding the EU's aviation emissions levy, which strongly resembles a BCA, testifies to that, as do a number of WTO disputes over unilateral extrajurisdictional action.[2] Using BCA leverage to demand participation in a multilateral agreement seems best suited to be negotiated as part of that agreement, not a unilateral measure.

Scope of applicability

The scope of a BCA's applicability determines which products, sectors, and countries the regime will cover. There are three key elements to this determination: which types of carbon regulations should be eligible for adjustment; which goods and sectors should actually be subject to adjustment; and from which countries or sectors should goods be exempt from coverage.

Policies eligible for adjustment

BCA should be used to protect only those sectors or products regulated with price-based climate policies, which offer a clear carbon price on which to base the adjustments. Sectors or products regulated with non-price-based policies should not be covered because it is impossible to transparently calculate the costs associated with the policy. Further, non-price-based policies do not require that producers pay for the remaining embodied carbon in their products, which is what BCAs are designed to adjust for.[3]

Identifying goods/sectors to be covered

Determining which products or sectors in the implementing country should be covered by a BCA scheme involves determining which products are actually at risk of leakage. In general, to prevent leakage, one would err on the side of broad coverage. For a number of reasons, however, we believe that narrower coverage is preferable. For example, applying BCA to sectors with low vulnerability will yield limited benefits relative to the administrative costs involved.[4] In addition, coverage that includes goods with long, complex supply chains will make it difficult to prevent trans-shipment, as noted above. Overly broad coverage also risks trade law violations as it constitutes support for domestic firms and sectors beyond what can be justified by environmental objectives.

To assess sectoral vulnerability, any workable regime must use a system that is simple enough to be operational and transparent, based on available data. Two criteria should be used simultaneously to avoid overly broad sectoral coverage:

- The first criterion should establish that the GHG regulations would result in substantially higher production costs for the sector in question, calculated as the tonnes of GHG emitted by the sector (including first-order indirect emissions from electricity) multiplied by the projected emissions tax or allowance price. These costs should then be evaluated relative to the economic size of the sector, as measured by value added. This ratio reflects the sector's GHG intensity.
- The second criterion should establish that attempts to pass increased costs to consumers would result in significant shifts of consumption to foreign sources. The ideal indicator for this criterion would be trade sensitivity – the degree to which cost increases would lead to a substitution to products sourced from abroad. Because reliable metrics for trade sensitivity are not generally available, one could use, as a proxy, trade exposure, measured as the value of imports and exports in the sector relative to total production plus imports.

Variants of these criteria have been used for preferential treatment in actual and proposed climate policies. The Waxman-Markey cap-and-trade bill used thresholds of 5 percent energy or CO_2 intensive (energy costs generally being a weaker hurdle) and 15 percent trade intensive; however, less trade-intensive sectors were also eligible if their energy intensity exceeded 20 percent. The result would make 44 of nearly 500 manufacturing industries presumptively eligible, mostly within the chemicals, paper, non-metallic minerals (e.g., cement and glass), or primary metals (e.g., aluminum and steel) sectors.[5] By way of comparison, the European Union used less stringent criteria for its benchmarking provisions, including a broader group of eligible sectors, with a resulting 201 products designated vulnerable to leakage.

Exemptions

Any national-level exemption raises two concerns. One is potential incompatibility with the GATT's Article I obligation for most-favored nation (MFN) treatment, which requires that no nation be favored above any other in the treatment of imported goods. However, the exemption might be justified under GATT's Article XX, which, as noted above, allows states to take otherwise-illegal measures that are aimed at, among other things, genuinely protecting the environment.

The other concern is trans-shipment problems; strong provisions would be required to ensure that any products coming from the exempted country have undergone a substantial transformation there. Otherwise, non-exempted countries could ship products to exempted countries for re-export to avoid coverage. Such provisions work best when the goods in question are wholly obtained in a single country or at least have a very simple supply chain. This argues for a high threshold for coverage of goods/sectors, which would, in effect, preclude all but a small handful of EITE goods such as cement, aluminum, iron and steel and some chemicals.

A number of possible exemptions are commonly considered for a BCA regime. We examine each exemption in light of the four criteria identified above.

Exempting countries that are party to a multilateral agreement on climate change to which the implementing state is party. If used for leverage, this exemption could backfire and make international agreement less likely. On the other hand, *not* employing this exemption may violate the principle of CBDR, as operationalized by the multilateral agreement in question. As such, this exemption may actually reduce international political friction.

Exempting countries that implement a national emissions cap. An effective national cap theoretically precludes leakage (even if emissions rise in EITE sectors). However, many formulations of emissions caps would be ineffective, allowing for leakage (examples include generous provisions for offsets, low price ceilings, caps that are not economy-wide, or lacking enforcement). A powerful case is made for exemption under GATT's Article XX because of the strong relationship between the defining national characteristic (an emissions cap) and the environmental objective (preventing leakage).

Exempting countries that take "adequate" national actions other than national caps. Any national climate regime other than a hard cap is susceptible to leakage. This exemption poses the challenge of defining *ex ante* what constitutes adequate action. Note that adjustments can be made for price-based mechanisms (like a carbon tax or allowance price) in the calculation of the BCA (see "Modification to Adjustment Levels" below); in contrast, exemptions are a much blunter instrument. Meanwhile, non-price-based mechanisms (including renewable energy standards and subsidies) can reduce emissions but the effectiveness and cost burdens are non-transparent. Still, one could use this exemption to bring the BCA regime into greater coherence with the principle of CBDR (and special and differential treatment [S&DT], the trade law equivalent).

Exempting sectors from countries that implement a sectoral cap. If a country effectively caps a given sector's emissions, no sector-level leakage will occur.

This exemption poses no trade law problem with non-discrimination, since the discrimination is based on sector, rather than country, characteristics.[6] Adjustments for sectoral carbon pricing (or export taxes) can also be made in the BCA calculations.

Exempting least developed countries (LDCs) and low-income countries (LICs). An exemption for LDCs and LICs would help the measure align with the UNFCCC principle of CBDR, and the WTO principle of S&DT, and definitions of such countries are standardized by the World Bank. However, very few of these countries export the type of EITE goods targeted by BCA. Moreover, this exemption creates problems with MFN treatment, and it is unlikely to be carved out by the WTO's Enabling Clause, which exempts some forms of developing country tariff treatment (only those intended to aid development in the target countries) from MFN obligations.[7]

Exempting countries by means of administrative flexibility. This would enable the implementing government to exempt certain countries from coverage, presumably based on broader public policy objectives. This exemption, which would conflict with the GATT's MFN provisions, lacks the predictability that should be the hallmark of any scheme.

Based on the considerations described above, we recommend the following exemptions for a BCA regime:

- Exemptions for countries adhering to a multilateral agreement on climate change to which the implementing state is also party.
- Exemptions for countries with an effective national emissions cap and for sectors with an effective sectoral cap.
- Exemptions for LDCs and LICs *if allowable under the WTO's Enabling Clause.*
- Calibrated credit (as opposed to outright exemptions) for national or sectoral actions in the case of price-based regimes, but no credit or exemptions for non-price-based actions (see "Modification to Adjustment Levels" below).

Determining the level and type of adjustment

Any BCA regime will need to elaborate how it calculates the adjustment it will assess on covered products. This requires first determining (or estimating) the amount of embodied carbon in a given product. Second, it requires calculating the level of adjustment, applying any necessary exemptions and deciding what form of adjustment to use, as well as how to price the embodied carbon in the adjustment.

Assessing the carbon content

Assessing carbon content involves setting system boundaries, determining which benchmarks to use in place of actual emissions data where necessary, and using accurate data reported to agreed-upon protocols.

The system boundary

The system boundary (i.e., what to include in calculating a product's carbon footprint) should always include *direct emissions* from a production process (scope 1 emissions[8]). Whether one should further hold the exporter responsible for emissions associated with other components of the product's life cycle depends on how significant the inclusion of those GHG emissions would be, whether the GHG emissions are already accounted for within another sector, and the practicality of collecting robust data.

Indirect emissions – emissions resulting from production but occurring at sources not owned by the producer – can be divided into energy-related emissions (scope 2) and other indirect emissions (scope 3). Scope 2 emissions (electricity, steam or heat generated off-site and purchased) sometimes represent the majority of emissions from processes such as metal smelting and represent a material share of total GHG emissions from sectors like steel and cement. In addition, such emissions are a significant portion of many national emissions inventories, so should probably be covered by any national climate policy; this would, in turn, necessitate coverage under a BCA regime to prevent leakage. As such, we recommend the inclusion of scope 2 emissions within the system boundary. Scope 3 emissions (any indirect production-related emissions not covered by scope 2) should be excluded from the system boundaries because the calculations would be complex and would require nonexistent data or benchmarks. Moreover, scope 3 does not tend to be a significant source of emissions.

The reasoning is similar for excluding emissions from the transportation of products to market, or from the consumption and disposal of products, which are further outside the scope. The calculations are excessively complex, the share of emissions in the industries of concern is relatively small, and there is no consensus on the extent to which responsibility to reduce these supply-chain emissions should lie with the manufacturers. Finally, system boundaries applied to exporting country producers cannot exceed those applied to implementing country producers.

The benchmarks

Ideally, actual emissions would be observed and verified by a third party subject to a protocol or standard. However, if exporters are unable or unwilling to provide such data, benchmarks should be used, aiming to capture carbon content as accurately as possible without being punitive. One question is how many benchmarks to have for a given product. Trade law considers identical commodities produced in different ways as "like," and so deserving of similar treatment, i.e., a single benchmark.[9] However, multiple benchmarks might be needed where multiple production processes can be used to produce a single product. For example, steel can be made from iron ore using a process starting with a blast furnace, or from scrap steel using an electric arc furnace; the two have vastly different GHG intensity profiles and thereby different leakage potential.

The second question is the level of the benchmark. A BCA only gives exporting firms an incentive to improve their performance if doing so lowers their adjustment. With benchmarks setting the default adjustment, incentives to provide data and establish superior performance exist only for firms that can outperform the benchmark. A low benchmark – like best available technology, or average emissions of a clean importer – provides little or no incentive for improvement, while a high benchmark – like worst available technology – encourages improvements but risks seeming punitive and, because it encourages submission of actual data, entails the additional costs of third-party certification.

The third question is whether the benchmark should differentiate among exporters. Exporting country data may not be readily available or verifiable, and gathering such data may be arduous. Country-specific benchmarks may conflict with GATT's MFN provisions and would require provisions to prevent trans-shipment from countries assigned higher-intensity benchmarks. Uniform benchmarks, such as those based on importing country characteristics, are calculated simply and avoid problems of discrimination, but they are less able to accurately reflect the actual emissions intensity.

Balancing these trade-offs of accuracy, incentives, and legal issues, we recommend a hybrid benchmark. While scope 1 emissions tend to be roughly similar for EITE sectors globally, scope 2 emissions tend to vary considerably based on country characteristics. A hybrid benchmark would assess a uniform benchmark – such as average implementing country practice – to scope 1 emissions, and subject scope 2 emissions to a benchmark that reflected exporting country practice, such as average GHG intensity. Such a system performs well at preventing leakage since it allows the more stringent benchmark to focus only on those areas (scope 2) with the most potential for regional variation – as well as the most potential for mitigation. It also avoids the data challenges of firm-specific scope 1 emissions. A hybrid scheme of this type would still face GATT MFN problems, but would probably be more defensible than a pure exporting country benchmark.

In any case, producers should be able to provide third-party-verified firm-level data on emissions intensity, using the same system boundaries used for implementing country producers. This increases the odds that any scheme will be found legal under WTO rules and provides incentives to producers to improve their processes. Using international standards and protocols where available, both for data submission and benchmark creation, would help ensure compatibility with WTO rules and may reduce administrative burdens.

To counter the negative impact of compliance costs, implementing states should offer support – in the form of financial and technical assistance in accounting, reporting, and verification – to assist foreign covered exporters in submitting verified individual data.

Modifications to the adjustment level

Since a BCA system asks imports to pay comparable emissions prices, it must recognize carbon payments made by exporting firms complying with market-based climate policies in their home countries. Such credit would probably be critical to allowing any BCA regime to pass the legal threshold presented by GATT Article XX.[10] Using modifications to the level of adjustment, rather than more blunt country-based exemptions, helps ensure that the BCA regime levied adjustments only to the extent necessary to offset the differential between the foreign climate policies and the domestic climate policies; this is the ideal.

For the same reasons that BCA should not be allowed for sectors or products regulated with non-price-based policies, such measures should not generate adjustments to the BCA: no comparable carbon price is transparent, nor are producers required to pay for the remaining embodied carbon in their products.

A BCA system must also recognize any free allowances or other compensatory mechanisms that shelter domestic firms and offer comparable benefits to covered imports. In some cases, the level of BCA may be adjusted down to zero.

Special benchmarks or calibrations could be developed for non-exempt LDCs to respect the principle of CBDR. The importing country could assume, for example, that all imports from LDCs have used BAT. Such special treatment would require trans-shipment provisions.

Type and price of adjustment

Adjustments may be in the form of levies, particularly in concert with a carbon tax, or allowance purchase requirements, when implemented with cap-and-trade. One alternative would be to allow importers or foreign producers to purchase international carbon offsets up to the determined value of adjustment.

If implementing country firms are regulated with a carbon tax, the carbon price they pay should be the basis for the price charged to exporting country firms. To avoid unpredictable swings in price, the price should be set on a regular and infrequent basis – say, annually – based on a rolling average of previous periods of measurement.

If the implementing country uses cap-and-trade, exporting country firms should be regulated such that they come as close as possible to purchasing allowances or offsets on the same terms offered to their competitors in the implementing country.

The application of BCA to exports

True destination-based carbon pricing would relieve exports from implementing countries of the burden of carbon payments associated with their production. Adjustment for exports would avoid the equivalent of double taxation for products being shipped to a destination state that also applied BCA to its imports. Export adjustment would also help avoid leakage from loss of market share in third country markets, making exports from regulating countries less disadvantaged in those markets relative to products from non-regulating countries.

To date, policy makers have preferred to focus on adjustment for imports only, probably at least in part because of the unclear legal status of BCA for exports under WTO law. Many legal scholars agree that it is not clear whether an export adjustment would constitute a prohibited subsidy under the WTO's Agreement on Subsidies and Countervailing Measures (De Cendra 2006; Tamiotti *et al.* 2009). But there does seem to be a "gentlemen's agreement" (i.e., without legal force) within the WTO not to rebate taxes levied on inputs that are consumed in the production process.[11]

In addition, empirical evidence shows that most of the benefits of a BCA regime, in terms of preventing leakage, can be captured by a scheme that contains only adjustment on imports (Bohringer, Rutherford and Balistreri 2012). This finding, however, holds only for countries that are heavy net importers of covered goods.

Border adjustment for exports is difficult to reconcile with an approach, like the one recommended here, that advocates exemptions from import adjustment. To illustrate, BCA should not be applied to the exports of countries with national emissions caps, because there is no risk of leakage to such countries. Any rebates to implementing country exports destined for such countries would constitute unfair subsidies that, if the destination country were not practicing BCA, would induce leakage to the implementing jurisdiction. But country-level exemptions are practically impossible on the export side, as any exempted goods could easily be trans-shipped from the destination country to other countries that do not qualify for an exemption.

Given these problems and the potential clash with trade law, we do not recommend the use of export adjustment in BCA regimes.

Use of revenues from import adjustments

Revenues collected from BCA for imports could be earmarked for various purposes. The better options would move the regime as a whole toward better respect for the principles of CBDR and S&DT. For example, collections could be refunded to the exporting country, either directly or to subsidize clean technology transfer. They could be contributed to internationally administered funds for climate change mitigation and/or adaptation. Or they could be disbursed by the implementing country in ways that benefit developing countries (e.g., finance for mitigation and adaptation projects). Any of these three options could improve a BCA regime's chance of success in a WTO dispute settlement by helping to demonstrate that the BCA regime was aimed at achieving environmental objectives.

We recommend against dedicating BCA collections to general revenues in the implementing country, though we recognize that any use of this revenue will have to take place within the context of domestic fiscal realities, and some jurisdictions discourage or prohibit hypothecation of tax revenues to specific purposes. Ensuring that the revenues are not retained by the implementing country removes incentives to use BCA to enhance domestic welfare by manipulating the terms of the adjustment.

Other design guidance

Best practice in institutions and governance for BCA can be drawn from a rich tradition of norms and principles found in trade and administrative law, industry practice, and economics.

For example, exporting countries should be notified of BCA proposals at an early stage (when amendments can still be introduced), with draft text distributed to them on request and ample opportunity for comment. Entry into force of any BCA regime should give exporters and exporting country governments enough lead time to adjust their policies and practices. An official point of contact should be designated to respond to questions and requests for documents from exporting countries and firms.

The decision-making process should be predictable and transparent, and methodologies for determining vulnerable sectors, level of adjustment, and country-level applicability, for example, should be public information. Calculations should be regularly reviewed and revised where necessary, and the parameters of the regime should be reviewed at least annually, to assess their effectiveness in meeting stated objectives. Mechanisms within the BCA regime should allow exporting countries and firms to appeal decisions and calculations that concern them.

Finally, the measures should be time limited and should have clear conditions for phase-out. BCA should be intended to offer only a temporary effect during a period of transition to a low-carbon economy and broader international cooperation. At a minimum, the continued application of BCA should be contingent on explicit criteria related to the state of progress toward these goals – or the lack thereof.

BCA as an administrative challenge

While this chapter's aim is to highlight best practice in the elaboration and application of BCA, it also serves to warn policy makers who might be considering BCA exactly how difficult it is to get right. Several elements of the design described here – for example the setting of product benchmarks and the negotiation of equivalency agreements for crediting climate action in exporting states – are long-term, complex undertakings to which numerous interests will lobby for access and influence. Even after being established, and ignoring the need for review and monitoring, the administrative costs of such a regime would be considerable (Persson 2010). These considerations, while not constituting deciding arguments, will need to be part of the equation when policy makers weigh BCA against their other options for addressing leakage.

Acknowledgements

This chapter draws on a group exercise in policy analysis, the result of which is published guidance on the elaboration and implementation of border carbon adjustment (Cosbey *et al.* 2012). That document, and this chapter, were produced with the support of the ENTWINED network of Sweden (MISTRA Foundation) and the Foreign Affairs Ministry of Norway.

Notes

1 We use the term "carbon" to include the carbon equivalent of other GHGs.
2 In one set of cases, disputes stemmed from US regulations that banned the import of shrimp caught in ways that killed endangered sea turtles. The WTO Appellate Body eventually agreed that the US had the right to take this sort of unilateral measure based on how the product was produced. But the divisive rift that this caused in the trade community still echoes today.
3 For example, performance standards may set a maximum emissions intensity, but they do not charge for all emissions. They are a less efficient version of tradable performance standards, which implicitly combine a price on emissions with a subsidy to output, equal to the value of the per-unit emissions allocation, the performance standard. This treatment is similar to output-based allocation under a cap-and-trade scheme, which is a substitute for BCA. At a minimum, the same (implicit or explicit) allocation would have to be afforded imports.
4 While a significant portion of total carbon trade globally is embedded in manufactured products, the value of carbon in those products as a percentage of value added tends to be low relative to the same calculation for EITE commodities. As a result, the potential for leakage is low for manufactured products, given the lower relative cost impacts from carbon pricing.
5 As defined at the six digit level of the North American Industrial Classification System (NAICS). Source: US Environmental Protection Agency 2009.
6 One could still argue that high-carbon and low-carbon goods are like in trade law terms. But compared to country-based discrimination, this exemption arguably stands a better chance of passing Article XX's strictures.
7 The Enabling Clause (1979): Differential and more favorable treatment reciprocity and fuller participation of developing countries, Decision of the GATT Contracting Parties of November 28 (L/4903), para. 3(a).
8 The definitions of scope 1, 2, and 3 emissions used here are taken directly from the GHG Protocol. See www.ghgprotocol.org.
9 This violation of GATT's non-discrimination provisions might be saved by Article XX if one can successfully argue that the environmental objectives of the regime (preventing leakage) are frustrated by a single benchmark per product. On the other hand, a single benchmark might be seen as unnecessarily trade-distorting and arbitrary.
10 A critical part of any Article XX defence would consist of proving (in the context of the Article's chapeau) that the regime was indeed aimed at environmental objectives, and was not arbitrary in its application. A BCA regime that did not credit efforts equivalent to the importing country's regime, or that did not give partial credit for significant climate policies, would be suspect on both these grounds.
11 Letter from Donald M. Phillips, Assistant US Trade Representative for Industry, to Abraham Katz, President, US Council for International Business (5 January 1994), printed in *Inside US Trade* (1994), "US secures agreement not to allow energy tax rebate," 28 January, p. 20.

References

Böhringer, C., T.F. Rutherford and E.J. Balistreri (2012). The role of border carbon adjustment in unilateral climate policy: insights from an EMF model comparison. *Energy Policy* 34, S97–S110.

Cosbey, A., S. Droege, C. Fischer, J. Reinaud, J. Stephenson, L. Weischer and P. Wooders (2012). *A Guide for the Concerned: guidance on the elaboration and implementation of border carbon adjustment.* Stockholm: ENTWINED Network.

De Cendra, J. (2006). Can emission trading schemes be coupled with border tax adjustments? An analysis vis-à-vis WTO law. *RECIEL* 15(2), 131–45.

Metz, B., O.R. Davidson, P.R. Bosch, R. Dave and L.A. Meyer (eds.) (2007). *Contribution of Working Group III to the Fourth Assessment Report of the Intergovernmental Panel on Climate Change*. Cambridge, UK, and New York: Cambridge University Press.

Persson, S. (2010). Practical aspects of border carbon adjustment measures: using a trade facilitation perspective to assess trade costs. ICTSD Programme on Competitiveness and Sustainable Development, Issue Paper No.13, International Centre for Trade and Sustainable Development, Geneva.

Tamiotti, L., R. Teh, V. Kulaçoğlu, A. Olhof, B. Simmons and H. Abaza (2009). *Trade and Climate Change.* Geneva: United Nations Environment Programme and World Trade Organization.

16 The effect of enforcement in the presence of strong reciprocity

An application of agent-based modeling

Håkon Sælen

Introduction and literature review

To what extent are enforcement mechanisms necessary for multilateral environmental agreements (MEAs) to be effective? International Relations scholars disagree on this question. Meanwhile, economic laboratory experiments show that enforcement mechanisms have a large and significant positive effect on cooperation among individuals. Notably, the observed effect violates predictions based on the standard assumptions of rational choice theory. The effect is consistent with *strong reciprocity*: a cooperation-enhancing force defined as "a predisposition to cooperate with others, and to punish (at personal cost, if necessary) those who violate the norms of cooperation, even when it is implausible to expect that these costs will be recovered at a later date" (Gintis *et al.* 2005: 8). In other words, strong reciprocators are willing to sacrifice resources both to be kind to those who are being kind (strong positive reciprocity) and to punish those who are being unkind (strong negative reciprocity) (Fehr *et al.* 2002). Concerning institutional design of MEAs, a recent experiment (Aakre *et al.* forthcoming – henceforth referred to as AHH) suggests that enforcement of compliance is effective if, and only if, participation is also incentivized. This chapter develops an agent-based model (ABM) of strong reciprocity that explains the collective outcomes observed in public good games with punishment. This model is then used to test the robustness of the conclusions drawn from the experiments. This analysis indicates some conditions under which international environmental cooperation will likely occur, given that some states act according to the principle of strong reciprocity.

This section reviews several strands of literature: on the debate concerning the need for enforcement in MEAs, on the effect of punishment in controlled behavioral experiments, and on the theory of strong reciprocity. The second section presents the model, while the third section describes and discusses the simulation results, focusing on the conditions under which enforcement is necessary and sufficient for cooperation. The fourth section concludes.

Do multilateral environmental agreements need enforcement?

Scholars disagree considerably on whether enforcement is needed to sustain meaningful cooperation among states. While it is widely accepted that "most states comply with most of their treaty obligations most of the time" (Chayes and Chayes 1991: 311), and that this compliance has been achieved with little attention to enforcement, there is a debate over what lessons can be learned from this observation. On the one hand, the so-called managerial school (see e.g., Chayes and Chayes 1991, 1993) contends that enforcement mechanisms are unnecessary and ultimately even damaging to international cooperation. These scholars argue that the compliance problems that do exist are best addressed as management problems rather than enforcement problems, and that a facilitative approach holds the key to the design of future treaties. On the other hand, the enforcement school argues that selection problems hamper the empirical findings because most treaties are shallow, in the sense that they represent modest departures from the non-cooperative outcome, and are therefore unlikely to provide the model for the future that managerialists claim they do. Deep cooperation is realistic only if compliance is enforced, through either centralized or decentralized mechanisms (Downs *et al.* 1996). A limitation of the debate is that it has tended to look at compliance enforcement in isolation from participation enforcement. Barrett argues (Barrett 1999b, 2003) that compliance and participation are linked problems that must be considered jointly. He argues that, while the enforcement school is right that compliance must be enforced, participation is the binding constraint on international cooperation.

In the absence of a global entity that can enforce treaties, much will be left to decentralized enforcement conducted by the parties themselves. Theorists suggest that agreements could specify that non-compliance or non-participation will be punished by increased emissions from the other countries. To ensure simultaneously broad participation and deep commitments by way of such enforcement is difficult because the punishment does not specifically target the defector (see e.g., Barrett 2003), although some very interesting theoretical designs have recently been proposed (Asheim *et al.* 2006; Froyn and Hovi 2008; Heitzig *et al.* 2011; Kratzsch *et al.* 2011). This chapter focuses on punishment through means such as trade-related measures that can be targeted specifically at the defector. Standard trade theory holds that unilateral tariffs hurt the country imposing them. Thus, in economic terms, such measures are costly to the country imposing them as well as to the defector. Threats to punish free-riders are therefore not credible in a model assuming as common knowledge that all actors are rational and self-interested. Consequently, access to such measures does not enhance cooperation in the standard model.[1] The model presented below departs from the standard assumptions by including agents who are willing to punish even if it is costly to them. The purpose of the model is to identify the conditions under which enforcement in general is necessary and sufficient to induce cooperation. Whether states are in fact willing to punish free-riders is beyond the scope of this study, although there is strong evidence for such preferences among individuals and some anecdotal evidence among larger groups, as will be discussed below.

Cooperation and punishment in behavioral experiments

The public good game is among the most widely used games to analyze cooperation, and has been applied across disciplines including social psychology (see, e.g., Yamagishi 1986, 1988), political science (see e.g., Ostrom 2000), and economics (see e.g., Fehr and Gächter 2000; Fischbacher and Gächter 2010). It has been applied specifically to the topic of MEAs, both theoretically (e.g., Barrett 2003; Heitzig *et al.* 2011) and empirically (e.g., Cherry and McEvoy 2012; Barrett and Dannenberg 2012; Aakre *et al.* forthcoming).

In the basic public good game, N subjects endowed with z units of a good simultaneously decide how much to contribute to a common project. Contributions are multiplied by a factor α and divided equally among the N subjects. If $\alpha < N$, the game constitutes a prisoners' dilemma in payoffs. The unique Nash equilibrium in a one-shot or finitely repeated such game (assuming rationality, self-interest, and complete information) is that no agent contributes anything. This outcome is Pareto inferior to the collective optimum in which all players contribute their entire endowment. Allowing costly punishment does not alter the equilibrium, because self-interested agents will not incur costs to punish others.

The experimental literature on repeated games finds that most subjects diverge from what a model assuming self-interest predicts. Average contributions typically start at around half the endowment, but decay over time, until in the final round most players are behaving as self-interested.

Another experimental finding is that when the structure of the game allows it, subjects incur personal costs to punish those who contribute less than the average. This finding has been replicated so many times that it can be considered a core fact (Ostrom 2000). In turn, the threat of punishment leads to a large and sustained rise in the cooperation level relative to a game with no punishment opportunity (Fehr and Gächter 2000; Fehr and Schmidt 1999; Fehr *et al.* 2002).

A recent laboratory experiment by AHH, using a public good game, investigates when enforcement matters for MEAs, and will provide the empirical grounding for the simulations presented here. The study concludes that the managerial school and the enforcement school are both right – but under different circumstances. It supports the argument that enforcing only compliance or only participation will likely have little or no effect on contribution to international public goods. This is because compliance enforcement without participation enforcement causes free-riding to take the form of non-participation (external free-riding), while participation enforcement without compliance enforcement causes free-riding to take the form of non-compliance (internal free-riding). Hence, if participation is voluntary, as is the case with most MEAs, enforcing compliance is likely to be ineffective, as the managerialists claim. However, if participation is incentivized, compliance enforcement is likely to be essential, as the enforcement school claims.

Inferring from behavioral experiments to international relations implies attributing features of personal psychology to organizations. Bureaucracies and domestic politics are among factors that may mute the effect of individual preferences on outcomes of international politics. Four reasons why determinants of cooperation among individuals can nevertheless shed light on actual decisions in international politics are provided by Hafner-Burton *et al.* (forthcoming): (1) negotiators have substantial autonomy; (2) evidence shows that individuals can have large effects on agendas and decision-making procedures; (3) studies of group reasoning show that the behavioral traits of group members can have large effects on outcomes when decision making requires "judgmental" tasks – as many decisions in international politics do; and (4) key decisions concerning international agreements often take place in small groups where a few decision makers can have inordinate influence.

The presence of strong reciprocity

The empirical patterns outlined above are consistent with the theory that some agents have preferences for strong reciprocity. The subjects' own explanation for decaying cooperation in the absence of punishment is that cooperators become angry with others who contribute less than they themselves do and retaliate against free-riders in the only way available to them – by lowering their own contributions. In contrast, when punishment is available, the strong reciprocators can force the selfish individuals to cooperate (Fehr *et al.* 2002). The argument is generalized and proven by Fehr and Schmidt (1999). The important contribution that the theory of strong reciprocity can make towards explaining empirical observations has been emphasized by several authors (Fischbacher and Gächter 2010; Ostrom 2000; Ye *et al.* 2011; Fehr and Fischbacher 2005a), and its neural underpinnings have been identified by brain imaging (de Quervain *et al.* 2004). Notably, strong reciprocity is not driven by the expectation of future economic benefits, which sets it apart from standard models of altruism in economics and political science (Gintis *et al.* 2005).

Several experiments have established that individuals can be categorized into distinct, stable types, and that strong reciprocators (30–80 percent[2]) and selfish types (10–35 percent) are the most common. Apart from parsimony, Fehr and Fischbacher (2005a) present three additional reasons to concentrate on strongly reciprocal and selfish types. First, in the domain of punishing behavior, empirical evidence shows that negative strong reciprocity is the dominant motive. Second, in the domain of rewarding behavior, positive strong reciprocity plays an important role, although other motives play a role too. Finally, since the presence of strong reciprocators can change the pecuniary incentives of selfish types, strong reciprocators often significantly influence aggregate outcomes.

Field experiments show that reciprocal preferences for contributing to public goods also exist outside the laboratory (for a review, see Gächter 2007). Whether they generalize to the behavior of states in international negotiations has been less investigated, but Hoel (2012) provides an initial discussion in the context of climate change. He finds mixed evidence as to whether groups behave more selfishly than individuals, and that there are no grounds for rejecting the observations of individual strong reciprocity as irrelevant for the behavior of larger groups. There is, furthermore, anecdotal evidence that strong reciprocity comes into play between nations and ethnic groups, for example in the form of blood revenge (Hoel 2012). Domestic politics is not incorporated into the current study, but could give rise to preferences for imposing unilateral trade sanctions, because national decision makers are lobbied more intensively by the domestic winners, those who benefit from trade restrictions, than by the diffuse set of losers (Helm *et al.* 2012). If so, rational self-interest at the domestic level could translate into strong reciprocity in a model where states are seen as unitary actors, even if strong reciprocity among individuals does not aggregate up to the national level. In any case, the Waxman-Markey Climate Bill proposed to the US Congress required importers to purchase emission allowances, a requirement also likely to be a core part of any US action on climate change (Helm *et al.* 2012).

Modeling strong reciprocity

Building analytically tractable models of strong reciprocity is extremely difficult. Fehr and Fischbacher (2005a) argue that when analyzing economic or social problems, one should routinely try to derive the implications of the assumption that, in addition to selfish types, 40–50 percent of people act as strong reciprocators. To derive such implications, a precise mathematical model of strong reciprocity is desirable. Several conceptual models exist (for a review, see Fehr and Fischbacher 2005b) and have been useful for sharpening the concept, but they quickly become very complex and difficult to operationalize (Fehr and Fischbacher 2005a). As a second-best approach, models that mimic strong reciprocity in many circumstances, but do not explicitly model strong reciprocity, have been developed by Fehr and Schmidt (1999) and Bolton and Ockenfels (2000). However, these models fail to explain the decline in contributions consistently observed in repeated public good games (Hoel 2012), as well as certain outcomes in which intentions seem to play a role – for example, that the desire to retaliate is much more important for reciprocating individuals than the desire to reduce inequality (Fehr and Fischbacher 2005b). A simple, precise, and operational model incorporating strong reciprocity is therefore currently missing. Agent-based simulations are a hopeful candidate for filling this gap. This chapter presents a model specific to the public good game. In this model, conditional contribution and punishment of low contributions are motivations per se. Future work should aim for more models that apply to a wider set of situations. Achieving that may require specifying an underlying motive from which strongly reciprocal behavior arises, drawing on the extant conceptual models.

Method

The advantage of simulations is that they are not limited by analytical tractability. Unlike numerical simulations, ABMs explicitly simulate the interactions between agents. Such models are therefore well suited for capturing that strong reciprocators behave as if their positive or negative valuation of the reference agent's payoff depends on the actions of that agent. The purpose here is to develop a simple and intuitive model of individual behavior that generates the collective outcomes observed by AHH, and that is grounded in the wider experimental literature. The outcomes of the model should be driven by its structure, rather than by assumptions about input parameters. One aim is that the model should be as simple as possible, while still being able to capture the essential features of the empirical observations (Janssen *et al.* 2009). The model will then be used to develop hypotheses about when enforcement is needed and when it is effective, by simulating conditions that have not been observed in reality and cannot be feasibly implemented in the laboratory. The empirical patterns that the model should explain are the following:

- Participation is near full when participation is enforced, and tapers off towards zero when it is not.
- Without punishment, contributions start off around 50 percent or lower and gradually taper off (Fehr and Schmidt 1999).
- When non-participation is not possible (forced participation), adding punishment opportunities leads to a large and sustained rise in the cooperation level relative to a game with no punishment opportunity (see also Fehr and Gächter 2000; Fehr and Schmidt 1999; Fehr *et al.* 2002).
- Punishing either non-participation or non-compliance, but not both, does not substantially increase the level of contributions.
- Punishing both non-participation and non-compliance has a very substantial effect. Contributions approach the level observed with forced participation and punishment.

The approach taken is to develop a baseline model whose output will be compared to the empirical patterns, and then to vary assumptions systematically in order to investigate under which conditions the observed patterns will likely hold.

Game structure

Following several experiments, a punishment stage is added to the standard public good game, allowing agents to hand out punishment points (*PP*) that detract b units from the recipient and one unit from the sender. Following AHH, the model also includes an initial participation stage where all agents simultaneously decide whether to join an agreement. Participating reduces a player's endowment from z to $z(1-d)$ units, where $0<d<1$; hence, participation is costly. We denote players who participate "insiders" and players who do not participate "outsiders." Let M be the number of insiders.

If player i is an insider, its payoff U_i^I is:

$$U_i^I = z(1-d) - c_i + \frac{\alpha}{N}\sum_{j=1}^{M} c_j - \sum_{j=1}^{N} PP_{ij} - b\sum_{j=1}^{M} PP_{ji}$$

Similarly, if player i is an outsider, its payoff U_i^O is:

$$U_i^O = z + \frac{\alpha}{N}\sum_{j=1}^{M} c_j - b\sum_{j=1}^{M} PP_{ji}$$

Five different punishment treatments are considered:

1 No punishment ("No enforcement").
2 Forced participation and insiders can punish other insiders ("Forced participation with punishment").
3 Insiders can punish outsiders ("Participation enforcement").
4 Insiders can punish other insiders ("Compliance enforcement").
5 Insiders can punish both insiders and outsiders ("Double enforcement").

Treatment 1 is included because it is the game used in most experiments that include punishment. Here there is no participation stage, and it is not possible to free-ride externally. The other treatments are more realistic models of real-world cooperation among sovereign states that cannot be forced into any treaty. Enforcement here is decentralized, consisting of granting individuals the option of allocating punishment points to others. However, the game may also be used to say something about when enforcement institutions in a more general sense matter under a MEA.

Model structure

The strong reciprocators operate according to very simple rules. They will join the agreement if they expect a certain number of others to do so, as in the influential model of collective action by Granovetter (1978). Once they have joined, they will contribute approximately the average amount they expect the other agents to contribute, and they will punish free riders if they have the opportunity. Applying the principle of moral subjectivism (Richter 2011), they are willing to punish those that contribute less than they do themselves and those that do not participate in the agreement. They punish all such agents equally. This specification implies that the expected level of punishment per defector decreases with the number of punishable defectors.

Selfish types will join if, and only if, they expect that the punishment for not doing so is greater than the participation fee. At the compliance stage, they seek to minimize the sum of the net cost of the contribution and the expected punishment, that is, $\left(1-\frac{\alpha}{N}\right)c_i + b\sum_{j\neq i} PP_{ji}$. They do so by copying the contribution level of the agent with the lowest value for this expression in the previous round.

Agents' beliefs are mainly backwards looking. Strong reciprocators form their expectations about others' participation and contributions by looking at what happened in the previous round. Selfish agents form their expectation about the amount of punishment for non-participation by looking at the average received in the last round where someone played this option. Similarly, as stated above, they set their belief about the strategic contribution level equal to the level yielding the highest payoff in the previous round. Figure 16.1 outlines the model's stages. The author programmed the model in Netlogo 5, and will make the code available upon request.

Model parameterization

Table 16.1 lists the baseline model's parameter values. Most are set rather arbitrarily at the mid-point of the range of possible values, because little empirical direction exists. Varying these values is the focus of a later section. The top six rows concern the game's structure and are taken directly from AHH's experiment. Next is the share of strong reciprocators versus selfish agents, which is set somewhat conservatively at one half. Moving down the list, it is assumed that strong reciprocators devote half their remaining endowment to punishment. This is captured in the parameter *vengeance.* Strong reciprocators' preferences for participation are drawn from a uniform distribution of the possible values, introducing a stochastic element in the participation across individuals. A stochastic error term also enters all contribution decisions. The term is drawn from a normal distribution with mean zero and standard deviation unity. The remaining assumptions concern beliefs in the first round. Because cooperation in the presence of strong reciprocity is belief dependent, and because the outcome of the first period affects the next one and so forth, initial beliefs are potentially important. Hence they will all be subject to sensitivity analysis in "Results and discussion" below. To include a stochastic element also in the participation decisions by selfish agents, their initial beliefs about the punishment received for non-participation are drawn from a uniform distribution between zero and two punishment points.

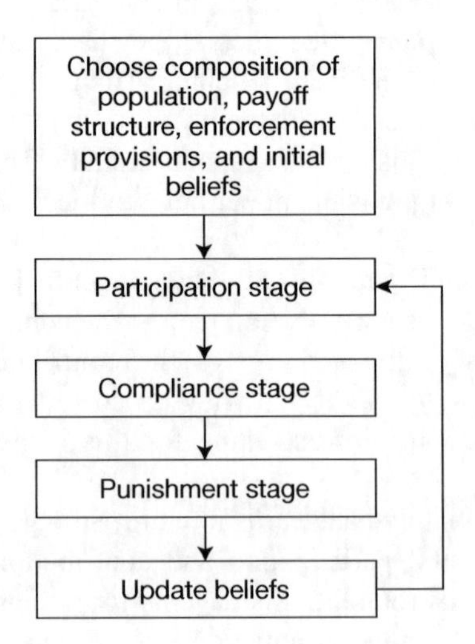

Figure 16.1 Model structure

Table 16.1 Parameterization of baseline model

Input parameter	*Baseline value*
N	4
z	22
α	1.6
b	3
d	1/11
Iterations	10
Share of selfish agents	0.5
Vengeance of strong reciprocators	0.5
Threshold of participation required by strong reciprocators to join	$\in [0,1]$
Standard deviation of error term	1 token
Initial beliefs about	
– The level of participation by others	0.5
– The level of contribution by others	10 tokens
– The amount of punishment received for non-participation	$\in [0,2]$ punishment points
– The payoff-maximizing contribution level	0 without compliance enforcement, 10 with compliance enforcement

Results and discussion

Baseline model

In the baseline model, the game's structure is exactly as in AHH's experiment and the game is played for ten rounds in groups of four. Assumptions about model parameters are as specified in Table 16.1. One hundred repetitions with different draws of the random variables are run for each enforcement regime. As Figure 16.2 illustrates, the simulations reproduce the qualitative pattern of contributions across enforcement regimes. Table 16.2 compares the observed and simulated patterns also in terms of participation and punishment. For all three outcome variables, the simulations reproduce the experiment's qualitative patterns. Quantitatively, there is some convergence from rounds 1 to 10. As the model's parameters were set rather arbitrarily, the fit occurs either by chance or because the model's results are driven by its structure, rather than by assumptions about parameters. The sensitivity analysis that follows investigates this question further.

Sensitivity analysis

The sensitivity analysis is a key contribution of this chapter. It seeks to analyze under which conditions the model reproduces the patterns observed in the

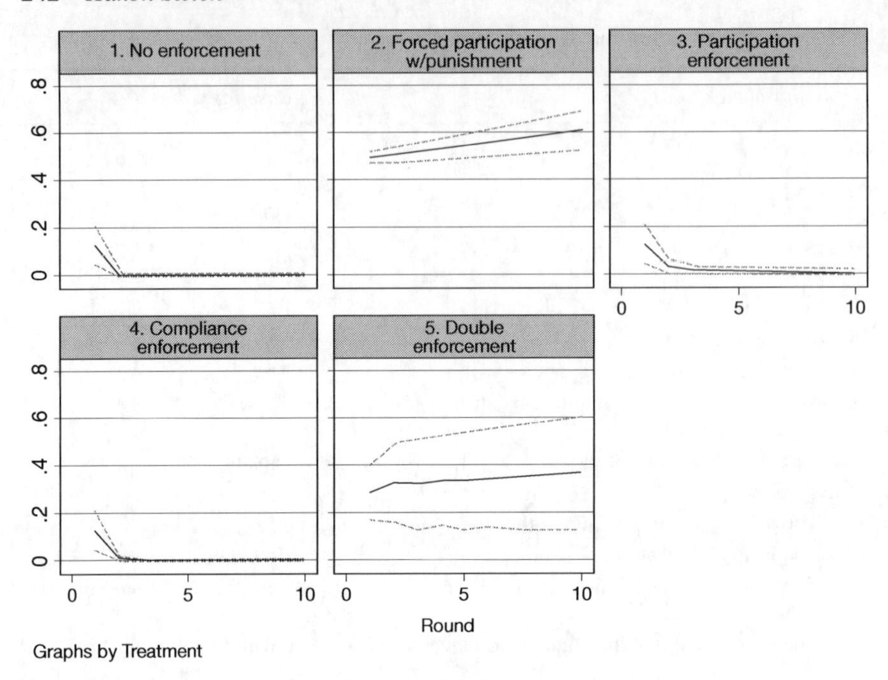

Figure 16.2 Contributions during 10 periods under different enforcement regimes

Note: Solid lines are means over 100 simulations and dashed lines are means +/– one standard deviation. N = 4.

Table 16.2 Comparison between simulated and observed patterns for participation, contributions, and punishment

		Participation		*Contribution*		*Punishment*	
		Period 1	*Period 10*	*Period 1*	*Period 10*	*Period 1*	*Period 10*
No enforcement	Exp.	0.7	0.5	6	2	n.a.	n.a.
	Sim.	0.3	0.0	3	0	n.a.	n.a.
Forced w/ punishment	Exp.	1.0	1.0	12	13	2.3	2.4
	Sim.	1.0	1.0	10	12	7.1	5.3
Participation enforcement	Exp.	0.9	0.9	7	4	1.0	0.4
	Sim.	0.6	0.8	3	0	4.5	0.9
Compliance enforcement	Exp.	0.7	0.2	3	3	1.4	0.3
	Sim.	0.3	0.0	3	0	0.9	0
Double enforcement	Exp.	0.8	0.8	9	11	2.7	2.3
	Sim.	0.6	0.8	6	7	4.9	5.3

experiments. The first analysis concerns the sensitivity of the results to the more-or-less arbitrary assumptions about model parameter values used in the baseline implementation. Next, some parameters of the game will be changed relative to what has been implemented in the laboratory. Finally, the model's structure will be altered to analyze alternative specifications for the behavioral rules of agents.

Model parameters

Figure 16.2 shows that group contributions vary considerably within the double enforcement regime. Two distinct sets of outcomes occur: either full participation with positive contributions, or no participation. This observation draws attention to the two stochastic elements affecting participation decisions: strong reciprocators' reciprocation threshold, and initial beliefs about punishment for non-participation. Because strong reciprocators compare their reciprocation threshold with their initial belief about participation, Figure 16.3 analyzes the interaction between the latter parameter and the two mentioned above. One might have expected tipping points at two-thirds punishment points (at which free-riders are indifferent about joining) and where the mean reciprocation threshold is equal to the belief about participation (at which strong reciprocators are indifferent about joining). The observed picture is somewhat more complex. One reason is that free-iders' participation affects strong reciprocators' motivation, and vice versa. Another is that heterogeneity in reciprocation thresholds increases the likelihood of cooperation, because early signatories affect other strong reciprocators' motivation (Granovetter 1978). As will be seen below, the latter effect means that when N increases, the outcomes with zero participation stop occurring.

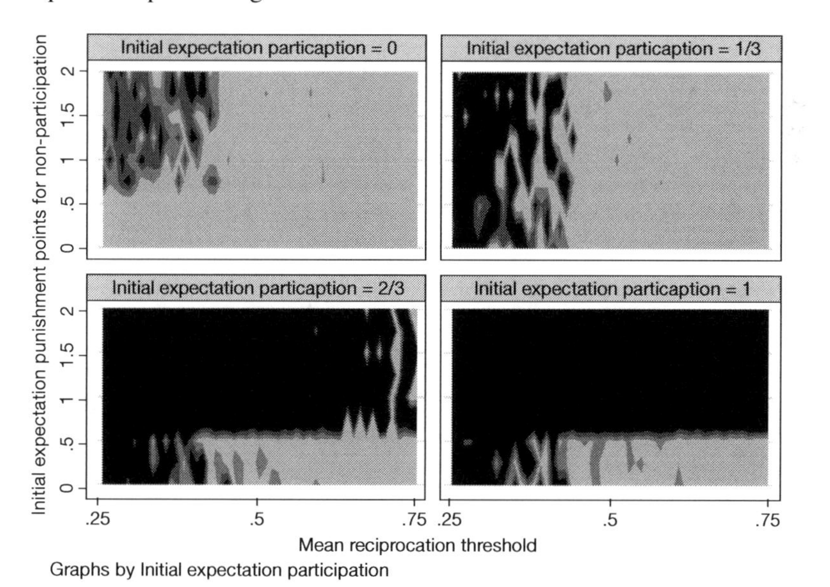

Figure 16.3 Mean contributions at period 10 under the double enforcement regime

Note: Darker colour signifies high contributions. Twenty-five repetitions for each vector of parameters.

Sensitivity to each remaining model parameter is analyzed individually in Figure 16.4. The conclusion that both participation and compliance must be enforced is robust across almost the entire range of each parameter. Focusing on the two effective enforcement regimes, particularly notable is that high levels of cooperation occur even with few strong reciprocators or little vengeance. Up to shares of around 60 percent, the presence of selfish agents does not reduce the level of cooperation, because strong reciprocators successfully incentivize selfish types to cooperate rather than free-ride. In small numbers, selfish types actually pull up cooperation, because to avoid punishment they must contribute more than the strong reciprocators would do on average, in the absence of selfish types. On their own, strong reciprocators would just try to coordinate their contributions at any level.[3]

The initial beliefs regarding contributions affect cooperation somewhat in the first 10 rounds, but this effect fades if more rounds are executed. The figure also shows that some random variation in contributions is important because observing the payoffs received from alternative decisions facilitates adaptation. Without randomness, the group can reach a stable state where actual and expected contributions are zero. It is, in other words, important that someone contributes even when expecting nobody else to do so. Once again, we observe that increasing variance increases cooperation. However, beyond certain levels of noise, the system becomes increasingly chaotic.

Overall, the sensitivity analysis demonstrates that the model's structure – rather than assumptions about parameters – drives the results, which is a desirable attribute for a model.

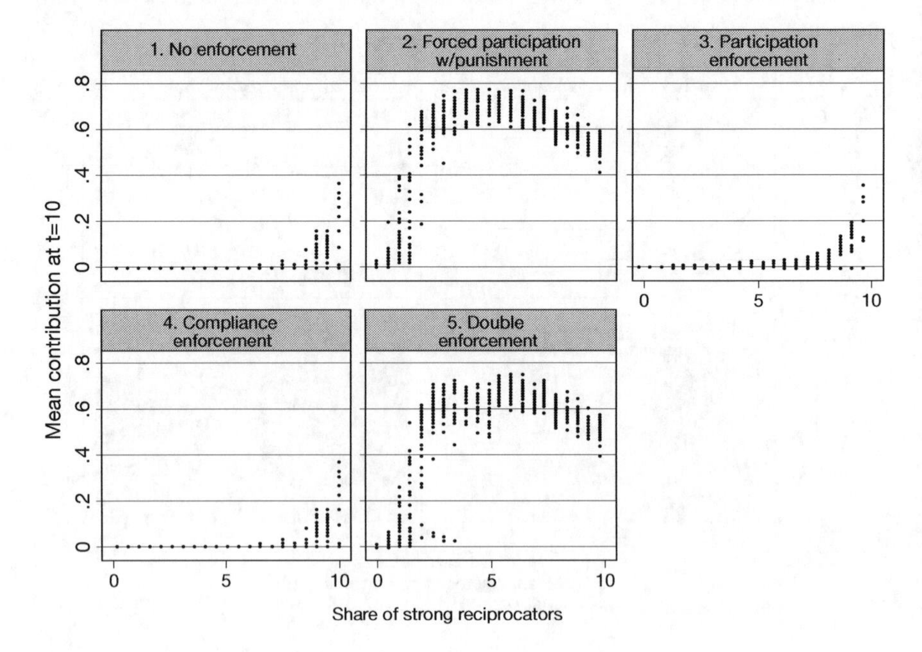

Figure 16.4 Sensitivity of results to different input parameters

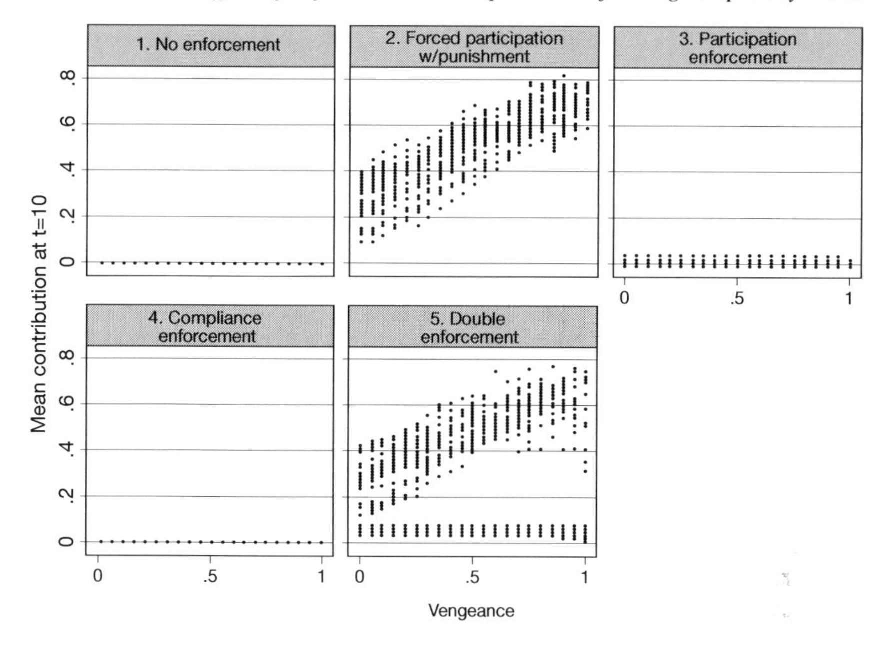

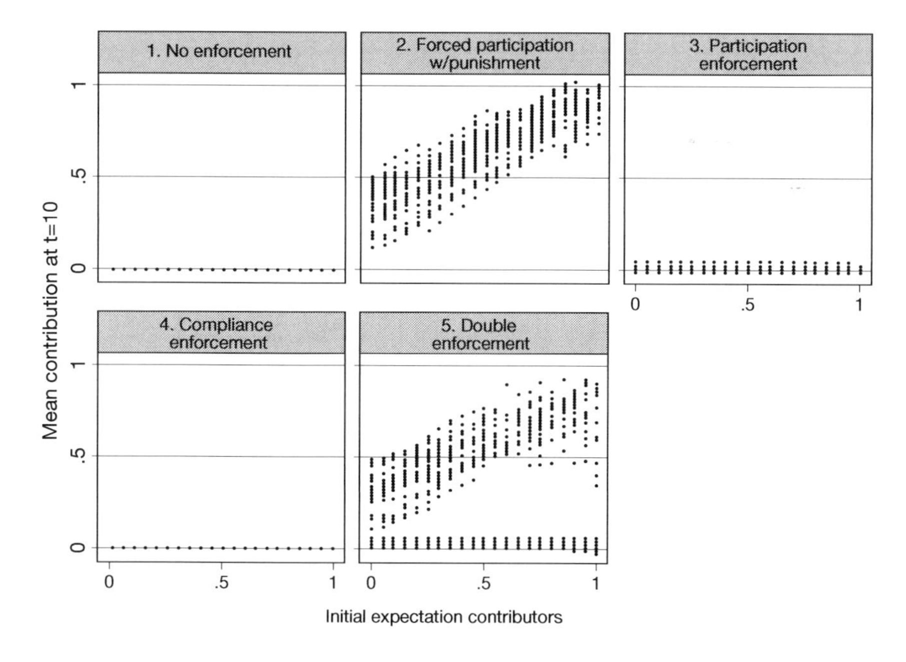

Figure 16.4 continued

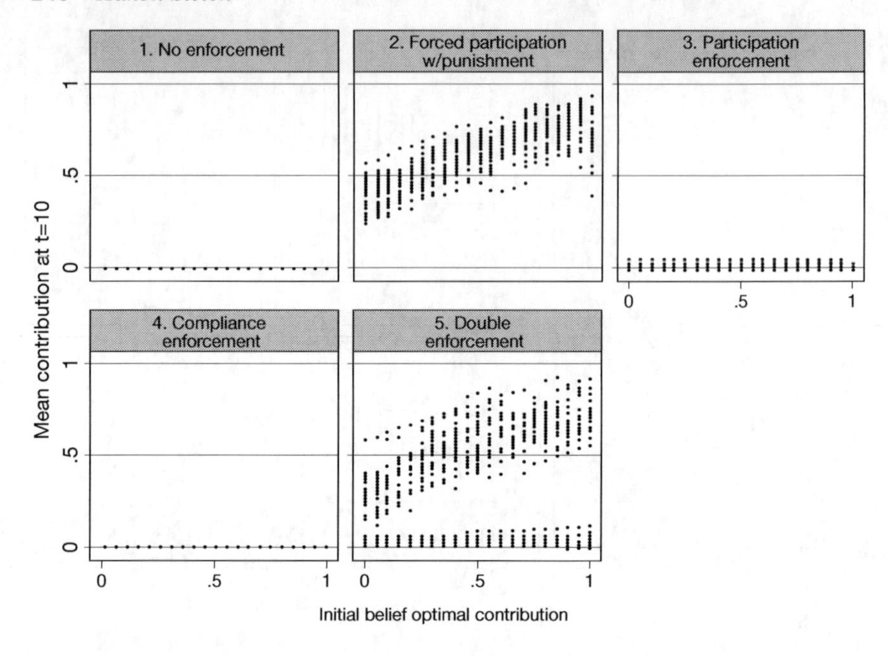

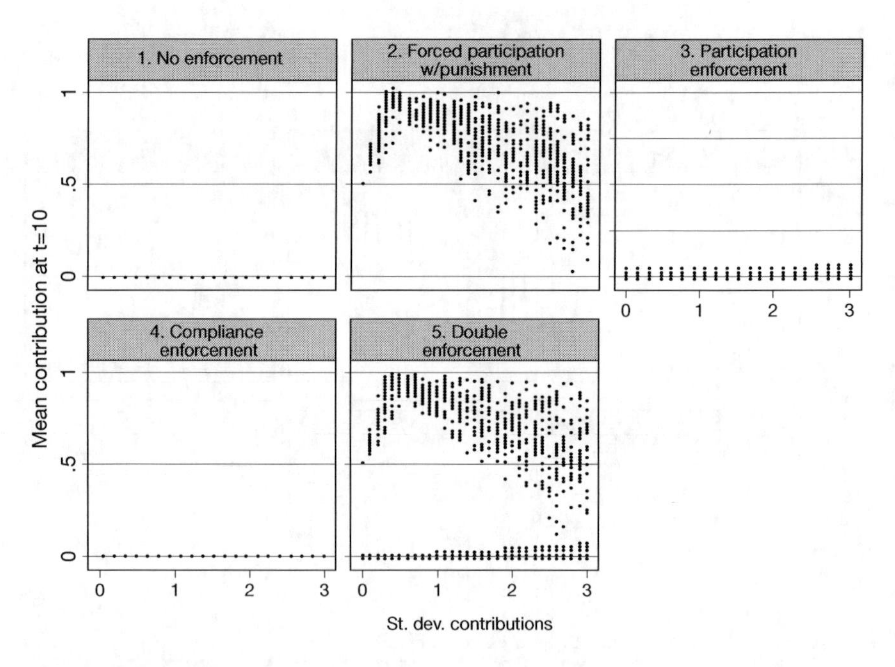

Figure 16.4 continued

Note: Each sensitivity analysis is carried out for N=4 and t=10, with 25 repetitions for each value of the independent variable. Exceptions are that N=20 for the analysis of the share of strong reciprocators to allow a finer resolution, and that t=100 for the analysis of noise because more noise means early outcomes are highly variable.

Increasing group size and time length

Two challenging variations to implement in the laboratory are increases in the size of the groups and increases in the number of iterations. Figure 16.5 shows the simulated outcomes for group sizes of 2, 4, 8, etc., up to 256. Larger groups display less noise, because the problem of getting cooperation started decreases. Cooperation initiates more easily in larger groups, *ceteris paribus*, because there are often enough agents with low thresholds to produce cascading effects. This dynamic again underscores the importance of variation in thresholds.

If the groups (still with 20 members) interact for 100 rounds, cooperation continues to increase under the effective enforcement regimes until it fluctuates near the maximum level. The reason that it keeps increasing is the following positive feedback effect: when contributions increase, more agents can punish someone who contributes a given level, and selfish agents must therefore continually increase their contributions to avoid excessive punishment.

Behavioral rules

Experiments that classify individuals as belonging to different types find that reciprocators on average prefer to contribute slightly *less* than the average of others' contributions. Such preferences imply that group contributions decline to zero over time, in the absence of enforcement, even if all members are strong reciprocators (Fischbacher and Gächter 2010). Figure 16.7 investigates the effect

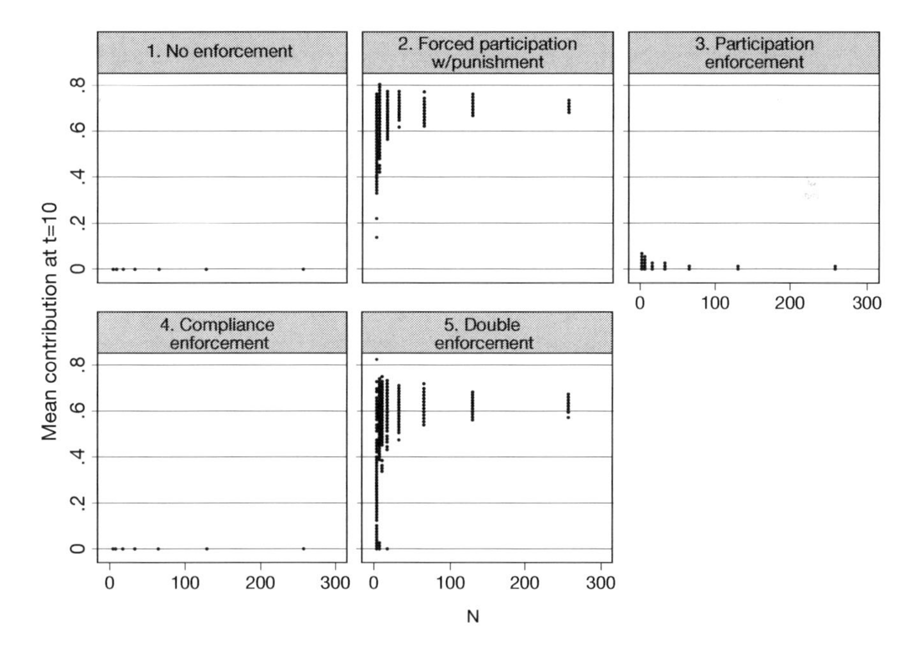

Figure 16.5 Sensitivity of results to group size

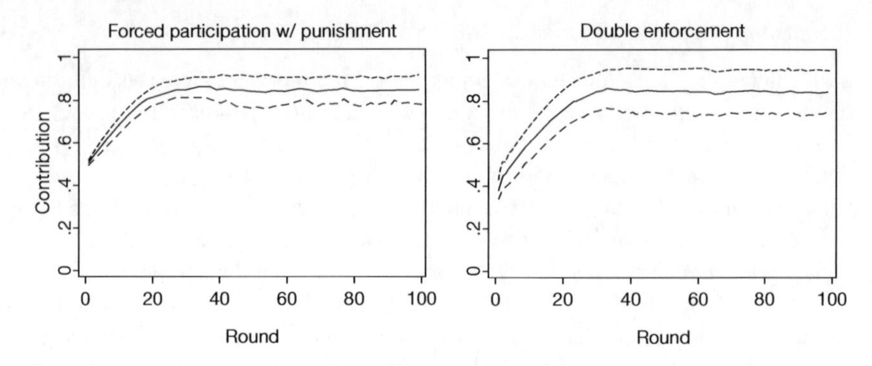

Figure 16.6 Contributions during 100 periods under the potent enforcement regimes

Note: Solid lines are means over 100 simulations and dashed lines are means +/– one standard deviation. N = 4.

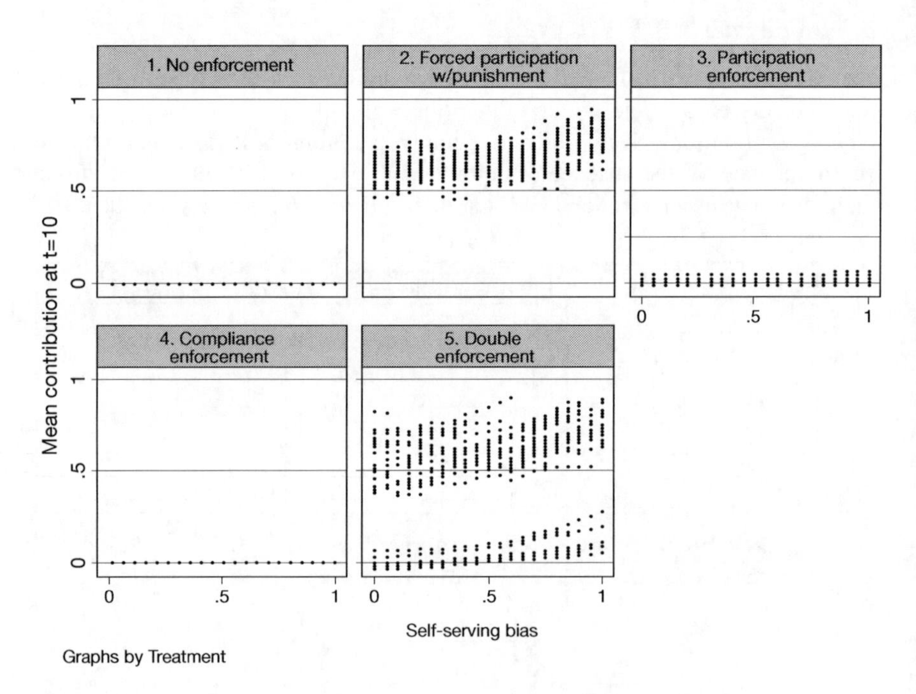

Figure 16.7 Sensitivity of results to self-serving bias among strong reciprocators

Note: N = 4 and T = 10, with 25 repetitions for each value of the independent variable.

of such a self-serving bias under different enforcement regimes. Strong reciprocators now contribute the weighted averages of their beliefs about others' contributions and the payoff-maximizing level (the level the selfish agents choose). The self-serving bias equals the relative weight given to the latter. Decision rules regarding participation and punishment of others are unchanged. Interestingly, contributions do not decline with an increasing self-serving bias, because the incentive of avoiding punishment is sufficient to achieve high levels of contributions. This result suggests that a sufficient requirement for achieving cooperation is that some agents will be willing to punish those who contribute less than they do themselves.

One assumption that does influence outcomes is that strong reciprocators apply the principle of moral subjectivism when punishing. Without this principle, the positive dynamic described earlier vanishes. For example, restricting punishment to those that contribute less than the average causes the contribution level to become a random walk, because all agents seek to contribute the average and deviations occur with equal probability in each direction. Hence the expected average contribution among insiders falls to 50 percent, as in the absence of selfish agents. The implication for the design of enforcement institutions is that such a limitation substantially reduces the effectiveness of punishment. The positive effect of moral subjectivism on cooperation is also highlighted by Richter (2011).

The analysis so far has not considered that selfish agents may find it in their interest to contribute to the public good when facing strong reciprocators, even when punishment is impossible, because of their own contributions' effect on the contributions of strong reciprocators. Sometimes, positive strong reciprocity is sufficient to induce selfish agents to cooperate. Consider, for example, a sequential two-person prisoner's dilemma, where a selfish agent moves first and knows that the second mover will reciprocate cooperation by cooperation and defection by defection. Then cooperation by both is the unique equilibrium. Furthermore, in certain games aggregate payoff is larger when costly punishment is not used (Ohtsuki *et al.* 2009). However, it will now be shown analytically that strong positive reciprocity is only sufficient to foster cooperation in N-player public good games under rather limited conditions. Consequently, this chapter's conclusions do not change when the model includes strategic contributions by selfish agents motivated by returns in the form of increased contributions by strong reciprocators.

It is shown in the Appendix that selfish agents contribute their entire endowment to the public good for strategic reasons, in the absence of punishment in an infinitely repeated game, if, and only if,

$$\frac{n\alpha\omega}{N^2-N}\frac{1}{1-\omega\dfrac{n-1}{N-1}}>1-\frac{\alpha}{N}$$

where n is the number of strong reciprocators and ω is the discount factor. Otherwise they contribute nothing. It is assumed that selfish agents know n and have linear utility functions. The condition in the case of finitely repeated games or bounded ability to reason ahead is given in the Appendix. It is worth noting that strong positive reciprocity can sustain cooperation even in finitely repeated games. Cooperation does not unravel through backward induction – as under standard assumptions – because strong reciprocators are willing to contribute even in the final round. This willingness derives directly from the definition of strong reciprocity. However, there is an end-game effect, because selfish agents no longer expect sufficient future returns from their contributions, and in turn this fall in contributions affects strong reciprocators' decisions. Such an effect is consistently observed in experiments (Ostrom 2000).

A more intuitive condition can be derived by holding constant the private return from contributions to the public good: $\beta = \frac{\alpha}{N}$ and the share of other agents who are strong reciprocators $\eta - \frac{n}{N-1}$:

$$\beta + \omega\eta > 1 + \frac{\omega(1-\beta)}{N-1}$$

In the absence of discounting, letting N approach infinity implies that the expression approaches the condition that the sum of the private marginal return from contributions to the public good and the share of other agents who are strong reciprocators must be greater than unity. This condition is not satisfied in the baseline model, but holds if the number of strong reciprocators is increased to three. In general, the condition is quite limiting. If the number of agents is increased while keeping constant the marginal *social* return from contributions to the public good – as was done in the sensitivity analyses – the *private* marginal return decreases and the condition becomes tougher to fulfill. Hence, none of the simulation results change.

This analysis shows that positive reciprocity is sufficient for inducing deep cooperation only when the private return from the public good and/or the share of strong reciprocators are rather large – in other words, when the social dilemma is blunted. The relevance of this dynamic to MEAs should not be overstated, especially given the number of actors involved. The discussion therefore does not fundamentally alter the conclusion that enforcement is necessary for effective MEAs.

Concluding remarks

This chapter has shown that a model of interactions between strong reciprocators and selfish agents robustly reproduces the empirical patterns from economic experiments allowing costly punishment. It supports the conclusion that achieving effective international environmental cooperation requires enforcement of both participation and compliance (Barrett 2008; Aakre *et al.* forthcoming).

When potent punishment is possible, strong reciprocators effectively induce cooperation even in large groups, when in the minority, when not particularly vengeful, and when exhibiting self-serving bias. Cooperation is higher when countries use their own – rather than the group average – contribution to the public good as the benchmark for good behavior. Notably, cooperation persists in the presence of numerous selfish agents and is actually enhanced by a small number of such agents. Noise and heterogeneity among agents sometimes also enhance cooperation. In contrast, when punishment is impossible, strong positive reciprocity is likely sufficient to sustain cooperation only in small groups.

That a modest number of countries behaving according to strong reciprocity can induce a larger number of selfishly motivated countries to cooperate seems theoretically plausible from this study's results. However, it is essential that countries be able to punish not only external free-riders but also signatories that contribute less than they do themselves.

The model presented is highly simplistic and artificial. Because empirical observations are sparse, many models could be formulated that would be consistent with the observations we have. Hence the model is currently only suggestive about what goes on in reality. Further work should therefore derive additional predictions from the current model, and test them in a new laboratory experiment. Another limitation is that the model's domain is restricted to public good games. Theoretical work should strive for a more general model of strong reciprocity, without sacrificing too much of the current model's simplicity and ability to be operationalized.

Acknowledgements

I am grateful for comments and suggestions from Lan Marie Berg, Jon Hovi, Andries Richter, Nils Weidmann, two anonymous referees, and participants at a workshop for authors of this volume.

Notes

1 An exception occurs if trade leakage is severe (Barrett 1999a), which arguably applies to ozone-depleting substances (Barrett 1997).
2 Large variations were found in a cross-country comparison (Kocher *et al.* 2008).
3 However, allowing communication might have enabled the latter group to coordinate on full cooperation as well.

References

Aakre, S., L. Helland and J. Hovi (forthcoming). When does enforcement matter? An experimental study.

Asheim, G.B., C.B. Froyn, J. Hovi and F. C. Menz (2006). Regional versus global co-operation for climate control. *Journal of Environmental Economics and Management* 51(1), 93–109.

Barrett, S. (1997). The strategy of trade sanctions in international environmental agreements. *Resource and Energy Economics* 19(4), 345–61.

——(1999a). The credibility of trade sanctions in international environmental agreements. In P.G. Fredriksson (ed.), *Trade, Global Policy, and the Environment.* Washington, DC: World Bank.

——(1999b). A theory of full international cooperation. *Journal of Theoretical Politics* 11(4), 519–41.

——(2003). *Environment and Statecraft: the strategy of environmental treaty-making,* Oxford: Oxford University Press.

——(2008). Climate treaties and the imperative of enforcement. *Oxford Review of Economic Policy* 24(2), 239–58.

Barrett, S. and A. Dannenberg (2012). Climate negotiations under scientific uncertainty. *Proceedings of the National Academy of Sciences* 109(43), 17372–76.

Bolton, G.E. and A. Ockenfels (2000). ERC: a theory of equity, reciprocity, and competition. *American Economic Review* 90(1), 166–93.

Chayes, A. and A.H. Chayes (1991). Compliance without enforcement: state behavior under regulatory treaties. *Negotiation Journal* 7(3), 311–30.

——(1993). On compliance. *International Organization* 47(2), 175–205.

Cherry, T.L. and D.M. McEvoy (2012). Enforcing compliance with environmental agreements in the absence of strong institutions: an experimental analysis. *Environmental and Resource Economics* 54(1), 1–15.

de Quervain, D.J.F., U. Fischbacher, V. Treyer, *et al.* (2004). The neural basis of altruistic punishment. *Science* 305(5688), 1254–58.

Downs, G.W., D.M. Rocke and P.N. Barsoom (1996). Is the good news about compliance good news about cooperation? *International Organization* 50(3), 379–406.

Fehr, E. and U. Fischbacher (2005a). The economics of strong reciprocity. In H. Gintis, S. Bowles, R. Boyd and E. Fehr (eds.), *Moral Sentiments and Material Interests: the foundations of cooperation in economic life.* Cambridge, MA, and London: MIT Press, pp. 151–92.

——(2005b) Modeling strong reciprocity. In A. Falk and U. Fischbacher (eds.), *Moral Sentiments and Material Interests: the foundations of cooperation in economic life.* Cambridge, MA, and London: MIT Press, pp. 193–214.

Fehr, E. and S. Gächter (2000). Cooperation and punishment in public goods experiments. *The American Economic Review* 90(4), 980–94.

Fehr, E. and K.M. Schmidt (1999). A theory of fairness, competition, and cooperation. *The Quarterly Journal of Economics* 114(3), 817–68.

Fehr, E., U. Fischbacher and S. Gächter (2002). Strong reciprocity, human cooperation, and the enforcement of social norms. *Human Nature* 13(1), 1–25.

Fischbacher, U. and S. Gächter (2010). Social preferences, beliefs, and the dynamics of free riding in public goods experiments. *The American Economic Review* 100(1), 541–56.

Froyn, C.B. and J. Hovi (2008). A climate agreement with full participation. *Economics Letters* 99(2), 317–19.

Gächter, S. (2007). Conditional cooperation: behavioral regularities from the lab and the field and their policy implications. *Psychology and Economics: a promising new cross-disciplinary field (CESifo Seminar Series),* 19–50.

Gintis, H., S. Bowles, R. Boyd and E. Fehr (2005). Moral sentiments and material interests: origins, evidence, and consequences. In H. Gintis, S. Bowles, R. Boyd and E. Fehr (eds.), *Moral Sentiments and Material Interests: the foundations of cooperation in economic life.* Cambridge, MA, and London: MIT Press, pp. 3–40.

Granovetter, M. (1978). Threshold models of collective behavior. *American Journal of Sociology* 83(6), 1420–43.

Hafner-Burton, E.M., B.L. LeVeck, D.G. Victor and J.H. Fowler (forthcoming). *Decision Makers' Preferences for International Legal Cooperation.*

Heitzig, J., K. Lessmann and Y. Zou (2011). Self-enforcing strategies to deter free-riding in the climate change mitigation game and other repeated public good games. *Proceedings of the National Academy of Sciences* 108(38),15739–44.

Helm, D., C. Hepburn and G. Ruta (2012). Trade, climate change and the political game theory of border carbon adjustments. *Oxford Review of Economic Policy* 28(2), 368–94.

Hoel, M. (2012). Klimapolitikk og lederskap – hvilken rolle kan et lite land spille? Vista Analyse.

Janssen, M.A., N.P. Radtke and A. Lee (2009). Pattern-oriented modeling of commons dilemma experiments. *Adaptive Behavior* 17(6), 508–23.

Kocher, M.G., T. Cherry, S. Kroll, *et al.* (2008). Conditional cooperation on three continents. *Economics Letters* 101(3), 175–78.

Kratzsch, U., G. Sieg and U. Stegemann (2011). An international agreement with full participation to tackle the stock of greenhouse gases. *Economics Letters* 115(3), 473–76.

Ohtsuki, H., Y. Iwasa and M.A. Nowak (2009). Indirect reciprocity provides only a narrow margin of efficiency for costly punishment. *Nature* 457(7225), 79–82.

Ostrom, E. (2000). Collective action and the evolution of social norms. *The Journal of Economic Perspectives* 14(3), 137–58.

Richter, A.P. (2011). The coevolution of renewable resources and institutions – implications for policy design. PhD thesis. Wagenigen: Wagenigen University.

Yamagishi, T. (1986). The provision of a sanctioning system as a public good. *Journal of Personality and Social Psychology* 51(1), 110–16.

——(1988). The provision of a sanctioning system in the United States and Japan. *Social Psychology Quarterly* 51(3), 265–71.

Ye, H., F. Tan, M. Ding, Y. Jia and Y. Chen (2011). Sympathy and punishment: evolution of cooperation in public goods game. *Journal of Artificial Societies and Social Simulation* 14(4), 20.

17 EU emissions trading

Achievements, challenges, solutions

Jon Birger Skjærseth

Introduction[1]

The EU ETS is a grand climate policy experiment – the first-ever international cap-and-trade system to target industry. The ultimate aim is to create a global carbon market by encouraging other major emitters, not least the USA and China, to follow suit. However, the economic recession that developed from 2008 has put the EU ETS to the test. A significant supply–demand imbalance of allowances has accumulated, followed by a low carbon price. This chapter explores possible solutions for the EU, as well as presenting a brief analysis of the evolution and achievements of the system.

The EU ETS is the EU's key climate policy instrument. It is a cap-and-trade system that covers about 11,000 industrial installations, representing close to 50 percent of CO_2 emissions in the EU. These installations are mainly owned and operated by the electricity-producing and energy-intensive companies. The intention behind the system is twofold. First, it aims at reducing emissions in a cost-effective manner in order to help achieve a 20 percent reduction in emissions by 2020 compared to 1990 levels. Second, it seeks to provide incentives for companies to invest in low-carbon solutions for the future. The EU aims at 80–95 percent reduction of emissions by 2050 so as to limit the rise in global temperature to 2°C. This long-term task is daunting indeed, and here the EU ETS is envisaged as driving a wide range of low-carbon technologies into the market (Commission 2011a). This will necessitate speeding up the development, commercialization and diffusion of new breakthrough technologies by a carbon price above a certain level.

Coal exemplifies the short- and long-term challenges to the ETS. Since 2010, cheap coal imports to Europe from the USA have increased, particularly to Germany. Moreover, with the carbon price at a very low level, it has become more economical for some power producers to burn coal and pay for the permits instead of shifting to renewables or other high-cost energy sources. In the long term, the challenge is to avoid continued lock-in of old energy technologies. Europe's coal-fired power plants are on average 34 years old (Ecoprog 2012). Between 2012 and 2020, approximately 80 power-plant units will be newly constructed or replaced, and others shut down. An important question here is whether these new and modernized plants will be equipped with Carbon Capture and Storage (CCS) or other low-carbon technologies. CCS is set to play a key role in meeting the EU's long-term climate targets. The ETS can hardly drive CCS by itself, but it can contribute to "pull" development and commercialization, together with other EU and national R&D policies "pushing" technological development.

Evolution

The 1997 Kyoto Protocol established emissions trading between countries as an optional flexible mechanism, in line with the preferences expressed by the USA. The EU was initially skeptical. And yet, only five years later, the EU member states and the European Parliament agreed on the world's first international cap-and-trade system for large industrial power-producing and energy-intensive emitters. There were several reasons behind this turnabout, including the failed EU carbon/energy tax, the need for a climate policy instrument to implement the EU's Kyoto commitment, and the efforts of a handful of economists acting as policy entrepreneurs within the European Commission's Directorate-General for the Environment. These entrepreneurs took the initiative to the EU ETS, built up knowledge based on US experience in other issue areas, and crafted support among reluctant stakeholders. These included parts of industry, green groups, the European Parliament, and most EU member states (Skjærseth and Wettestad 2008; Skjærseth and Eikeland 2013).

In March 2000, Directorate General (DG) Environment in the European Commission issued a Green Paper on the ETS, outlining specific design proposals. In order to sell the idea to reluctant stakeholders, the Commission set up a working group under the European Climate Change Programme (ECCP), where representatives from industry, green groups and the Commission met regularly. The Commission presented the ETS as an instrument that could provide economic opportunities to sell allowances for decreasing industrial emitters and bring emissions costs effectively down to the cap set: this framing proved effective in reducing resistance from reluctant stakeholders and building support. Cap-and-trade stood out as a good idea that would meet less industrial resistance than a tax, and – if properly designed – would perform better than voluntary agreements to change behavior. Importantly, fewer allowances should be handed out to industry than were expected to be needed, to create scarcity in the market and a better carbon price.

The Commission's plans were significantly advanced when newly-elected US President George W. Bush decided to withdraw from the Kyoto Protocol in March 2001. The US exit served to unify the positions within the EU in support of the EU ETS as a key measure for strengthening the EU's credibility in ensuring the entry into force of the Kyoto Protocol – which was crowned with success in 2005 (Skjærseth and Wettestad 2008).

In October 2001, the Commission put forward a proposal for the Emissions Trading (ET) Directive. The Commission had taken into account the interests of energy-intensive industries and the various member states, and its final proposal included decentralized allocation of mainly free allowances at the member-state level. The final ET Directive was agreed by the member states in late 2002, approved by the European Parliament and formally adopted in July 2003. The system was launched on 1 January 2005. A first pilot phase ran from 2005 to 2007, followed by the Kyoto commitment phase, 2008–2012. In 2004, the EU adopted the associated Linking Directive connecting the EU ETS to the Kyoto Protocol's flexible project mechanisms, mainly the Clean Development Mechanism (CDM). Within certain limits, external credits from investing in more affordable projects abroad could be used by companies to comply with the EU ETS.

The ET Directive was implemented in the form of National Allocation Plans (NAPs). The Commission was given the authority to assess the NAPs, and could reject any not in accordance with the relevant provisions of the Directive. When the first verified ETS emission figures for 2005 were announced in mid-May 2006, they indicated strongly that allocations for the first pilot phase had distributed large quantities of excess allowances, with the market price for allowances plunging (Skjærseth and Wettestad 2008). The decentralized nature of cap-setting had resulted in a "race to the bottom," where each member state had incentives for allocating allowances generously, so as to protect its own industries. To avoid the repetition of such a calamity, for the second phase (2008–2012) national emissions from the ETS sectors were set at an average around 6.5 percent below 2005 emission levels. The penalty for non-compliance (when an operator does not surrender sufficient allowances each year to cover its emissions) was also increased from €40 to €100 for each tonne of CO_2 equivalent – significantly higher than the actual and foreseen carbon price. This more stringent system led industry to expect higher carbon prices.

On the basis of experience gained, in January 2008 the Commission put forward a proposal for a revised EU ETS for a third phase, 2013–2020. The changes agreed in December 2008, and formally adopted in 2009, were significant. First, the revised ETS introduced a single EU-wide cap on allowances, to be reduced in a linear manner by 1.74 percent annually from 2013 onwards, scrapping the system of NAPs. The EU ETS has no deadline and is intended to continue beyond 2020 with subsequent phases. Second, allowances are now to be allocated mainly by payment (auctioning) for the power sector on the basis of fully EU-harmonized rules. Industrial sectors or sub-sectors particularly exposed to global competition, and hence in danger of "carbon leakage," will get allowances for free, based on harmonized performance benchmark rules. Allocation of allowances based on benchmarks is intended to provide a competitive advantage to the most energy-efficient installations. A further mandatory element in the revised ETS is the use of 300 million allowances from the New Entrants Reserve to support up to 12 CCS demonstration projects and projects involving renewable energy technologies (NER300). Third, the rules on the use of external credits give companies roughly similar access to such credits as before. Transfer of CDM credits is permitted from the second to the third trading period. In January 2012, international aviation was included in the ETS, but implementation has been postponed due to opposition from airlines in the USA, China, and India in particular.

This EU ETS revision formed part of a broader post-2012 energy and climate strategy. The European Council adopted targets to reduce emissions of greenhouse gases (GHGs), increase energy efficiency and the share of renewable energy in total energy consumption – all by 20 percent by 2020 compared to 1990. The ETS reform formed the core of a climate and energy package of policy instruments adopted in December 2008 that also included: (1) a decision on effort-sharing among the member states in the form of differentiated targets for sectors not covered by the ETS, such as transport and agriculture; (2) a directive on the promotion of renewable energy sources; and (3) a directive on safe carbon capture and storage (CCS). New legislation was also adopted for reducing CO_2 emissions from new cars and for fuel quality, from "well to wheel."

Action on climate change was put at the center of the new European energy policy for making the use of energy more efficient, reducing the need for imported hydrocarbons, and reducing vulnerability to volatility in oil and gas prices. Action on energy policy was to contribute to climate change mitigation, more effective application of the ETS, and create "green" jobs. This was to be achieved by strengthening policies on renewables, by energy efficiency, liberalization of European energy markets, and energy-technological innovation. A European energy technology plan was proposed to lower the cost of clean energy and put the EU at the forefront of the low-carbon technology sector. Energy and climate policies were initially designed to be mutually reinforcing and synergistic (Commission 2007a, 2007b). Largely swept under the carpet were potential conflicts like downward pressures in carbon prices caused by renewables, and energy-efficiency measures in the ETS sectors that would reduce the demand for allowances (Skjærseth 2013).

In 2009, the EU declared that it would work for an OECD-wide carbon market by 2015, intended as an agreement for the upcoming Copenhagen summit. However, when Copenhagen-2009 failed and proposed cap-and-trade plans stalled in the US Congress, an OECD-wide carbon market became totally unrealistic (Skjærseth *et al.* 2013). An additional challenge to the ETS came from the financial crisis, which caused a significant drop in emissions and thus a plunge in demand for allowances and a drop in the carbon price, particularly from 2011 (see below).

Achievements

Is the EU ETS worth saving? Part of the answer can be found in how the system performed in practice as regards achieving EU goals, before the carbon price dropped. Opinions and interests in Brussels have varied. Some actors would prefer to see the EU ETS experiment fail, so as to provide stronger signals for low-carbon investment. Others argue that the system should be strengthened as key driver for decarbonization in Europe (van Renssen 2012).

First-generation studies on ETS consequences focused on expected (*ex ante*) effects of the system. These studies sought to predict whether and how the ETS would influence the sectors covered by the system. They generally asserted that the initial incentives for innovation under the system were rather weak. The next generation of studies centered on *ex post* evaluations after the first 2005–2007 trading phase. These studies concluded that, although emissions trading had entered the boardrooms and affected the corporate agenda, the system had not induced fundamental shifts in corporate strategy or corporate investment decisions. What are perhaps the most extensive accounts of actual short-term CO_2 abatement conclude that there has been abatement in both the energy-intensive and power sectors since the introduction of the EU ETS (Ellerman and Buchner 2008; Ellerman *et al.* 2010). Other scholars have contested such findings, holding that it is difficult to attribute changes in emissions to the ETS as such (see e.g., McIlveen and Helm 2010).

Since 2009, several studies have focused on longer-term effects of the EU ETS, with greater emphasis on technological innovation and long-term strategy changes. One lesson drawn is that the EU ETS has exerted an impact on innovation especially in the power sector, but due to design flaws and teething problems, the effects have been limited and varying. An extensive study on the ETS and investment in innovation has documented that companies generally expect(ed) tighter emissions caps for the third phase and significantly higher carbon prices in the future, averaging €40 per ton in the post-2012 phase (Martin *et al.* 2011).[2] Moreover, 70 percent of the firms studied were found to be engaged in R&D on cutting emissions or improving energy efficiency. This 2011 study also found a significant positive correlation between company expectations as to the future stringency of their cap, and "clean" innovation.

These and other valuable studies have improved our understanding of the consequences of the EU ETS, but the dominant focus has been on economic incentives as the only mechanism and system, or on the consequences for specific sectors (particularly in the German power sector). In the following, we turn to a recent study that concentrates on corporate responses to the EU ETS in the context of company internal and wider external factors within all the major sectors covered: power, oil, cement, steel, and pulp and paper (Skjærseth and Eikeland 2013). The study, covering the period up to 2011, is based on three models of corporate response to regulation based on different behavioral assumptions and mechanisms linking regulation to response. According to these models, companies can be expected to: (1) resist the system and comply only shortsightedly and reluctantly; (2) support the system and search for innovative low-carbon business opportunities; or (3) "crowd in" or "crowd out" corporate norms of social responsibility. The key findings of this new study can be summarized in four main points, as presented below.

First, since the introduction or anticipation of the ETS, all companies and sectors studied have, to varying degrees, adopted more proactive strategies with "innovative" elements. The EU ETS was the first mandatory climate instrument that directly affected most of the sectors and companies. Combined with short- and long-term climate targets, the ETS has contributed to shaping expectations of the need to decrease emissions. Cooperation on low-carbon innovation has also increased.

Second, the most significant aggregate changes can be found in the electric power industry, which has embarked on a strategy for decarbonizing power supply in Europe by 2050. Most of the large power companies have supported the ETS, and have increased investments in low-carbon R&D and renewables. Emissions have gone down, as they also did before the financial crisis.[3] The changes observed in energy-intensive industries appear less significant, but there are interesting examples of collaborative and innovative initiatives – for instance, in the steel sector through the Ultra-low CO_2 Steelmaking (ULCOS) program. In the oil industry, the main importance of the ETS lies in the longer-term strategic consequences of a political agreement on carbon pricing, sending a price signal that may be copied in other parts of the world. As a response to the adoption of the revised ETS and the EU climate and energy package in 2008, many major oil companies announced that they would step up or implement new strategies in renewables, biofuels, and CCS. The expected increase in regulatory pressure also affected the position of Europia, the association for the oil-refining industry, in a more offensive direction, at least temporarily. The European refining industry now paints a dark picture of the future, due to decreasing demand for petroleum products, a mismatch between supply and demand for diesel and petrol, and increasing regulatory pressures.

Third, the EU ETS has significantly affected these changes, particularly in the electric power sector, which has faced the strongest direct regulatory pressure from the system (and the lowest exposure to international competition). Still, the strength of the causal relationship between the ETS and corporate climate strategies is extremely hard to determine precisely because of other international, EU, and national policies that have co-evolved with the ETS. In addition, various company characteristics, like carbon intensity and management structure, affect how companies respond to regulation. The ETS has impacted on corporate climate strategies through a wide range of mechanisms. These include regulatory pressures that create incentives for cost-cutting in periods when the carbon price was expected to increase, and by triggering attention, experimenting, learning and investment with respect to low-carbon solutions beyond business-as-usual for the companies.

And finally, the Skjærseth and Eikeland study finds no strong evidence of a causal link between the ETS and norms of responsibility. What we can say is that the ETS has not reduced voluntary social responsibility activities, independent of the motivation.

The main conclusion is that companies have responded to the EU ETS by adopting more proactive climate strategies, with elements of low-carbon innovation. The question is now: will this development end? Or will the EU ETS drive industry responses further toward a low-carbon economy, when significant changes in economic circumstances occur?

Challenges

As a market-based policy instrument, the EU ETS has proven vulnerable to market changes. Companies expected the EU ETS to become ambitious from the start of the second trading period in 2008. This expectation was reinforced by the revision of the system, to apply from 2013. But the financial crisis has lowered expectations of an ambitious system. International credits could also be used – the use of international credits for compliance increases the surplus of allowances, as an international credit frees up one allowance that does not need to be used.

As shown in Table 17.1, from 2008 the number of allowances has increased every year alongside supply and use of international credits, most notably in 2011 (Commission 2012a, 2012b). By early 2012, a surplus of 955 million allowances had accumulated. Of these, 549 million came from the use of international credits.

Increasing supply of allowances combined with low demand is, according to the Commission, partially reflected in the evolution of the carbon price since 2008. The carbon price has dropped from nearly €30 in spring 2008 to just above €5 in spring 2012, with a significant reduction in the second half of 2011 coinciding with an accelerated build-up of allowances and international credits. The low carbon price means scant incentives for companies to invest in low-carbon solutions for the future. The surplus is expected to build up in 2012 and 2013, and may reach a structural surplus of 2 billion allowances in phase 3 due to various elements related to the transition to phase 3.[3] The amount of surplus by 2020 will depend on the speed of economic recovery and various energy factors, such as the development of renewable energy, and energy efficiency. Increasing diffusion of renewable energy and energy efficiency in the EU ETS sectors will reduce the demand for allowances, further fuelling increases in the imbalance between supply and demand.

Solutions

The drop in emissions due to the financial crisis will contribute significantly to meeting the 20 percent emissions reduction target. However, a low carbon price cannot stimulate investment in low-carbon technologies for the long term. Accordingly, the Commission has proposed a range of solutions for dealing with the structural surplus problem presented above.

In March 2011, the Commission published a Roadmap for moving towards a competitive low-carbon economy by 2050 (Commission 2011a). The analysis presented shows that GHG emissions will have to be reduced by 25 percent below 1990 levels in 2020, 40 percent in 2030, and 60 percent in 2040, in order to

Table 17.1 Supply–demand balance 2008–2011 (in Mt)

Year	*2008*	*2009*	*2010*	*2011*	*Total*
Supply: Issued allowances and used international credits	2076	2105	2204	2336	8720
Demand: Reported emissions	2100	1860	1919	1886	7765

Source: European Commission 2012a.

achieve 80 percent reduction by 2050. The Roadmap encourages decision makers (here the European Parliament and the Council) to deal with the structural surplus problem of the EU ETS by revising the 1.74 percent linear reduction factor. For the short term, allowances for the third phase (2013–2020) should be set aside to provide scarcity in the market and raise carbon prices. In two rounds, 26 of the 27 EU member states – Poland was the blocking state on both occasions – agreed on watered-down versions of the Roadmap. The Parliament voted in favor of the Roadmap and the strengthening of the EU ETS (European Parliament 2012). Poland also vetoed the Energy Roadmap 2050, which was released the same year (Commission 2011b). The Energy Roadmap argues that greater energy efficiency and renewables to achieve 80–95 percent reduction by 2050 will cost about the same as continued heavy reliance on nuclear power and fossil fuels. Poland's opposition can be explained by its reliance on coal and a concern for higher energy prices (Polish Chamber of Commerce 2012). Poland, the largest coal producer in the EU, covers almost all its electricity demand with domestic coal. It does not see market opportunities in renewables.

Failure to agree on a more ambitious 2020 climate target or stepwise targets toward 2050 has pitted EU climate policy and energy policy instruments against each other. Initially, these policies were designed to be synergistic. The conflict of interaction between policy instruments has been exacerbated by the adoption of a new Energy Efficiency Directive (EED) in October 2012. The EED explicitly refers to the need to strengthen the EU ETS to make it more effective. Because a more ambitious climate target has proved politically unfeasible, energy policies on renewables and energy efficiency have been strengthened at the expense of climate policy and the EU ETS.

In addition to the Roadmap, the Commission has proposed various measures for remedying the problems with the EU ETS.[4] In the short term, it has proposed a Regulation to postpone or "backload" 900 million allowances from the beginning to the end of the 2013–2020 period (Commission 2012c). Backloading is expected to stabilize the carbon price – but it cannot solve the surplus problem, since it does not remove the allowances permanently from the system. Most energy-intensive industries have resisted "fixing" the ETS. In June 2012, a legal opinion prepared for the alliance of Energy Intensive Industries in the EU by the German law firm Luther argued that the ET Directive does not permit the Commission to intervene if the carbon price is too low (Luther 2012). This interpretation has been supported by Business Europe (BusinessEurope 2012). The Commission has responded by proposing a decision to clarify its own competence on this point. Backloading can be combined with the permanent retiring of allowances for phase 3 at a later stage. This measure can thus prove an effective way of dealing with the overall supply–demand imbalance.

Various solutions have also been proposed for tackling the growing imbalance between supply and demand of allowances in the longer term (Commission 2012a). One option is to revise the 1.74 percent annual linear factor earlier than foreseen in the ET Directive (which set the year by 2025). This measure could solve both the imbalance problem and raise ambition levels after 2020. The current annual linear factor needs strengthening to be in line with the EU's ambition of 80–95 percent reduction by 2050. Other options include extending the scope of the EU ETS to sectors less heavily influenced by economic cycles (e.g. fuel consumption in other sectors), a price floor, or restricted access to international credits for phase 4. In retrospect, access to international credits has proven too generous, as international credits account for a significant share of the currently expected surplus of allowances. Ever since the inception of the system, industry has lobbied to maximize access to such credits so as to lower the costs of abatement.

The upshot? There are several design options available for dealing with the problem in the short and long term. Whether these options will be realized hinges largely on political feasibility. The political feasibility of strengthening the EU ETS depends first on EU-internal factors: the positions of "veto players," decision-making procedures, and coordination with other parts of the climate and energy package. "Veto players" are actors that will have to agree, if policy is to be changed. Poland vetoed the low-carbon Roadmap, but it will not necessarily be in a position to block specific legislative proposals, because the decision-making procedures required for the different options vary. That also affects the probability and speed of adoption. Whereas most options require a qualified majority decision among the EU member states and support from the European Parliament, a carbon price floor is a fiscal measure – and that will require unanimity among the member states. The retiring of allowances in phase 3 can in theory be adopted relatively quickly by a separate decision of the Council and Parliament, whereas other options will require a full revision of the ET Directive (see below).

Various options can also affect other parts of the climate and energy package. For example, strengthening the EU reduction target in 2020 will also affect the targets adopted under the ESD Directive for sectors outside the ETS. The reason for this is that the ETS sectors are to contribute with 21 percent reduction and the non-ETS sector with 10 percent reduction, in order to achieve the target of 20 percent reduction from 1990 levels by 2020. Package deals can promote agreement by combining differently valued issues, overcoming distributional obstacles, and promoting synergies. However, such packages may act to impede revision if circumstances change, since amending one component could have repercussions for the package as a whole.

The Commission has put forward a Green Paper on a framework for climate and energy policies towards 2030 (Commission 2013a). The Green Paper launches a public consultation on the content of the 2030 framework, including new targets, coherence between policy instruments, contribution to competitiveness, and how different capacities of the member states can be taken into account. The Commission has also published a Consultative Communication on the future of CCS in Europe in response to slow deployment (Commission 2013b). One of the key barriers identified is the lack of a long-term business case and the cost of CCS technology. At the current low carbon price, CCS investment does not make sense for operators. This situation is unlikely to change in the short term. In April 2013, the European Parliament rejected the short-term fix to the EU ETS by voting against the backloading of 900 million tonnes of allowances from the EU ETS.

Political feasibility also depends on factors external to the EU. The development of EU climate policy has always been linked to the EU's ambitions to climate leadership and has been based on the assumption that a new comprehensive, ambitious, and binding climate treaty will be adopted. The costs of EU climate policy could be cut by bringing the major world emitters into a global effort (Commission 2005). The 2011 Energy Roadmap explicitly states that the decarbonization of Europe should not develop in isolation, but should take into account international developments so as to minimize carbon leakage and effects on competitiveness. As yet, ambitions for including other major emitters have failed, since the USA, China, and India have refused to go along. Moreover, as part of its global effort, the EU has sought to facilitate an OECD-wide carbon market. This effort to level the playing field for energy-intensive industries in Europe has also failed. In fact, there has been active resistance to EU climate policies that would affect third countries, such as including international aviation in the EU ETS. As a result, some member states and industry have increasingly questioned the EU ambition of "leadership by example." The 2011 Durban Platform for Enhanced Action aims at finalizing a new 2020 climate agreement "with legal force" by 2015. However, even with the Doha 2012 outcome of including a new commitment period under the Kyoto Protocol, it is highly uncertain whether a new climate agreement can prove sufficient to align EU actors in support of a more stringent climate target and a more stringent EU ETS. Slow progress on a new international climate treaty has so far impeded any swift solutions to the challenges facing the EU ETS.

Concluding remarks

Three main conclusions can be drawn. First, practical experience has shown that the EU ETS has significant potential as a climate policy instrument, through influencing corporate climate strategies in a low-carbon direction. Second, the recent economic recession has brought a significant drop in emissions from the ETS sectors, leading to a substantial surplus of allowances that is expected to increase. This has in turn led to a drop in the carbon price, which has meant scant incentives for companies to invest in low-carbon solutions for the future. Finally, solving this problem of supply–demand imbalance is mainly a question of political feasibility.

What will be at stake if the EU proves unable to fix the EU ETS? First, the *legitimacy* of the system may be at risk if it fails to deliver results. The mandatory "cap" part of the system will ensure reduction of emissions in the ETS sectors according to the EU 2020 target, but the low carbon price will not provide industry with incentives for investing in abatement. If this situation continues, the system may also lose support from those who have favored emissions trading for dealing with the problem of climate change.

Second, the system has a key role in securing long-term, low-carbon innovation when circumstances change and decision makers might be tempted to renege on earlier commitments. One option for coping with this time inconsistency problem is to "tie hands" by adopting pre-commitment strategies. As Poland has blocked stepwise commitments to achieve the EU's 2050 decarbonization ambitions, a failed ETS may place the EU's long-term ambition at risk. Increasing the shares of renewables and more energy-effective production will not guarantee a reduction in emissions if energy consumption grows. This means there is a need for gradually decreasing the cap in the ETS sectors beyond 2020, as well as for sectors not covered by the ETS. EU imports of cheap coal from the USA illustrate this point. Another option for promoting low-carbon solutions over the long term is to eliminate or significantly reduce emissions from fossil fuels by developing new low-carbon technologies that, once commercially viable, may become irreversible. CCS is one option that is needed to realize the 2050 target. Development and commercialization of CCS technologies will suffer when the carbon price is low.

Finally, the system cannot function as a good climate policy model for other countries unless its deficiencies are remedied. On the other hand, the importance of this point should not be exaggerated: after all, although the EU ETS has been a source of empirical insight, among the emerging trading systems in North America (regional), Australia, New Zealand, Japan (regional), South Korea and China (regional), there are few that have sought to directly incorporate lessons from the EU ETS in designing their own systems, and the diversity of these systems may hamper any market convergence (Tuerk *et al.* 2013). The dynamic is now located outside Europe – and the performance of these systems may prove more important for other countries than what unfolds in Europe.

Notes

1 The author would like to thank Jon Hovi and Torbjørg Jevnaker for constructive comments.
2 The study by Martin, Muûls and Wagner was based on telephone interviews with almost 800 manufacturing firms in Belgium, France, Germany, Hungary, Poland and the UK. It included randomly selected firms that are both ETS and non-ETS companies.
3 Cheap coal imports to Europe from the USA have increased, particularly to the German power sector.
4 These include forward-selling to generate funds for the NER300 program, early auctioning to meet power-sector hedging demand, and the sales of left-over allowances in phase 2 New Entrants Reserves.

References

BusinessEurope (2012). Letter to Mr. José Manuel Barroso on the EU ETS. Brussels: BusinessEurope, June 27, 2012.

Commission of the European Communities (2005). Winning the battle against global climate change. *SEC(2005)*180, February 9, 2005.

——(2007a). Limiting global climate change to 2 degrees Celsius: The Way Ahead for 2020 and Beyond. *COM(2007)* 2 final, January 10, 2007.

——(2007b). An energy policy for Europe. *COM(2007)* 1 final, January 10, 2007.

——(2011a). A roadmap for moving to a competitive low carbon economy in 2050. *COM(2011)* 112 final, March 8, 2011.

——(2011b). Energy roadmap 2050. *COM(2011)* 885 final, December 15, 2011.
——(2012a). The state of the European carbon market in 2012. *COM(2012)*652 (undated).
——(2012b). Commission staff working document. Proportionate impact assessment. Brussels: draft, 2012 (undated).
——(2012c). Commission submits draft amendment to back-load 900 million allowances to the years 2019–2020. http://ec.europa.eu/clima/news/articles/news_2012111203_en.htm (accessed 10 December, 2012).
——(2013a). Green Paper: A 2030 framework for climate and energy policies. COM(2013)169 final. Brussels, March 27, 2013.
——(2013b). On the future of carbon capture and storage in Europe. Brussels, March 27, 2013. COM(2013) 180 final.
Ecoprog (2012). The market for coal power plants in Europe (extract). Cologne: Ecoprog GmbH, March 2012.
Ellerman, A.D. and B.K. Buchner (2008). Over-allocation or abatement? A preliminary analysis of the EU ETS based on the 2005–06 emissions data. *Environmental and Resource Economics* 41(2), 267–87.
Ellerman, A.D., F. Convery and C. de Perthuis (2010). *Pricing Carbon: the European Union Emissions Trading Scheme.* Cambridge: Cambridge University Press.
European Parliament (2012). Parliament calls for low-carbon economy by 2050. Press release, Brussels: PR\40876.
Luther (2012). Luther's legal opinion: Brussels' plans for CO_2 set-aside violate EU law. Press release, Cologne, June 21, 2012.
Martin, R., M. Muûls and U. Wagner (2011). *Climate Change, Investment and Carbon Markets and Prices: evidence from manager interviews.* Carbon Pricing for Low-Carbon Investment Project, Climate Strategies, Berlin: Climate Policy Initiative.
McIlveen, R. and D. Helm (2010). *Greener, Cheaper.* London: Policy Exchange.
Polish Chamber of Commerce (2012). *Assessment of the Impact of the Emission Reduction Goals Set in the EC Document 'Roadmap 2050' on the Energy System, Economic Growth, Industry and Households in Poland.* Warsaw: EnergSys, February 15, 2012.
Skjærseth, J.B. (2013). The EU climate and energy package: Causes, content, consequences. Lysaker: FNI Report–2, The Fridtjof Nansens Institute.
Skjærseth, J.B. and P.O. Eikeland (2013). *Corporate Responses to EU Emissions Trading.* Aldershot: Ashgate.
Skjærseth, J.B. and J. Wettestad (2008). *EU Emissions Trading: initiation, decision-making and implementation.* Aldershot: Ashgate.
Skjærseth, J.B., G. Bang and M. Schreurs (2013) Explaining growing climate policy differences in the European Union and the United States. *Global Environmental Politics* (forthcoming). Global Environmental Politics, 13:4, November 2013: 61–80.
Tuerk, A., M. Mehling, S. Klinsky and X. Wang (2013). Emerging Carbon Markets: experiences, trends and challenges. London: Climate Strategies, Working Paper January 2013.
van Renssen, S. (2012). The fate of the EU carbon market hangs in the balance. *European Energy Review*, April 12, 2012.

18 The EU's quest for linked carbon markets

Turbulence and headwind

Jørgen Wettestad and Torbjørg Jevnaker

Introduction

The Emissions Trading System (ETS) of the European Union (EU) involves the largest carbon trading market in the world, with a turnover of some 90 billion euro in 2010 (Skjærseth and Wettestad 2008, 2010). But the challenge of climate change is fundamentally global, with most emissions emanating outside the EU. Hence mitigation measures should ideally have a global reach and create a level global economic playing field. Over time, the EU ETS has been complemented by similar initiatives around the globe. Linkage between emissions trading systems can enable participants in one system to use units from another system for compliance purposes (Commission and Combet 2012). This makes the international linking of carbon markets a key challenge in global climate policy (see e.g., Ellis and Tirpak 2006; Jaffe and Stavins 2008; Tuerk *et al.* 2009; Flachsland *et al.* 2009; Metcalf and Weisbach 2010; Zetterberg 2012).

The EU has put forward the ambition of having linked carbon markets OECD-wide by 2015 (Commission 2009). This chapter assesses the EU's progress in reaching this goal, inquires whether the main explanatory factors are found within the EU itself or in the external environment, and discusses the central prospects ahead. As the functioning of the EU ETS has proven troublesome, with recent serious problems of oversupply (Wettestad 2012 forthcoming 2014), a key question becomes: have the internal push and the external pull for linkage decreased apace with the mounting ETS turbulence?

While previous research has mapped and explained the establishment of national and regional emissions trading systems (see e.g., Skjærseth and Wettestad 2008; Tuerk *et al.* 2013), or pointed out differences in design or plans that might obstruct linking (Jaffe *et al.* 2009; Metcalf and Weisbach 2010; Tuerk *et al.* 2009; Zetterberg 2012), the actual establishment of links between systems remains understudied. Although some studies have noted some barriers to linking (e.g., Tuerk *et al.* 2009), few if any have mapped the perceptions of various groups of relevant actors. It is not only differing system designs that can be problematic; differing perceptions and preferences within each of the national/regional actors may also constitute serious barriers to linking. This study contributes to mapping one of the core actors in global carbon-market politics: the EU. Although the

European Commission (hereafter Commission) plays a central role in the finalization of links, the member states within the Council of Ministers (hereafter Council) and the European Parliament (hereafter Parliament) make the final decisions.

Why are "linking dynamics" so important to understand? On the one hand, there are in theory a number of potential benefits. Linking can contribute to lower overall costs of emissions reductions; it can reduce price volatility, and improve market liquidity. It can add extra demand for allowances and help counteract the oversupply and low carbon price problems recently experienced by the EU ETS. Successful linking can also assuage fears among EU industries about an uneven playing field in global regulation and related "carbon leakage" (Flachsland *et al.* 2009: 363). Furthermore, successful linking could increase the general attractiveness of emission trading, and add further credibility to the view of a "bottom-up approach" as a feasible route to a well-functioning global climate regime (see Victor 2007; Jaffe and Stavins 2008; Metcalf and Weisbach 2010).

On the other hand, there are several serious potential problems. Linking can lead to the export and import of problems experienced by one of the systems. Linking to another system that is oversupplied can worsen the grave problems of an oversupplied EU market with a low and volatile carbon price. Not least, linking also gives rise to tricky issues of distribution of transnational powers and competences. Take a simple example: design choices in system A influence the operation of system B. Should system B then be consulted and have co-decision power if system A is about to undertake changes? Substantial attention needs to be paid to the issues and challenges of political legitimacy and feasibility.

As regards method and evidence, this chapter presents a qualitative case study, seeking to base its main assessments and conclusions on a broad base of evidence (triangulation). We draw on relevant earlier studies (see references above), main EU documents, and articles from news services such as *ENDS Europe* and *Point Carbon,* as well as interviews with central policy actors and observers (see list of interviews).

The EU's linking progress so far: rather moderate?

How did the EU's main linking goals come about? The first Emissions Trading Directive (hereafter ETS Directive), adopted in 2003, established rules for the pilot phase 2005–2007 and the Kyoto commitment phase 2008–2012. Article 25 of the ETS Directive dealt with the issue of linkage by stating the general goal of entering into agreements with countries that had ratified the Kyoto Protocol for "mutual recognition" of allowances from their ETS (EU 2003). The intention of linking the EU ETS to other systems was confirmed during sessions of the Council in its Environment constellation during the years 2004 to 2006 (see e.g., Council 2006).

In 2006 the EU started the process of revising the ETS and developing rules for the post-2012 phase. Linking was also discussed in this process. In the Communication on "building a global carbon market" published in November

2006, linkage to third countries was explicitly mentioned as one of four main topics to be further reviewed by stakeholder working groups (Commission 2006: 6–7).

In a 2007 Communication on the EU's potential contributions to limiting climate change, the Commission noted linkage to "compatible mandatory schemes" as one measure that should be considered, citing California and Australia as examples (Commission 2007: 6). Plans for the US Regional Greenhouse Gas Initiative (RGGI) had been launched in August 2006. But in 2007 EU officials also stated that no links between the EU ETS and the RGGI and California initiatives could be expected before 2013 at the earliest (Point Carbon 2007).

In October 2007 the International Carbon Action Partnership (ICAP) was launched, with the EU Commission and the EU ETS as a cornerstone, but also including several other partners: US and Canadian members of the Western Climate Initiative; RGGI representatives; California representatives; EU member states Germany, France, Portugal, the Netherlands and the UK; New Zealand; and Norway (ICAP 2007). ICAP is intended to serve as an international forum where governments and public authorities adopting mandatory greenhouse gas (GHG) emissions cap-and-trade systems can share and discuss design experiences and best practices (ICAP 2007). The first formal link was established in 2008, to the European Economic Area (EEA) countries Norway, Iceland and Liechtenstein. With Norway, this marked the conclusion of a long history of hooking up with the EU in this field (Sæverud and Wettestad 2006). The EEA context facilitated the link, as this was mainly a matter of these countries taking over the EU ETS design, with some minor adjustments.

Early in 2008, the Commission put forward a proposal for a revised ETS post-2012 (Commission 2008b). Here, linkage to other emission trading systems was seen as a step towards establishing a "global carbon market" (Commission 2008b: 10). According to Stavros Dimas, Environment Commissioner at the time, "the ETS is going to be the prototype for the world to imitate" (ENDS Europe 2008). In the impact assessment accompanying the proposed directive, internal harmonization within the EU ETS was seen as an important step towards external harmonization, i.e., linkage (Commission 2008a: 136). The revised ETS Directive, formally adopted in 2009, retained Article 25's goal of linkage to other carbon trading systems (EU 2009).

What countries (or "sub-federal or regional entities" within these) might the EU ETS be linked up to? According to Article 25.1 these "should" be third countries listed in the Kyoto Protocol's Annex B, and "may" be third countries with "compatible mandatory greenhouse gas emissions trading systems with absolute emissions caps" (EU 2009). In other words, the pool of available partners was expanded as compared to the 2003 directive, which had limited linkage to Annex B countries.[1]

Importantly, later the same year, the Commission put forward the ambition of linkage within the OECD area by 2015: "The EU should reach out to other countries to ensure an OECD-wide market by 2015 and an even broader market by 2020" (Commission 2009: 2).[2] Shortly after, the Council restated the 2015 ambition, adding that this should be done by linking systems that were "comparable in ambition and compatible in design" (Council 2009: 8). Moreover, the broader market to be built by 2020 was further specified as consisting of "economically more advanced developing countries" (Council 2009: 8). Importantly, the Council provided positive backing for linking, and the ongoing cooperation within ICAP was highlighted (Council 2009: 8). Given this Council statement and hence general endorsement by member states, we find it reasonable to regard the 2015/2020 ambitions as EU goals whose progress and achievement are meaningful to assess.

Between 2009 and spring 2012 no new and additional formal links were established, but the EU certainly carried out substantial relevant "linking diplomacy," both within Europe and with respect to the USA and Asia. As to further linking within Europe, negotiations with Switzerland started in 2010, with the goal of having the link in place by 2013 (FOEN 2010), although 2015 has since emerged as a more likely date (Point Carbon 2012).

As regards the USA, to the disappointment of many, President Obama failed to get a US ETS adopted in 2010 (ENDS Report 2010). But at the regional level a climate policy dynamic continued. In 2011 it was decided to establish a Californian emissions trading system, to start in January 2013 (Zetterberg 2012). In April 2011 EU Climate Commissioner Hedegaard announced the intention to establish a link with the California ETS, on the basis of a meeting with Californian representatives (Guardian 2011).

In September 2011 a meeting in Canberra between Commission President Barroso and Australian Prime Minister Gillard resulted in an announcement of the intention to start discussing linking of trading systems between the two entities (Reuters Planetark 2011). This was followed up in August 2012, when the Commission announced the intention to link the ETS to an Australian system by July 2018. A partial link, enabling Australian firms to use EU allowances for compliance, is to be put in place in 2015. Australia and the Commission agreed on a "pathway" towards full linkage between the two systems. This included two adjustments to be made to the Australian system: the scrapping of a planned price floor, and a new limit to the use of Kyoto offsets (Commission and Combet 2012; ENDS Europe 2012). In January 2013 the Commission then put forward a recommendation to the Council that formal negotiations on the link to Australia should be opened (Commission 2013).

Summing up, the EU has now forged emissions-trading links with EEA countries Norway, Iceland and Liechtenstein and is currently negotiating with Switzerland. It has announced a general ambition of linking up with the California ETS, and has concluded a more specific deal on linking up with the Australian system partially in 2015 and fully by 2018. The EU ETS covers 23 of the OECD's 34 member countries, including two EEA countries. If negotiations proceed according to schedule, this means that 2015 will see an interconnected ETS covering 25 of the 34 OECD countries. That will mean progress – but no complete OECD-wide carbon market will exist by 2015, because concrete plans for links to the remaining OECD countries have yet to materialize. The question then becomes: is this due to a low and/or decreasing internal push towards linking? Or does it reflect a lacking and/or decreasing external pull, with few other trading systems operative and "linkable?" Or perhaps a combination of these factors?

Liberal intergovernmentalist lenses: member states split and decreasingly interested?

The first perspective on an internal push is a "Liberal Intergovernmentalist" one (see e.g., Moravcsik 1998; Moravcsik and Schimmelfennig 2009). As with most complex theories, weight can be given to different theory sub-dimensions. Here we will highlight how this theory emphasizes the key role that member state positions and the distribution of power backing different positions play in understanding EU policy development. What was the main distribution of member state positions on linking in 2007–2009, when the EU's climate and energy package was negotiated and the 2015 goal was established? Have member state positions changed since then – perhaps in connection with the financial crisis and a reassessment of climate policy priorities?

What do we know about the development of member state positions on linking? We may first note that a general quest for linking was expressed at Council meetings throughout 2007/2008. For instance, the Council meeting in February 2007 expressed the wish "to link the EU ETS with other compatible emission trading schemes with comparable levels of ambition" (Council 2007a: 9). This was regarded as a step towards establishing a global carbon market (Council 2007b: 14).

The linking issue was also the theme of a meeting on ETS revision held in 2007, with national representatives and stakeholders. At this meeting, internal harmonization of EU ETS was noted as important for external harmonization, i.e., linkage. Flexibility and stepwise development were emphasized. Linkage would also be aided by similar price levels on carbon credits within the various trading systems. Compatibility between systems was emphasized, as a way of building trust in linkage, although the option of using a gateway was also mentioned (ECCP 2007).

While the goal of a global carbon market was generally supported – and was, moreover, seen as reducing compliance costs and increasing efficiency – views differed within the working group. A UK representative stressed "the importance for the international cooperation of linking" (ECCP 2007: 3).

In March 2008, the Environment Council, as well as the European Council, emphasized that the revised ETS Directive then under discussion should enable linkage to "other mandatory emissions trading systems capping absolute emissions" (Council 2008a: 13; 2008c: 9). In the preparatory documents to the June 2008 Council meeting, the UK explicitly stated its support for linking different emission trading systems (Council 2008b: 7). This stands in contrast to another key member state, Germany, which focused on internal ETS coverage, supporting "in principle the planned expansion of the trading scheme to cover all major industrial greenhouse-gas emitters" (Council 2008b: 5). Hence the UK stood out as something of a "linking frontrunner," consistent with the country's general embrace of the emissions trading instrument (Interviews in Brussels, November 2012). From Poland and other Eastern European member states there was little attention to the external linking issue.

As noted, the Commission's proposed goal of linked OECD markets by 2015 was supported by the Council meeting in its Environment configuration in March 2009. Thus we see that the member states were initially not particularly split on the issue of external linking, but interest was fairly lukewarm, with the exception of the UK in particular.

The Copenhagen climate summit of December 2009 was a general disappointment, with a decidedly meager outcome (see e.g., Dimitrov 2010). However, we have not been able to find any specific conclusions drawn about the implications for further ETS linking dynamics. As indicated above, the next EU linking milestone was Climate Commissioner Hedegaard's announcement in the spring of 2011 of the intention to establish a link to the California ETS. To our knowledge, the push for this initiative did not come from the member state side, nor did it spark any specific comments or enthusiasm from member states at the time.

Despite some scattered references to the issue of linking, no specific initiatives were taken by member states at this point.[3] For instance, as will be elaborated on below, the initiative for the Australian link announced in 2012 did not come from the EU member states. Neither did the Australian link seem to trigger any public reactions from member states. Here we should note that the ETS reform process was then moving up the agenda, with the Commission announcing its plans for "backloading" (i.e., postponing allowance auctioning) and more structural ETS reform measures in July 2012 (Commission 2012). Furthermore, from 2010 on, the carbon price was overall moderate and over time decreasing, even dipping below €5 in 2013. This removed the possible need for external linking in order to ease the internal regulatory burden.

On the whole, the member states seem to have been generally positive towards linking, although only the UK stands out as paying substantial attention to this issue. It may perhaps be argued that the rather tepid member state interest has been a function of loose ambitions about linking, with more attention expected once concrete measures could be placed on the agenda. Yet the pattern of uninterested support seems to have been repeated, even when plans for specific links were announced. On the other hand, since no bilateral agreements have been submitted to the Council for ratification (with the exception of Norway), little has been at stake for the member states thus far.

Supranationalist lenses: decreasing attention to linking from Commission and Parliament?

A second perspective on the internal push is based on literature on Multi-level Governance and supranationalism (see e.g., Stone Sweet and Sandholtz 1997, Hooghe and Marks 2001). A common element in these analytical perspectives is the far greater attention and weight given to the role and positions of EU bodies such as the Commission and the Parliament than to Liberal Intergovernmentalism. Hence, such lenses are helpful in zooming in on characteristics such as the internal agreement within these bodies and the role the bodies play in building broader networks and alliances. We know that both the Commission and the Parliament have been internally divided on other climate policy issues (Boasson and Wettestad 2013). Has this also been the case as regards linking? Furthermore, has the interest in and attention paid to the issue waned within these bodies, alongside the mounting EU ETS problems?

Central statements on linking from the Commission were described above. The Commission has certainly directed attention as well as resources to the issue. It seems clear that the initial main linking vision of the Commission was a linked EU–US carbon market (interviews in Brussels 2012). This vision was upheld for a while, into 2010, but suffered a serious blow when in mid-2010 it became clear that a US ETS would not be established. This external influence is further described below.

In 2011, the Commission put forward a "roadmap" for EU climate policy up to 2050. This document touched on linking only in general terms, emphasizing the important need to "work towards a gradual development of global carbon markets" (Commission 2011: 13). The Commission also continued a reasonably active linking diplomacy in these years. The impetus for Hedegaard's California initiative in 2011 came from the Directorate General (DG) Climate Action (hereafter Clima). Furthermore, Hedegaard re-stated the EU's linking ambitions towards Asia on a trip to South Korea in spring 2011, referring to South Korean, Chinese and Australian carbon market developments (Point Carbon 2011b). And when the Californian ETS was adopted in the fall of 2011, Commission officials stated: "Brussels is actively working with [California] state authorities to make the Californian and European ETS schemes compatible" (ENDS Europe 2011).

However, it also seems very likely that increasing EU ETS problems from 2011 on, including allegations of fraud and, not least, an increasingly volatile and sinking carbon price related to oversupply of allowances, served to reduce the attention paid by Commission personnel to this issue. With regard to the Australian link announced in 2012, it now seems clear that the initiative did not come from the EU side (interviews in Brussels 2012).

As regards internal unity within the EU bodies, the impression is that the linking issue has been driven mainly by DG Clima, and that internal Commission disagreement has not been an issue here. Furthermore, as to the role of the Parliament, this body has touched on the linking issue, for instance during the debate on the proposed revisions to the 2003 ETS Directive (EP 2008), but this particular ETS sub-issue has not figured centrally on the Parliament's agenda (interviews in Brussels 2012).

Thus it seems likely that the Commission's "linking campaign" has been somewhat affected over time by the mounting internal EU ETS problems, those of the most recent years in particular. Still, it is indisputable that the Commission has paid continuous attention to linking, as demonstrated by documents issued over the course of this period, as well as its meetings with potential linking partners. The Commission has been interested in strengthening the legitimacy of the EU's own ETS by highlighting emissions trading as a tool used outside of Europe as well (interviews in 2012). With the many and repeated initiatives, the picture that emerges regarding linkage is one of surprising continuity through difficult times.

External interaction lenses: slow global progress and little to link up to?

Studies of EU environmental policy have begun to pay greater attention to how the external environment, the climate regime in particular, affects EU policy-making (see e.g., Oberthür and Pallemaerts 2010; Boasson and Wettestad 2013). Three developments in the external environment will be in focus here. The first factor highlights the development of trading systems outside of the EU. If such systems do not emerge or are seriously delayed, that will make the fulfilling of the EU's own linking ambitions increasingly difficult. The second factor highlights the interest in linking up with the EU ETS among potential partners. Have they become more interested, or less, in linking up with an increasingly crisis-ridden European ETS – or have these problems not played a role? The third factor concerns the development of the global climate negotiations. Has the lackluster development of these negotiations led to greater pressure on the EU to speed up the carbon-market linking process? Or has it decreased pressure, as disappointment with global progress has led the EU to focus inwards?

As of 2013, the main emissions trading systems in operation outside of the EU are in New Zealand, California, Quebec and Japan (two regional systems). In addition, pilot activities have been started in several cities and provinces in China. Furthermore, various other actors (such as South Korea and Australia) are about to establish trading systems or (like Russia) are seriously considering doing so (Tuerk *et al.* 2013; Moscow Times 2012). But the development of emissions trading globally has generally progressed slowly, and mostly without the cap-and-trade design preferred by the EU. That has meant fewer candidates for the EU to approach seriously with a view to establishing linkages. The slow growth of ETS outside the EU constitutes an important part of the explanation of why the EU's ambition of linked markets in the OECD area is unlikely to come about.

What about external interest in linking with the EU ETS? We may first note that there was external interest in the EU ETS from quite early on. For instance, an EU delegation of ETS experts visited the US Senate in March 2007. Republican Senator Pete Domenici then stated: "I do not think the majority of US legislators think the EU is doing quite well – but at least you are trying something" (Europolitics Environment 2007). Moreover, a California delegation visited Brussels to learn about the EU ETS. The delegation expressed hopes that California could be the first non-European region to link up to the ETS, from 2013 on (Reuters Planetark 2007).

It is clear that the main initial linking vision of the Commission was a linked EU–US carbon market. For instance, in the spring of 2009 a Commission representative stated, "a key [to building a broad carbon market by 2015] is a transatlantic carbon market between the EU and US" (quoted in Point Carbon

2009; Reuters Planetark 2009). This "plan A" was upheld for a while into 2010, with DG Clima official Peter Zapfel stating in May 2010 that "the US is the prime [linking] candidate. The development of the market there would facilitate the development of markets elsewhere such as Canada, Australia and Japan" (EU Energy 2010: 13). But this plan suffered a serious blow in mid-2010 when the US ETS bill was not adopted, after months of hefty opposition from the fossil-fuel lobby as well as from Republicans and Democrats in coal-dependent areas; *ENDS Report* termed it "a major setback for the European Commission" (ENDS Report 2010: 57).

From 2011 on, the EU ETS was seen as increasingly in troubled waters. This development may have contributed to muting the interest of central potential partners in linking up to the EU ETS. For instance, Commissioner Hedegaard, still keen on establishing a transatlantic link, met with Californian representatives in April 2011 (Zetterberg 2012: 41). But the Californian representatives downplayed the importance of this meeting, dismissing it as mainly a "courtesy call" (Point Carbon 2011a). Indeed, the Californian authorities explicitly discussed linking with other ETS (Western Climate Initiative, Quebec), without mentioning the EU ETS as an option (Zetterberg 2012: 31).

In early 2013, *EurActiv* reported that the continuing instability of the EU ETS had "unsettled" other countries and regions inspired by the EU's market-based model of countering carbon emissions, such as California, South Korea, and China. According to the EU's chief climate negotiator, Artur Runge-Metzger, "one worry is that the Chinese might think that nobody in Europe takes climate change seriously any longer" (EurActiv 2013).

But this picture is not clear-cut, not least as shown by the Australian process. Although a meeting between Commission President Barroso and Australian Prime Minister Gillard was held in September 2011 and a linking intention declared, central EU officials have indicated that they were just as surprised as everyone else when concrete follow-up processes were initiated in 2012. The initiative came mainly from the Australian side, with the ETS link mentioned as being a factor in facilitating the troubled Australian road towards an effective carbon pricing policy (interviews in Brussels 2012; see also Bailey *et al.* 2012). Moreover, linking with the EU ETS was seen as a way of preserving the Australian ETS, because it would limit the possibility of fallback following Australian elections (interviews in 2012).

The slow progress of the international negotiations does not appear to have led the EU to focus only inwards. We have noted the interest in the Californian link. Further, the EU has been actively involved in assisting China to develop its pilot systems, including a specific collaboration agreement between the two parties concluded in September 2012 (see Biedenkopf and Torney 2013).

Conclusions

Since 2009 the EU has forged ETS links with EEA countries Norway, Iceland and Liechtenstein, and is currently negotiating a link with Switzerland. Moreover, it has announced the general ambition of linking up with the California ETS, and has concluded a specific deal linking up with the Australian system partially in 2015 and fully by 2018. But 2015 is not that far away, and there have been numerous problems involved in linking efforts – so the conclusion here must be that the EU is not on schedule to meet its declared goal of linked carbon markets within the OECD area by 2015.

Why is this so? Has it been caused by low or decreasing internal push, or by low or decreasing external pull? Using a Liberal Intergovernmentalist approach, we first examined the EU member states' interest in linking. On the whole, the push from member states for external linking emerges as rather moderate, and other ETS concerns seem to have dominated their agendas. Furthermore, an overall moderate and decreasing carbon price has removed the possible need for external linking in order to ease the internal regulatory burden.

Putting on supranationalist lenses, we then zoomed in on the role of the Commission and the Parliament. The latter has not been a key actor in this process. The linking issue has been mainly driven by DG Clima, and internal Commission disagreement has not been a problem here. As mentioned, the linkage initiatives have been many and repeated, yielding a picture of surprising continuity through difficult times.

As regards external interaction, the external pull has most likely decreased. Crucial events in 2009 and 2010, particularly the failure to establish a US ETS, shot down the Commission's linking plan A. The US failure meant that a more difficult and complex plan B needed to be implemented. As the development of emissions trading globally has progressed slowly, there have simply not been that many candidates for the EU to approach for linking purposes. Furthermore, with the EU ETS increasingly finding itself in troubled waters, the interest among key players in emerging carbon markets around the world for linking with the EU may well have lessened. But it should be noted that US interest has been quite moderate all along, and the effect has seemingly also been uneven, with actors like Australia showing a stable interest in linking up with the ETS.

With regard to the main analytical implications for further studies of the linking of carbon markets, greater attention must be paid to questions of political dynamics, instead of focusing solely on the economic aspects and implications. For instance, in order to understand the dynamics of the EU–Australian linking process, it seems crucial to give weight to considerations of domestic politics. Establishing links and then making them function effectively requires support from a wide-ranging group of societal actors who are also potential veto-players. Good analyses of the "politics of linking" can make important contributions to the linking processes ahead.

Finally, what of the prospects for establishing a new, more effective global climate agreement? Potentially, the gradual linking of trading systems offers an alternative route to creating a global system for cutting emissions. But experience thus far indicates that such a route is a long-term project – certainly not a "quick fix" for a stalled global climate regime.

Appendix: list of interviews

Egenhofer, Christian, CEPS, November 19, 2012.
Eickhout, Bas, European Parliament (by telephone), November 29, 2012.
Ervik, Leif K. and Ingrid Hoff, Norwegian Ministry of Environment, April 24, 2013.
Lind, Øystein, Statoil Brussels office, November 19, 2012.
Scott, Jesse, EURELECTRIC, November 21, 2012.
Svarstad, Dag and Elen Richter Alstadheim, Norwegian Ministry of the Environment, September 7, 2012.
Wyns, Tomas, CCAP, November 19, 2012.
Zapfel, Peter, European Commission, November 20, 2012.

Acknowledgements

Our thanks to Jon Hovi, an anonymous reviewer, and participants at a panel on "diffusion or decline of emissions trading" at the ISA Annual Convention, San Francisco, April 3–6, 2013, for helpful comments. Thanks also to Susan Høivik for language polishing.

Notes

1 In addition to EU member states, this includes Australia, Canada, Croatia, Iceland, Japan, Liechtenstein, Monaco, New Zealand, Norway, Russia, Switzerland, Ukraine, and the USA (UNFCCC 2012).
2 The Organization for Economic Co-operation and Development (OECD) has 34 members: in addition to 21 EU countries, these are Australia, Canada, Chile, Iceland, Israel, Japan, Korea, Mexico, New Zealand, Norway, Switzerland, Turkey, and the United States.
3 For instance, Poland expressed general support for linking in 2012. See Ministerstwo Srodowiska (2012: 9).

References

Bailey, I., I. MacGill, R. Passey and H. Compston (2012). The fall (and rise) of carbon pricing in Australia: a political strategy analysis of the carbon pollution reduction scheme. *Environmental Politics* 21(5), 691–711.

Biedenkopf, K. and D. Torney (2013). EU–China environmental cooperation: the case of emissions trading. Paper presented at the panel *Diffusion or Decline of Emissions Trading?* ISA Annual Convention, San Francisco, April 3–6, 2013.

Boasson, E.L. and J. Wettestad (2013). *EU Climate Policy: industry, policy interaction and external environment*. Aldershot: Ashgate.

Commission (2006). Building a global carbon market – Report pursuant to Article 30 of Directive 2003/87/EC. Brussels.

——(2007). Limiting Global Climate Change to 2 degrees Celsius: The way ahead for 2020 and beyond (COM (2007) 2 final). Brussels.

——(2008a). Accompanying document to the Proposal for a Directive of the European Parliament and of the Council amending Directive 2003/87/EC so as to improve and extend the EU greenhouse gas emission allowance trading system: Impact Assessment (SEC(2007) 52). Brussels.

——(2008b). Proposal for a Directive of the European Parliament and of the Council amending Directive 2003/87/EC so as to improve and extend the greenhouse gas emission allowance trading system of the Community (COM (2008) 16 final). Brussels.

——(2009). Towards a comprehensive climate change agreement in Copenhagen (COM (2009) 39 final). Brussels.

——(2011) A Roadmap for Moving to a Competitive Low Carbon Economy in 2050.

——(2012). The state of the European carbon market in 2012 (COM (2012) 652 final). Brussels.

——(2013). Linking EU ETS with Australia: Commission recommends opening formal negotiations. Press release, Brussels, January 24.

Communication from the Commission to the European Parliament, the Council, the European Economic and Social Committee and the Committee of the Regions, COM (2011) 112 Final, 8 March. Brussels.

Commission and Combet [Hon. Greg Combet MP] (2012). Australia and European Commission agree on pathway towards fully linking Emissions Trading systems. Brussels.

Council (2006). Press release: 2757th Council Meeting (October) (13989/06 (Presse 287)). Luxembourg.

——(2007a). Press release: 2785th Council Meeting (February) (6272/07 (Presse 25)). Brussels.

——(2007b). Press release: 2826th Council Meeting (October) (14178/07 (Presse 247)). Luxembourg.

——(2008a). Brussels European Council 13/14 March 2008 Presidency Conclusions (7652/1/08 REV 1). Brussels.

——(2008b). Climate–Energy Legislative Package: Progress report, policy debate (10236/08 ADD 1). Brussels.

——(2008c). Press release: 2865th Council Meeting (March) (6847/08 (Presse 50)). Brussels.

——(2009). Climate Change – Contribution to the Spring European Council (19 and 20 March 2009): Further development of the EU position on a comprehensive post-2012 climate agreement – Council conclusions (7128/09). Brussels.

Dimitrov, R. (2010). Inside Copenhagen: the state of climate governance. *Global Environmental Politics* 10(2), 18–24.

ECCP (2007). Final Report of the 4th meeting of the ECCP working group on emissions trading on the review of the EU ETS on linking with emissions trading schemes of third countries. Brussels.

ENDS Europe (2008). Global carbon market 'unlikely by 2020'. February 25.

——(2011). California adopts emissions trading scheme. October 21.

——(2012). EU and Australia agree emissions trading link. August 28.

ENDS Report (2010). Stillborn cap-and-trade bill forces policy rethink, no. 427, August 2010, p. 57.

EP (2008). Greenhouse gas emission allowance trading system (debate). CRE December 16, 2008–11. Available at: www.europarl.europa.eu/sides/getDoc.do?type=CRE&reference=20081216&secondRef=ITEM-011&language=EN&ring=A6-2008-0406 [accessed September 15, 2012].

EU (2003). Directive 2003/87/EC of the European Parliament and of the Council of 13 October 2003 establishing a scheme for greenhouse gas emission allowance trading within the Community and amending Council Directive 96/61/EC.

——(2009). Directive 2009/29/EC of the European Parliament and of the Council of 23 April 2009 amending Directive 2003/87/EC so as to improve and extend the greenhouse gas emission allowance trading scheme of the Community.

EU Energy (2010). EC pens final ETS auction rules, *EU Energy* 232, 11–13.

EurActiv (2013). MEPs throw lifeline to EU's ailing carbon market, February 19, 2013. Available at: www.euractiv.com/climate-environment/meps-vote-carbon-market-future-b-news-517898 [accessed February 20, 2013].

Europolitics Environment (2007). Commission briefs US Senate on EU emissions trading scheme, *Europolitics Environment* 721 (March 30, 2007), 6.

Ellis, J. and Tirpak, D. (2006). Linking GHG emissions trading schemes and markets, OECD: Paris (October).

Flachsland, C., R. Marschinski and O. Edenhofer (2009). To link or not to link: benefits and disadvantages of linking cap-and-trade systems. *Climate Policy* 9(4), 358–72.

FOEN (2010). Commission proposes opening negotiations with Switzerland on linking emission trading systems. Available at: www.bafu.admin.ch/emissionshandel/10923/10926/10927/index.html?lang=en [accessed February 20, 2013].

Guardian (2011). EU plans to link emissions trading scheme with California. *The Guardian*, April 7.

Hooghe, L. and G. Marks (2001). *Multi-level Governance and European Integration.* Lanham, MD: Rowman & Littlefield.

ICAP (2007). International Carbon Action Partnership Political Declaration. Lisbon.

Jaffe, J. and R.N. Stavins (2008). Linkage of tradable permit systems in international climate policy architecture. Faculty Research Working Papers Series, John F. Kennedy School of Government, Harvard University.

Jaffe, J., M. Ranson and R.N. Stavins (2009). Linking tradable permit systems: a key element of emerging international climate policy architecture. *Ecology Law Quarterly* 36(4), 789–808.

Metcalf, G.E. and D. Weisbach (2010). Linking policies when tastes differ: global climate policy in a heterogeneous world. Discussion Paper 10-38 (July), John F. Kennedy School of Government, Harvard University.

Ministerstwo Srodowiska (2012). Polish policy paper on the future of the ETS up to 2030, Warsaw.

Moravcsik, A. (1998). *The Choice for Europe*. Ithaca, NY: Cornell University Press.

Moravcsik, A. and F. Schimmelfennig (2009). Liberal Intergovernmentalism. In T. Diez and A. Wiener (eds.), *European Integration Theory*. Oxford: Oxford University Press, pp. 67–87.

Moscow Times (2012). Russia Mulls Internal Carbon Trading. *Moscow Times*, September 13.

Oberthür, S. and M. Pallemaerts (2010). The EU's internal and external climate policies: an historical overview. In S. Oberthür and M. Pallemaerts (eds.), *The New Climate Policies of the European Union.* Brussels: VUB Press, pp. 27–65.

Point Carbon (2007). EU emissions trading not seen linking with US until after 2013, 22 January.

——(2009). EU hopeful of building joint carbon market with US. February 4.

——(2011a). EU eyes California market link, but faces barriers. April 14.

——(2011b). Interview: Europe looks east for partners in carbon markets. May 3.

——(2012). Switzerland sees EU ETS link in 2014. Available at: www.pointcarbon.com/news/1.1861881 [accessed September 17, 2012].

Reuters Planetark (2007). California eyes joining EU emissions trading scheme. March 30.

——(2009). US gives cap and trade boost for climate treaty. March 2.

——(2011). EU, Australia to discuss linking carbon trading schemes. September 6.

Sæverud, I.A. and J. Wettestad (2006). Norway and emissions trading: from global front-runner to EU follower. *International Environmental Agreements* 6(1), 91–108.

Skjærseth, J.B. and J. Wettestad (2008). *EU Emissions Trading: initiation, decision-making and implementation*. Aldershot: Ashgate.

——(2010). Fixing the EU Emissions Trading System? Understanding the post-2012 changes. *Global Environmental Politics* 10(4), 101–23.

Stone Sweet, A. and W. Sandholtz (1997). European integration and supranational governance. *Journal of European Public Policy* 4(3), 297–317.

Tuerk, A., M. Mehling, C. Flachsland and W. Sterk (2009). Linking carbon markets: concepts, case studies and pathways. *Climate Policy* 9(4), 341–57.

Tuerk, A., M. Mehling, S. Klinsky and X. Wang (2013). Emerging carbon markets: experiences, trends, and challenges. Working Paper (January), London: Climate Strategies.

UNFCCC (2012). Kyoto Protocol. Available at: http://unfccc.int/kyoto_protocol/items/3145.php [accessed September 13, 2012].

Victor, D.G. (2007). Fragmented carbon markets and reluctant nations: implications for the design of effective architectures. In J. Aldy and R. Stavins (eds.), *Architectures for Agreement: addressing global climate change in the post-Kyoto world.* Cambridge, UK: Cambridge University Press, pp. 133–60.

Wettestad, J. (2012). The EU Emissions Trading System: repaired but still leaking? *Klima* no.2.

______(forthcoming 2014). Rescuing EU Emissions Trading: Mission Impossible?, *Global Environmental Politics* 14(2).

Zetterberg, L. (2012). Linking the Emissions Trading Systems in EU and California. FORES Study 6, IVL (Swedish Environmental Research Institute), Gothenburg.

Index

Please note that page references to Figures or Tables will be in *italics*, while the letter 'n' will follow references to Notes.